TRANSLATIONS OF
MATHEMATICAL
MONOGRAPHS
VOLUME 45

D. A. Suprunenko

Matrix Groups

American Mathematical Society

ГРУППЫ
МАТРИЦ

Д. А. СУПРУНЕНКО

Издательство "Наука"
Главная редакция
физико-математической литературы
Москва 1972

Translated from the Russian by
Israel Program for Scientific Translations

Translation edited by K. A. Hirsch

AMS (MOS) subject classifications (1970). Primary 20G15;
Secondary 20B05, 15A15, 20E15, 20F50, 20E25.

Library of Congress Cataloging in Publication Data

Suprunenko, Dmitriĭ Alekseevich.
Matrix groups.

(Translations of mathematical monographs ; v. 45)
Translation of Gruppy matrits.
Bibliography: p.
Includes index.
1. Matrix groups. I. Title. II. Series.
QA171.S77813 512'.22 75-45115
ISBN 0-8218-1595-4

10 9 8 7 6 5 4 3 2   02 01 00 99

PREFACE

In this book we give an account of the classical resutls on the structure of normal subgroups of the general linear group over a division ring, of Burnside's and Schur's theorems on periodic linear groups, and of the theorem on the normal structure of $SL(n, Z)$ for $n > 2$. We also discuss the theory of solvable, nilpotent, and locally nilpotent linear groups. To give a better idea of the content of the book, here is a brief chapter-by-chapter survey.

In the first chapter we deal with permutation groups (finite and infinite). After an account of the elementary facts we establish a connection between the theory of primitive solvable permutation groups (not necessarily of finite degree) and the theory of solvable linear groups over prime fields. Here we also investigate nilpotent and locally nilpotent permutation groups. For example, we give a complete classification of the maximal nilpotent subgroups of the symmetric groups of finite degree. We prove, in particular, that the maximal transitive nilpotent subgroups of a finite symmetric group are conjugate.

A substantial part of the second chapter is devoted tô the theory of determinants of matrices over a division ring and the structure of normal subgroups of the general linear group over a division ring; the exposition is based on Dieudonné [1] and Artin [1]. We also consider properties of endomorphisms of a module over an associative ring. In particular, we establish the permutability properties of transvections of a module.

In the third chapter we describe the normal subgroups of the stable general linear group over an arbitrary associative ring (Bass's theorem [1]) and the normal subgroups of $SL(n, Z)$, $n > 2$ (Bass, Lazard and Serre [1], Mennicke [1]). The stable (or limiting) general linear group $GL(\Delta)$ over a ring Δ is defined as the union of the following infinite ascending chain of groups:

$$GL(1, \Delta) \subset GL(2, \Delta) \subset \ldots \subset GL(n, \Delta) \subset \ldots$$

with a standard embedding of $GL(n, \Delta)$ in $GL(n + 1, \Delta)$.

In the fourth chapter we prove Mal′cev's local theorem on conditions for a faithful linear representation of a group and Clifford's theorem on the normal subgroups of an irreducible linear group. We also investigate properties of imprimitive linear groups and conditions for linear groups to be completely reducible.

In the fifth chapter we consider solvable matrix groups over a field. Attention is concentrated mainly on the structure of maximal primitive solvable subgroups of $GL(n, \Delta)$ for an arbitrary field Δ. We examine in detail the invariant series

$$G \supset V \supset A \supset F \supset (E_n), \qquad (\alpha)$$

where G is a maximal primitive solvable subgroup of $GL(n, \Delta)$, F a maximal abelian normal subgroup of G, V the centralizer of F in G, and A/F a subgroup of G/F that is maximal among the normal abelian subgroups of G/F contained in V/F. The series (α) uniquely determines G. In certain cases the study of (α) enables us to classify the maximal irreducible solvable subgroups of $GL(n, \Delta)$. From the properties of (α) it is easy to derive the theorems of Zassenhaus and Mal′cev on solvable linear groups. The construction of maximal solvable subgroups of the general linear group over an algebraically closed field is reduced to the construction of solvable subgroups of the symplectic group over a finite prime field. We list the maximal irreducible solvable groups of prime degree over an algebraically closed field and over a finite field. We establish conditions for complete reducibility of a solvable linear group. In particular, a solvable linear group over a field of characteristic zero is completely reducible if and only if all its cyclic subgroups are completely reducible.

The sixth chapter is devoted mainly to periodic linear groups over a field. We prove Burnside's criterion for the finiteness of a linear group, and Schur's theorems on the local finiteness of a periodic linear group over an arbitrary field and on the existence of an abelian normal subgroup of finite index in a complex periodic linear group. We study Sylow p-subgroups and Sylow Π-subgroups of $GL(n, \Delta)$. We present Platonov's structure theorems for periodic linear groups: the theorem that the Sylow p-subgroups of a periodic linear group are conjugate and the analog of the Schur-Zassenhaus theorem for periodic linear groups. We consider properties of a linear group, not necessarily periodic, that are connected with properties of the characteristic polynomials of its matrices.

In the seventh chapter we study nilpotent and locally nilpotent matrix groups over a field. Here we give a complete classification of all maximal locally nilpotent subgroups of the general linear group over an algebraically closed field. In particular, we prove that if Δ is an algebraically closed field such that char Δ divides n, then $GL(n, \Delta)$ contains no irreducible locally nilpotent subgroups; while if char $\Delta = 0$, or char $\Delta = p$ but n is not divisible by p, then $GL(n, \Delta)$ contains (up to conjugacy) a unique maximal irreducible locally nilpotent subgroup. For an algebraically closed ground field we also give a complete classification of the maximal irreducible nilpotent subgroups of class 2 of the general linear group. For an arbitrary ground field the classification of the maximal irreducible locally nilpotent subgroups of the general linear group is reduced to a complete description of the Sylow p-subgroups of the projective group $PGL(p^\alpha, \Delta)$. For a perfect field we give a canonical form of locally nilpotent reducible indecomposable linear groups.

The numbering of theorems, propositions, lemmas and formulas is independent within each section. References to the number of a theorem (lemma, proposition) are prefaced by the section number. For example, if Theorem 5 of §7 is referred to outside §7, it is called Theorem 7.5. Corollary 1 of Theorem 7.5 is referred to as Corollary 7.5.1. If this reference is in §7 itself, we write "Corollary 5.1".

I remember with gratitude the late A. I. Mal′cev, who encouraged me to begin work on this book. I would also like to thank the editor, F. I. Kizner, and my research students G. F. Žavrid, V. S. Konjuh, L. P. Lugovskaja, N. N. Metel′skiĭ, O. M. Neroslavskiĭ and V. P. Juferev, who helped me with the preparation of the manuscript.

TABLE OF CONTENTS

FIRST CHAPTER

ELEMENTS OF THE THEORY OF PERMUTATION GROUPS

§1. Properties of mappings

1. *Mappings.* Let X and Y be arbitrary nonempty sets. We say that a *mapping of X into Y* is defined if each element $x \in X$ is associated with a unique element $y \in Y$.

Mappings of a set X into a set Y are written as $f\colon X \longrightarrow Y$, $g\colon X \longrightarrow Y$, etc. The image of X under a mapping f is denoted by $f(X)$ or $\operatorname{Im} f$. The notation $x \longmapsto y$ means that $y = f(x)$ is the image of $x \in X$ under f. If $f\colon X \longrightarrow Y$ and $B \subset Y$,[1] then $f^{-1}(B) = \{x \mid x \in X; f(x) \in B\}$.

Two mappings $f\colon X \longrightarrow Y$ and $f_1\colon X_1 \longrightarrow Y_1$ are *equal* if $X = X_1$, $Y = Y_1$ and $f(x) = f_1(x)$ for every $x \in X$.

Let $\varnothing \neq A \subset X$. A mapping $\varphi\colon A \longrightarrow Y$ is called the *restriction of $f\colon X \longrightarrow Y$ to A* if $\varphi(a) = f(a)$ for every $a \in A$. The restriction of a mapping f to a set A is often denoted by $f|A$. If H is a set of mappings of X into Y, then $H|A$ denotes the set of all restrictions $h|A$, $h \in H$.

A mapping $f\colon X \longrightarrow Y$ is said to be *injective* if for any two elements x_1 and x_2 of X

$$f(x_1) = f(x_2) \Rightarrow x_1 = x_2$$

(as usual, the symbol $\Rightarrow$ denotes implication).

A mapping $f\colon X \longrightarrow Y$ is said to be *surjective* if $f^{-1}(y) \neq \varnothing$ for every $y \in Y$.

A mapping $f\colon X \longrightarrow Y$ is said to be *bijective*, or a *bijection*, if it is both surjective and injective.

2. *Multiplication of mappings.* Let $f\colon X \longrightarrow Y$ and $g\colon Z \longrightarrow U$, where $Y \subset Z$. The mapping $gf\colon X \longrightarrow U$, $x \longmapsto g(f(x))$, is called the *product* of g and f.

Under the action of gf an element $x \in X$ goes into $g(y)$, where $y = f(x)$.

The product gf of two mappings g and f is defined only if the second set of the second mapping is a subset of the first set of the first mapping.

---

(1) In this book the symbol $\subset$ indicates nonstrict inclusion; that is, the sets in question may be equal.

THEOREM 1. *Let f, g, h be mappings such that at least one of the two products $f(gh)$ or $(fg)h$ exists. Then the other product also exists, and*

$$f(gh) = (fg)h. \tag{1}$$

PROOF. Suppose, for example, that $f(gh)$ exists. Then gh exists and so g and h must be of the form $g\colon X \longrightarrow Y$, $h\colon Z \longrightarrow U$, $U \subset X$. Hence $gh\colon Z \longrightarrow Y$, $f\colon V \longrightarrow W$ and $Y \subset V$. Therefore fg exists and is of the form $fg\colon X \longrightarrow W$. Since $U \subset X$, the mapping $(fg)h$ also exists.

We now prove (1). Obviously $f(gh)\colon Z \longrightarrow W$ and $(fg)h\colon Z \longrightarrow W$. Let $z \in Z$, $a = f(gh)$ and $b = (fg)h$. Then $a(z) = f(gh)(z) = f(gh(z)) = f(g(h(z)))$ and $b(z) = (fg)h(z) = fg(h(z)) = f(g(h(z))) = a(z)$; thus $f(gh) = (fg)h$. ∎

Now let $\Sigma(X)$ be the set of all mappings of X into itself. Then, clearly, the product of any two mappings in $\Sigma(X)$ exists and, by Theorem 1, $\Sigma(X)$ *is a semigroup. This semigroup has a unit element, namely the identity mapping* $e\colon X \longrightarrow X$, $x \longmapsto x$.

We call $\Sigma(X)$ the *symmetric semigroup on* X, and the group $S(X)$ of all its invertible elements the *symmetric group on* X. Every subgroup of $S(X)$ is called a *permutation group on* X.

The cardinality card X of X is called the *degree* of a permutation group on X. If card $X = n < \infty$, we sometimes write S_n for $S(X)$.

THEOREM 2. *A mapping in* $\Sigma(X)$ *is invertible if and only if it is bijective.*

PROOF. Let $f\colon X \longrightarrow X$ be a bijection. If we let x range over the elements of X, taking each value only once, then $f(x) = y$ also ranges over the same elements, taking each value once. Thus there exists a $g \in \Sigma(X)$ such that $g(y) = x$. Obviously $fg(y) = f(x) = y$ and $gf(x) = g(y) = x$, so that $fg = gf = e$. Thus bijectivity of f implies its invertibility in $\Sigma(X)$.

Now let $f \in S(x)$, so that there exists a mapping g in $\Sigma(X)$ such that $fg = gf = e$. If y is any element of X, then $fg(y) = e(y) = y$ and $f(g(y)) = y$. Thus f is surjective. If $x_1, x_2 \in X$ and $f(x_1) = f(x_2)$, then $gf(x_1) = gf(x_2)$, $e(x_1) = e(x_2)$ and $x_1 = x_2$. Thus f is bijective. ∎

3. *Similarity.* Let $A \subset \Sigma(X)$ and $B \subset \Sigma(Y)$. We say that A is *similar* to B if there exists bijections $\gamma\colon X \longrightarrow Y$ and $\varphi\colon A \longrightarrow B$, such that for any $a \in A$ and $x \in X$

$$\varphi(a)(\gamma(x)) = \gamma(a(x)). \tag{2}$$

If

$$\varphi(a) = b, \qquad \gamma(x) = y, \tag{3}$$

then $b(y) = \gamma(a(x)) = \gamma a \gamma^{-1}(y)$, whence

$$b = \gamma a \gamma^{-1}. \tag{4}$$

It is clear that (3) and (4) together imply (2), so that *a set A is similar to a set B if and only if there exists a bijection γ: $X \longrightarrow Y$ such that ϕ: $A \longrightarrow B$, $a \mapsto \gamma a \gamma^{-1}$, is also a bijection. If A and B are similar, then $B = \gamma A \gamma^{-1}$.*

4. *Representation by mappings.* Let A be an arbitrary nonempty set. Any mapping

$$g\colon A \rightarrow \Sigma(X) \tag{5}$$

is called a *representation of A by mappings.*

Two representations g: $A \longrightarrow \Sigma(X)$ and h: $A \longrightarrow \Sigma(Y)$ are said to be *equivalent* if there exists a bijection γ: $X \longrightarrow Y$ such that for any $a \in A$ and $x \in X$

$$h(a)(\gamma(x)) = \gamma(g(a)(x)). \tag{6}$$

Setting $y = \gamma(x)$ we infer from (6) that

$$h(a)(y) = \gamma g(a)\gamma^{-1}(y), \qquad h(a) = \gamma g(a)\gamma^{-1}.$$

Thus *the sets* Im g *and* Im h *are similar.*

Below we consider two such representations of a semigroup (in particular, a group), one of which is a homomorphism, the other an antihomomorphism.

5. *The regular representation.* Let A be a semigroup with unit element 1. Each element a of A determines two mappings:

$$\sigma_a\colon A \rightarrow A, \qquad x \mapsto ax, \qquad x \in A,$$

and

$$\tau_a\colon A \rightarrow A, \qquad x \mapsto xa, \qquad x \in A.$$

We introduce two representations:

$$\lambda\colon A \rightarrow \Sigma(A), \qquad a \mapsto \sigma_a,$$

and

$$\rho\colon A \rightarrow \Sigma(A), \qquad a \mapsto \tau_a.$$

We call λ the *left regular representation of A* and ρ its *right regular representation.*

THEOREM 3. (i) *λ and ρ are injective mappings; λ is a homomorphism and ρ an antihomomorphism.*

(ii) Im ρ *is the centralizer of* Im λ *and* Im λ *the centralizer of* Im ρ *in* $\Sigma(A)$.

PROOF. (i) We consider, say, ρ. Let $\rho(a) = \rho(b)$, $a, b \in A$. Then $\tau_a = \tau_b$, $\tau_a(1) = \tau_b(1)$ and so $a = b$. Thus ρ is injective. Next,

$$\rho(ab) = \tau_{ab}, \qquad \tau_{ab}(x) = xab = \tau_a(x)\,b = \tau_b\tau_a(x).$$

Hence

$$\tau_{ab}=\tau_b\tau_a, \qquad \rho(ab)=\rho(b)\rho(a).$$

Similarly one proves that λ is injective and $\lambda(ab) = \lambda(a)\lambda(b)$.

(ii) We set $\operatorname{Im}\lambda = A_\lambda$ and $\operatorname{Im}\rho = A_\rho$. Let $Z(A_\lambda)$ be the centralizer of A_λ in $\Sigma(A)$. For $a, b \in A$

$$\sigma_a\tau_b(x)=axb=\tau_b\sigma_a(x),$$

and so $Z(A_\lambda) \supset A_\rho$. For $\varphi \in Z(A_\lambda)$

$$\varphi\sigma_x=\sigma_x\varphi, \qquad \varphi\sigma_x(1)=\sigma_x\varphi(1),$$

or $\varphi(x) = xc$, where $c = \varphi(1)$. Hence

$$\varphi(x)=\tau_c(x), \qquad \varphi=\tau_c, \qquad \varphi\in A_\rho, \qquad Z(A_\lambda)=A_\rho.$$

Similarly one proves that $Z(A_\rho) = A_\lambda$. ■

COROLLARY 1. *If A is a commutative semigroup with unit element, then* $\operatorname{Im}\lambda = \operatorname{Im}\rho$ *is a maximal commutative subsemigroup of* $\Sigma(A)$.

COROLLARY 2. *If A is an abelian group, then* $\operatorname{Im}\lambda = \operatorname{Im}\rho$ *is a maximal abelian subgroup of* $S(A)$.

THEOREM 4. *The left regular representation of a semigroup A with unit element is equivalent to its right regular representation if and only if A is commutative.*

PROOF. For a commutative semigroup A

$$\sigma_a(x)=ax=xa=\tau_a(x), \qquad \sigma_a=\tau_a.$$

Conversely, let $\gamma\colon A \longrightarrow A$ be a bijection such that $\tau_a = \gamma\sigma_a\gamma^{-1}$ for any $a \in A$. Then $\tau_a\gamma = \gamma\sigma_a$ and $\tau_a\gamma(1) = \gamma\sigma_a(1)$. Setting $\gamma(1) = c$ we get

$$\tau_a(c)=\gamma(a), \qquad \gamma(a)=ca, \qquad \gamma=\sigma_c.$$

Now $\sigma_a = \gamma^{-1}\tau_a\gamma = \gamma^{-1}\gamma\tau_a = \tau_a$. For $b \in A$ we have $\sigma_a(b) = ab$ and $\tau_a(b) = ba$. Thus $ab = ba$ and A is commutative. ■

§2. Transitivity

1. *Orbits.* Let Γ be a permutation group on a set X. We define a binary relation R on X by setting xRy for $x, y \in X$ if and only if there is an element g in Γ such that $g(x) = y$. Since Γ is a group, R is an equivalence relation. Now let

$$X=\bigcup X_\alpha, \qquad \alpha\in I, \tag{1}$$

be the partition of X determined by this relation. Then the sets X_α, $\alpha \in I$, are called the *orbits* of Γ. If Γ has only one orbit; that is, if xRy for any $x, y \in X$, we say that the group is *transitive*. If card $I > 1$, we say that Γ is *intransitive*. If $A \subset X$, $g \in \Gamma$ and $\Gamma(A) = A$, then $\Gamma|A$ will sometimes be treated as a subgroup of $S(A)$ that agrees with Γ on A, and $g|A$ as the element of $S(A)$ that agrees with g on A.

We say that two permutation groups G_1 and G_2 on X are *of the same orbital type* if there exists a bijection $\varphi: X \longrightarrow X$ that maps each orbit of G_1 onto an orbit of G_2.

Let $s, t \in S(X)$ and $x_1 \in X$. If $t(x_1) = x_2$, then $sts^{-1}(s(x_1)) = s(x_2)$. Hence it follows, in particular, that if (1) is the partition of the set X into orbits of a group Γ, then $X = \bigcup s(X_\alpha)$, $\alpha \in I$, is its partition into orbits of $s\Gamma s^{-1}$. Consequently, *conjugate subgroups of* $S(X)$ *have the same orbital type.*

2. *Permutation cycles. Cycle type.* Let $a \in S(X)$, and let X_α be an orbit of the cyclic group (a). Then the restriction $a_\alpha = a|X_\alpha$ is called a *cycle* of the permutation a. The cardinality card X_α is called the *length of the cycle* a_α.

It is clear that *any orbit of* (a) *is the set of all points* $x_k = a^k(x_0)$, $k = 0, \pm 1, \pm 2, \ldots, x_0 \in X$. It follows that *the length of any cycle is either finite or countable. In the latter case* $x_k \neq x_l$ *if* $k \neq l$.

If the group (a) is transitive, the permutation a is called a *complete cycle.*

It is obvious that $S(X)$ *contains complete cycles only if* X *is finite or countable.*

If only one orbit of (a) has cardinality greater than 1, the permutation a is said to be *cyclic.*

We say that two permutations a and b of $S(X)$ are *of the same cycle type* if the cyclic groups (a) and (b) are of the same orbital type.

THEOREM 1. *Two permutations of a set* X *are conjugate in* $S(X)$ *if and only if they have the same cycle type.*

PROOF. Let a and b be permutations in $S(X)$ and let $b = tat^{-1}$, $t \in S(X)$. Then $(b) = t(a)t^{-1}$. As we have mentioned above, conjugate subgroups have the same orbital type; hence a and b have the same cycle type, and we have proved the necessity.

To prove the sufficiency, let a and b be permutations in $S(X)$ of the same cycle type. Then there exists a permutation φ in $S(X)$ such that for any orbit X_α, $\alpha \in I$, of the group (a), the set $\varphi(X_\alpha)$ is an orbit Y_α of the group (b). Obviously card X_α = card Y_α, $X_\alpha = \{x_{k\alpha} | x_{k\alpha} = a^k(x_\alpha)\}$ and $Y_\alpha = \{y_{k\alpha} | y_{k\alpha} = b^k(y_\alpha)\}$. For every $\alpha \in I$ we define a bijection $t_\alpha: X_\alpha \longrightarrow Y_\alpha$, $x_{k\alpha} \longmapsto y_{k\alpha}$. We now define a bijection $t: X \longrightarrow X$ by setting $t|X_\alpha = t_\alpha$ for any $\alpha \in I$. Then $tat^{-1} = b$. Indeed, for any $y_{k\alpha} \in Y_\alpha$ we have

$$tat^{-1}(y_{ka}) = tat^{-1}t(x_{ka}) = ta(x_{ka}) = t(x_{k+1,a})$$
$$= y_{k+1,a} = b(y_{ka}). \blacksquare$$

3. *Multiple transitivity.* Let Γ be a permutation group on a set X and U_k the set of all points $(x_1, \dots, x_k)$ in the cartesian product $X^k = X \times \cdots \times X$ such that $x_i \neq x_j$ for $i \neq j$. In particular, if card $X = n$ is a natural number, then card $U_k = n(n-1)\cdots(n-k+1)$. We define a binary relation R_k on U_k by stipulating that aR_kb for elements $a = (x_1, \dots, x_k)$ and $b = (y_1, \dots, y_k)$ of U_k if and only if there is an element g in Γ such that $g(x_i) = y_i$, $i = 1, \dots, k$. When $k = 1$ the relation R_k is obviously the R defined above. It is also obvious that R_k is an equivalence relation.

We say that Γ is *k-ply transitive* if aR_kb for any two points $a, b \in U_k$.

A group Γ acting on a set X determines an isomorphic group $\mathfrak{G}$ of permutations $\bar{g}$ of the set U_k: each $g \in \Gamma$ is associated with a $\bar{g} \in S(U_k)$ by the formula

$$\bar{g}(a) = (g(x_1), \dots, g(x_k)), \qquad a = (x_1, \dots, x_k) \in U_k.$$

Obviously, Γ *is k-ply transitive if and only if* $\mathfrak{G}$ *is transitive.*

4. *Stabilizers.* Let Γ be a permutation group on a set X and a some point of X. The set Γ_a of all elements of Γ leaving a fixed is obviously a subgroup, which is called the *stabilizer of a in* Γ.

THEOREM 2. *Let X_α be an orbit of Γ, $a \in X_\alpha$, and Γ_a the stabilizer of a. For each point $x \in X_\alpha$, let g_x be a permutation in Γ such that $g_x(a) = x$. Then $\Gamma = \bigcup g_x\Gamma_a$, $x \in X_\alpha$, is the partition of Γ into left cosets of Γ_a.*

PROOF. Let g be an arbitrary element of Γ. Then $g(a) = x \in X_\alpha$. Clearly $g_x^{-1}g \in \Gamma_a$ and $g \in g_x\Gamma_a$. On the other hand the cosets $g_x\Gamma_a$ and $g_y\Gamma_a$ are distinct for different x and y in X_α, because $g_x\Gamma_a(a) = x$ and $g_y\Gamma_a(a) = y$. ■

COROLLARY 1. *The order of a permutation group on a finite set is divisible by the number of elements in any of its orbits. In particular, the order of a transitive permutation group on a finite set is divisible by its degree.*

COROLLARY 2. *The order of a k-ply transitive group of finite degree n is divisible by $n(n-1)\cdots(n-k+1)$.*

THEOREM 3. *Let X_α be an orbit of Γ and $\mathfrak{K}_\alpha$ the family of all stabilizers Γ_a for $a \in X_\alpha$. Then the system of all distinct subgroups in $\mathfrak{K}_\alpha$ is a class of conjugate subgroups of Γ.*

PROOF. Let $a \in X_\alpha$ and $g \in \Gamma$. Consider the group $g\Gamma_a g^{-1}$. Since X_α is an orbit of Γ, it follows that $g(a) = b \in X_\alpha$. Further, $g\Gamma_a g^{-1}(b) = b$, so that $g\Gamma_a g^{-1} \subset \Gamma_b$. Since $g^{-1}(b) = a$, we see that

$$g^{-1}\Gamma_b g \subset \Gamma_a, \quad \Gamma_b \subset g\Gamma_a g^{-1}, \quad \Gamma_b = g\Gamma_a g^{-1}.$$

Hence it follows that together with Γ_a the family $\mathfrak{K}_\alpha$ contains also $g\Gamma_a g^{-1}$, and any two subgroups in $\mathfrak{K}_\alpha$ are conjugate in Γ. ∎

COROLLARY 1. *If Γ is a transitive group and $\mathfrak{K}$ the family of its stabilizers, then the set of all distinct subgroups in this family is a class of conjugate subgroups of Γ.*

COROLLARY 2. *Let Γ be a permutation group whose subgroups are all normal, and let X_α be one of its orbits. Then $\Gamma_a = \Gamma_b$ for any $a, b \in X_\alpha$.*

A permutation group is said to be *regular* if it is transitive and its stabilizers are all trivial.

COROLLARY 3. *A transitive permutation group whose subgroups are all normal is regular.*

COROLLARY 4. *A transitive permutation group on a set X whose subgroups are all normal is not contained in any other permutation group on the same set whose subgroups are all normal. In particular, an abelian transitive subgroup of $S(X)$ is maximal among the abelian subgroups of $S(X)$.*

COROLLARY 5. *The order of a transitive permutation group on a finite set whose subgroups are all normal is equal to its degree.*

§3. Imprimitivity

1. *Systems of imprimitivity.* Let Γ be a transitive permutation group on a set X. Suppose that there exists a partition

$$X = \bigcup X_\alpha, \quad \alpha \in I, \tag{1}$$

of X with the following properties:

(i) card $I > 1$.

(ii) There exists an $\alpha \in I$ such that card $X_\alpha > 1$.

(iii) For any $\alpha \in I$ and $g \in \Gamma$ there exists a $\beta \in I$ such that $g(X_\alpha) \subset X_\beta$.

Then (1) is called a *partition of X into systems of imprimitivity for* Γ, and Γ is said to be *imprimitive.* If no such partition exists, Γ is said to be *primitive.*

LEMMA 1. *If (1) is a partition of a set X into systems of imprimitivity for a group Γ, then all the sets X_α have the same cardinality. For any $\alpha \in I$ and $g \in \Gamma$, there exists a $\beta \in I$ such that*

$$g(X_\alpha) = X_\beta. \tag{2}$$

PROOF. Consider $g(X_\alpha)$. By (iii), $g(X_\alpha) \subset X_\beta$, $\beta \in I$, and so $X_\alpha \subset g^{-1}(X_\beta)$.

If $x \in g(X_\alpha)$, then $g^{-1}(x) \in X_\alpha$. Hence, by (iii), $g^{-1}(X_\beta) \subset X_\alpha$. This proves (2). Since Γ is transitive, it follows that for any $\gamma, \delta \in I$ there is a permutation h in Γ such that $h(X_\gamma) \subset X_\delta$. By (2), $h(X_\gamma) = X_\delta$, and thus the sets X_γ and X_δ are equipotent. ■

COROLLARY 1. *A transitive permutation group of prime degree is primitive.*

THEOREM 1. *A doubly transitive permutation group is primitive.*

PROOF. Let Γ be an imprimitive permutation group on a set X, (1) a partition of X into systems of imprimitivity for Γ, $a, a_1 \in X_\alpha$, $b \in X_\beta$, $a \neq a_1$ and $\alpha \neq \beta$. If Γ were doubly transitive, we could find in it a permutation g such that

$$g(a) = a, \tag{3}$$
$$g(a_1) = b. \tag{4}$$

By (3) it would follow that $g(X_\alpha) = X_\alpha$, and by (4) that $g(X_\alpha) = X_\beta$. This contradiction proves the theorem.

Note that *a partition of a set X into systems of imprimitivity for a group* Γ *determines a group homomorphism.* For let us introduce the mapping

$$\varphi\colon \Gamma \to S(I), \quad g \mapsto \bar{g}, \quad \bar{g}(\alpha) = \beta, \quad \text{if} \quad g(X_\alpha) = X_\beta. \tag{5}$$

Obviously φ is a homomorphism; we call it *the homomorphism determined by the partition* (1). *Its kernel* Ker φ *consists of all permutations* $g_0 \in \Gamma$ *such that* $g_0(X_\alpha) = X_\alpha$ *for any* $\alpha \in I$. Since Γ is transitive, Im φ *is a transitive subgroup of* $S(I)$.

For a fixed $\alpha_1 \in I$ we associate two subgroups with the partition (1):

$$H_{\alpha_1} = \{g \mid g \in \Gamma,\ g(X_{\alpha_1}) = X_{\alpha_1}\}, \qquad H_{\alpha_1} \subset \Gamma,$$
$$F_{\alpha_1} = H_{\alpha_1} \mid X_{\alpha_1}, \qquad F_{\alpha_1} \subset S(X_{\alpha_1}).$$

Let $\{g_\alpha\}$, $\alpha \in I$, *be the set of elements of* Γ *such that* $g_\alpha(X_{\alpha_1}) = X_\alpha$. *Then* $\Gamma = \bigcup g_\alpha H_{\alpha_1}$, $\alpha \in I$, *is the partition of* Γ *into left cosets of* H_{α_1}.

Indeed, for $\alpha \neq \beta$ we have $g_\alpha H_{\alpha_1} \neq g_\beta H_{\alpha_1}$, because $g_\alpha H_{\alpha_1}(X_{\alpha_1}) = X_\alpha$ and $g_\beta H_{\alpha_1}(X_{\alpha_1}) = X_\beta$. Now if $g \in \Gamma$, then $g(X_{\alpha_1}) = X_\alpha$ for some $\alpha \in I$. Furthermore $g_\alpha^{-1} g(X_{\alpha_1}) = X_{\alpha_1}$. Thus $g_\alpha^{-1} g \in H_{\alpha_1}$ and $g \in g_\alpha H_{\alpha_1}$. ■

Note that F_{α_1} *is a transitive subgroup of* $S(X_{\alpha_1})$. This follows from the transitivity of Γ and the definition of F_{α_1}.

2. *A primitivity criterion.*

THEOREM 2. *A transitive group* Γ *is primitive if and only if any stabilizer of it is maximal.*

PROOF. Let Γ be a transitive permutation group on a set X, $a \in X$, Γ_a a stabilizer in Γ, and suppose that Γ_a is not maximal in Γ, so that there exists an intermediate subgroup U: $\Gamma \supset U \supset \Gamma_a$, $\Gamma \neq U \neq \Gamma_a$. We show that this makes Γ imprimitive. To prove this we write down the partition of Γ into left cosets of U:

$$\Gamma = \bigcup g_\alpha U = \bigcup U_\alpha, \quad U_\alpha = g_\alpha U, \quad \alpha \in I, \tag{6}$$

and we set $X_\alpha = U_\alpha(a)$. Since Γ is transitive,

$$X = \bigcup X_\alpha, \quad \alpha \in I. \tag{7}$$

We now show that (7) is a partition; in other words, $X_\alpha \cap X_\beta = \varnothing$ for $\alpha \neq \beta$. For let $u(a) = v(a)$, $u \in U_\alpha$, $v \in U_\beta$. Then $u^{-1}v(a) = a$, $u^{-1}v \in \Gamma_a$ and $v \in u\Gamma_a \subset U_\alpha$. This contradicts the fact that (6) is a partition. Thus (7) is a partition. We now show that (7) is a partition into systems of imprimitivity for Γ. Consider $g(X_\alpha)$, $g \in \Gamma$:

$$g(X_\alpha) = gU_\alpha(a) = gg_\alpha U(a) = g_\beta U(a) = U_\beta(a) = X_\beta;$$

that is,

$$g(X_\alpha) = X_\beta. \tag{8}$$

Since $U \neq \Gamma_a$, it follows that card $X_\alpha > 1$; but $U \neq \Gamma$, so that card $I > 1$. Hence and from (8) it follows that (7) is a partition of X into systems of imprimitivity for Γ.

Now for the converse. Let Γ be an imprimitive group. We prove the existence of an intermediate subgroup U:

$$\Gamma \supset U \supset \Gamma_a, \quad \Gamma \neq U, \quad U \neq \Gamma_a. \tag{9}$$

Let $a \in X_\alpha$, where X_α is one of the systems of imprimitivity for Γ. Let U be the set of all permutations $h \in \Gamma$ for which $h(X_\alpha) = X_\alpha$. Since Γ is transitive, $X_\alpha \neq \{a\}$ and $X_\alpha \neq X$, it follows that U is in between Γ and Γ_a, and (9) is proved. ∎

COROLLARY 1. *A regular permutation group is either imprimitive or of prime order.*

COROLLARY 2. *A transitive permutation group whose subgroups are all normal is primitive, except when the order and degree of the group are one and the same prime number.*

COROLLARY 3. *Any stabilizer in a symmetric group is maximal.*

The first part of the proof of Theorem 2 implies

COROLLARY 4. *Let Γ be a transitive subgroup of $S(X)$, $a \in X$, and Γ_a the stabilizer of a in Γ. If U is a subgroup of Γ such that $\Gamma \supset U \supset \Gamma_a$* and $\Gamma \neq U \neq \Gamma_a$, *then $U(a)$ is a system of imprimitivity for Γ.*

3. *Refinement of a partition into systems of imprimitivity.* Let Γ be an imprimitive permutation group on a set X, and let

$$X=\bigcup X_\alpha, \quad \alpha\in I, \quad X=\bigcup Y_\beta, \quad \beta\in B, \tag{10}$$

be two partitions of X into systems of imprimitivity for Γ. If each set Y_β is a proper subset of some X_α, the second partition is called a *refinement* of the first.

LEMMA 2. *If for some $\beta_1 \in B$ the inclusion*

$$Y_{\beta_1}\subset X_{\alpha_1}, \quad \alpha_1\in I, \tag{11}$$

is strict, then the second partition in (10) *is a refinement of the first.*

PROOF. Let $\beta \in B$. Since Γ is transitive, it contains a permutation g such that $g(Y_{\beta_1}) = Y_\beta$. It follows from (11) that

$$Y_\beta=g(Y_{\beta_1})\subset g(X_{\alpha_1})=X_\alpha, \quad \alpha\in I,$$

is a strict inclusion. ■

A partition of a set X into systems of imprimitivity for Γ that has no refinement is said to be *nonrefinable.*

By means of the group F_{α_1} defined at the beginning of the section we now state a condition for a partition into systems of imprimitivity to be nonrefinable.

LEMMA 3. *The first of the two partitions in* (10) *is nonrefinable if and only if the group F_{α_1} is primitive.*

PROOF. Suppose that the second partition in (10) is a refinement of the first. Then

$$X_{\alpha_1}\supset Y_{\beta_1}, \quad \beta_1\in B, \quad X_{\alpha_1}\neq Y_{\beta_1}. \tag{12}$$

We set $H = \{h|h \in \Gamma, h(Y_{\beta_1}) = Y_{\beta_1}\}$. Since X_{α_1} is a system of imprimitivity for Γ, it follows from (12) that

$$H\subset H_{\alpha_1}=\{g\,|\,g\in\Gamma,\ g(X_{\alpha_1})=X_{\alpha_1}\}.$$

The group F_{α_1} is transitive, and so $H|X_{\alpha_1} \subset F_{\alpha_1}$ is a strict inclusion. Now if $y_1 \in Y_{\beta_1}$ and $(F_{\alpha_1})_{y_1}$ is the stabilizer of y_1 in F_{α_1}, then $(F_{\alpha_1})_{y_1} \subset H|X_{\alpha_1}$ is also a strict inclusion. By the primitivity criterion F_{α_1} is imprimitive.

Conversely, if F_{α_1} is imprimitive, then the first partition in (10) is refinable. For let F_{α_1} be imprimitive and let Y_{β_1} be one of its systems of imprimitivity. We set $V = \{v|v \in F_{\alpha_1}, v(Y_{\beta_1}) = Y_{\beta_1}\}$. Then $F_{\alpha_1} \supset V \supset (F_{\alpha_1})_a$, where $(F_{\alpha_1})_a$ is the stabilizer of $a \in Y_{\beta_1}$ in F_{α_1}, and both inclusions are strict. Now we set

$$U=\{u\,|\,u\in\Gamma,\ u(X_{\alpha_1})=X_{\alpha_1},\ u\,|\,X_{\alpha_1}\in V\}.$$

For the stabilizer Γ_a of a in Γ we then have $\Gamma \supset U \supset \Gamma_a$, both inclusions being strict. Obviously $U(a) = Y_{\beta_1}$. Consequently Y_{β_1} is a system of imprimitivity for Γ. Since $Y_{\beta_1} \subset X_{\alpha_1}$ and the inclusion is strict, it follows by Lemma 2 that the first partition in (10) is refinable. ∎

It is clear that *if X is a finite set, then for any imprimitive subgroup of $S(X)$ there exists a nonrefinable partition of X into systems of imprimitivity.*

4. *Imprimitive groups and wreath products.* Among the imprimitive permutation groups, those obtained by means of wreath products of transitive groups are distinguished by their especially "good behavior". The definition of the wreath product of permutation groups, which is due to Kaloujnine [1], [2], [3], is as follows.

Let $X = A \times B$ be the cartesian product of two sets A and B, card $A > 1$, card $B > 1$; let F be a permutation group on A, G a permutation group on B, and Π the set of all mappings $\pi: B \longrightarrow F$. We identify each pair $h = \langle g, \pi\rangle$, $g \in G$, $\pi \in \Pi$, with the mapping

$$h: X \to X, \quad x = (a, b) \mapsto (f_b(a), g(b)), \quad f_b = \pi(b), \quad a \in A, \quad b \in B. \tag{13}$$

We define an additive operation $+$ on Π by setting for $\pi_1, \pi_2 \in \Pi$ and $b \in B$,

$$(\pi_1 + \pi_2)(b) = f_1 f_2, \qquad f_1 = \pi_1(b), \qquad f_2 = \pi_2(b).$$

Then $\langle \Pi, +\rangle$ *is a group isomorphic to the unrestricted direct product of v copies of F, where $v =$* card B. It is obvious that *for $g \in G$ and $\pi \in \Pi$, the product πg exists and $\pi g \in \Pi$.*

LEMMA 4. *Let $g, g_1 \in G$ and $\pi, \pi_1 \in \Pi$. Then*

$$\langle g, \pi\rangle \langle g_1, \pi_1\rangle = \langle g g_1, \pi g_1 + \pi_1\rangle.$$

PROOF. We set $h = \langle g, \pi\rangle$ and $h_1 = \langle g_1, \pi_1\rangle$. Then by (13)

$$\begin{aligned} hh_1(a, b) &= h(\pi_1(b)(a), g_1(b)) = (\pi g_1(b)\pi_1(b)(a), gg_1(b)) \\ &= ((\pi g_1 + \pi_1)(b)(a), gg_1(b)) = \langle gg_1, \pi g_1 + \pi_1\rangle(a, b). \end{aligned}$$ ∎

From Lemma 4 it is easy to deduce that *the set of all mappings $h = \langle g, \pi\rangle$ is a subgroup of $S(X)$.* For by Lemma 4 this set is a semigroup with unit element $\langle e, 0\rangle$, where e is the unit element of G and 0 the zero of $\langle \Pi, +\rangle$. Again by Lemma 4,

$$\langle g, \pi\rangle^{-1} = \langle g^{-1}, -\pi g^{-1}\rangle.$$

The group of all mappings $h = \langle g, \pi\rangle$ is called the *unrestricted wreath product* of the groups F and G and is denoted by $F \operatorname{Wr} G$.

LEMMA 5. *The group $F \operatorname{Wr} G$ has a normal subgroup N such that*

(i) $(F \operatorname{Wr} G)/N \cong G$, *and*

(ii) *N is the unrestricted direct product of ν copies of F, where* $\nu = \operatorname{card} B$.

PROOF. Consider the mapping γ: $F \operatorname{Wr} G \longrightarrow G$, $\langle g, \pi\rangle \longmapsto g$. By Lemma 4 the mapping γ is an epimorphism; that is, a surjective homomorphism. Obviously, $\operatorname{Ker} \gamma = N = \{\langle e, \pi\rangle\}$. Consequently $(F \operatorname{Wr} G)/N \cong G$. By Lemma 4,

$$\langle e, \pi\rangle\langle e, \pi_1\rangle = \langle e, \pi + \pi_1\rangle.$$

Thus $N \cong \langle \Pi, +\rangle$. ∎

COROLLARY 1. *If* $\operatorname{card} A = \mu$ *and* $\operatorname{card} B = \nu$, *where μ and ν are natural numbers, then $F \operatorname{Wr} G$ is a permutation group of order* $(\operatorname{card} G)(\operatorname{card} F)^\nu$ *and of degree $\mu\nu$.*

By means of (13) it is easy to verify

LEMMA 6. *The group $F \operatorname{Wr} G$ is transitive if and only if both F and G are transitive.*

Let F and G be transitive. For $b \in B$ we set $X_b = A \times b$. Then clearly $X = \bigcup X_b$, $b \in B$, is a partition of X into systems of imprimitivity for $F \operatorname{Wr} G$. In this case the group G describes how $F \operatorname{Wr} G$ permutes the systems of imprimitivity and F describes the action of $F \operatorname{Wr} G$ within each system of imprimitivity.

The next theorem shows that every imprimitive permutation group is contained in the unrestricted wreath product of two groups.

THEOREM 3. *Let Γ be an imprimitive subgroup of $S(Y)$, $Y = \bigcup Y_\alpha$, $\alpha \in I$, a partition of Y into systems of imprimitivity for Γ, and φ: $\Gamma \longrightarrow S(I)$ the homomorphism defined by this partition. For a fixed $\alpha_1 \in I$ let $H = \{h | h \in \Gamma, h(Y_{\alpha_1}) = Y_{\alpha_1}\}$, $F = H | Y_{\alpha_1}$ and $G = \operatorname{Im} \varphi$. Then Γ is a subgroup of $F \operatorname{Wr} G$.*

PROOF. We set $A = Y_{\alpha_1}$. For every $\alpha \in I$ we choose a permutation g_α in Γ such that $g_\alpha(Y_{\alpha_1}) = Y_\alpha$ and $g_{\alpha_1} = e$. Each $y_\alpha \in Y_\alpha$ can be expressed as $y_\alpha = g_\alpha(a)$, $a \in A$. We set $y_\alpha = (a, \alpha)$. Then $Y_\alpha = A \times \alpha$. Thus $Y = A \times I$. As we have seen above (at the beginning of the section), $\{g_\alpha | \alpha \in I\}$ is a left transversal to H in Γ. For $y \in Y$ and $v \in \Gamma$ we can write

$$v(y) = v(a, \alpha) = v g_\alpha(a, \alpha_1) = g_\beta h_\alpha(a, \alpha_1), \tag{14}$$

where $h_\alpha \in H$. We set $f_\alpha = h_\alpha | Y_{\alpha_1}$. Then by (14)

$$\begin{aligned} v(y) &= g_\beta h_\alpha(a, \alpha_1) = g_\beta f_\alpha(a, \alpha_1) = g_\beta(f_\alpha(a), \alpha_1) \\ &= (f_\alpha(a), \beta) = (f_\alpha(a), \varphi(v)(\alpha)), \qquad \varphi(v) \in G. \end{aligned}$$

Consequently $v(a, \alpha) = \langle g, \pi\rangle (a, \alpha)$, where $g = \varphi(v)$ and π: $I \longrightarrow F$, $a \longmapsto f_\alpha$. Thus $v = (g, \pi) \in F \operatorname{Wr} G$. ∎

5. *Normal subgroups of transitive groups.*

THEOREM 4. *Let Γ be a transitive permutation group on a set X and H a nontrivial intransitive normal subgroup of Γ. Then the partition of X into orbits of H is a partition of X into systems of imprimitivity for Γ.*

PROOF. Let $X = \bigcup X_\alpha$, $\alpha \in I$, be the partition of X into orbits of H. Since $H \neq (e)$, there exists an $\alpha \in I$ such that card $X_\alpha > 1$. Since H is intransitive, card $I > 1$. It remains to show that for $g \in \Gamma$

$$g(X_\alpha) \subset X_\beta, \qquad \alpha,\ \beta \in I. \tag{15}$$

If a_1, $b_1 \in g(X_\alpha)$, then $a_1 = g(a)$ and $b_1 = g(b)$, where a, $b \in X_\alpha$; also $b = h(a)$, where $h \in H$. Now $h_1 = ghg^{-1} \in H$, whence $h_1(a_1) = b_1$, and consequently a_1, $b_1 \in X_\beta$. This implies (15). ∎

COROLLARY 1. *Any nontrivial normal subgroup of a primitive permutation group is transitive.*

COROLLARY 2. *A nontrivial abelian normal subgroup H of a primitive permutation group is characteristically simple; that is, H contains no nontrivial characteristic subgroups.*

PROOF. Let F be a nontrivial characteristic subgroup of H. By Corollary 1, F is a transitive abelian group. By Corollary 2.3.4 F is maximal among the abelian subgroups of the symmetric group. This contradicts the strict inclusion $F \subset H$. ∎

COROLLARY 3. *The additive group of a division ring is characteristically simple.*

PROOF. Let T be a division ring. Let Γ be the subgroup of the symmetric group $S(T)$ consisting of all permutations f of the form $f(x) = \alpha x + \beta$, x, α, $\beta \in T$, $\alpha \neq 0$. Obviously Γ is doubly transitive. By Theorem 1, Γ is primitive. The permutations h of the form $h(x) = x + \tau$, x, $\tau \in T$, form an abelian normal subgroup H of Γ. By Corollary 2, H is characteristically simple. Finally, it is clear that H is isomorphic to the additive group of T. ∎

THEOREM 5. *A transitive permutation group Γ with nontrivial center is imprimitive except when the order and degree of Γ are equal to one and the same prime number.*

PROOF. Let Γ be a primitive group, Z its center, $Z \neq (e)$. By Corollary 4.1, the group Z is transitive. For any $g \in \Gamma$, $(g)Z$ is a transitive abelian group. By Corollary 2.3.4, $(g)Z = Z$, $g \in Z$ and $\Gamma = Z$. By Corollary 2.2, the order and degree of Γ are equal to one and the same prime number. ∎

COROLLARY 1. *If X contains at least three elements, then the center of $S(X)$ is trivial.*

Let Γ be a transitive subgroup of $S(X)$, H an intransitive normal subgroup of Γ, and $X = \bigcup X_\alpha$, $\alpha \in I$, the partition of X into orbits of H. We introduce transitive representations r_α of H:

$$r_\alpha\colon\ H \to S(X_\alpha), \qquad h \mapsto h_\alpha = h \mid X_\alpha. \tag{16}$$

We split the system of all orbits X_α of H into classes, putting X_α and X_β into the same class if and only if the representations r_α and r_β are equivalent. The unions of the orbits in each class form a partition of X:

$$X = \bigcup Y_\lambda, \qquad \lambda \in K, \tag{17}$$

where K is the set of all classes, $Y_\lambda = \bigcup X_\alpha$, $\alpha \in I_\lambda$, $I = \bigcup I_\lambda$, $\lambda \in K$, and r_α and r_β are equivalent if and only if α and β belong to the same I_λ. With this notation we have

THEOREM 6. (i) *The groups $H_\alpha = H|X_\alpha$, $\alpha \in I$, are pairwise similar.*

(ii) *If* card $K > 1$, *then* (17) *is a partition into systems of imprimitivity for* Γ.

PROOF. Let $\alpha, \beta \in I$. We prove that there exists a bijection $\gamma\colon X_\alpha \to X_\beta$ such that

$$H_\beta = \gamma H_\alpha \gamma^{-1}. \tag{18}$$

Since Γ is transitive, it contains a permutation g such that $g(X_\alpha) = X_\beta$. Consider the mapping $\gamma\colon X_\alpha \to X_\beta$, $x_\alpha \mapsto g(x) = x_\beta$. Clearly γ is a bijection. We write any point $x_\beta \in X_\beta$ as $x_\beta = \gamma(x_\alpha)$, $x_\alpha \in X_\alpha$. For $h_\alpha \in H_\alpha$ we can write

$$\gamma h_\alpha \gamma^{-1}(x_\beta) = \gamma h_\alpha (x_\alpha) = gh(x_\alpha) = \bar{h}\gamma(x_\alpha) = \bar{h}(x_\beta) = \bar{h}_\beta(x_\beta),$$

where $\bar{h} = ghg^{-1}$, $\bar{h}_\beta = \bar{h}|X_\beta \in H_\beta$. Therefore $\gamma h_\alpha \gamma^{-1} \in H_\beta$, so $\gamma H_\alpha \gamma^{-1} \subset H_\beta$. Interchanging α and β, γ and γ^{-1}, and g and g^{-1}, we find that $\gamma^{-1} H \gamma \subset H_\alpha$. This implies (18), which shows that the groups H_α and H_β are similar, proving (i).

We now prove (ii). Let r_α and r_β be equivalent. Then by the definition of equivalence there exists a bijection $\gamma\colon X_\alpha \to X_\beta$ such that $r_\beta(h) = \gamma r_\alpha(h)\gamma^{-1}$ for any $h \in H$; that is,

$$h_\beta = \gamma h_\alpha \gamma^{-1}, \tag{19}$$

where $h_\alpha = h|X_\alpha$ and $h_\beta = h|X_\beta$. Now let $g \in \Gamma$, $g(X_\alpha) = X_\sigma$ and $g(X_\beta) = X_\tau$. We claim that the representations r_σ and r_τ are equivalent. To prove this we construct a mapping

$$f\colon X_\sigma \to X_\tau, \quad f\colon g(x_\alpha) \mapsto g(\gamma(x_\alpha)), \tag{20}$$

where x_α is an arbitrary point of X_α. Obviously f is bijective. We set $x_\sigma = g(x_\alpha)$, $\bar{h} = g^{-1}hg$ for any $h \in H$, and $\bar{h}_\alpha = \bar{h}|X_\alpha$. Then it follows from (19) and (20) that for any $h \in H$

$$\begin{aligned} h_\tau f(x_\sigma) &= h_\tau f(g(x_\alpha)) = h_\tau g(\gamma(x_\alpha)) = hg(\gamma(x_\alpha)) = g\bar{h}(\gamma(x_\alpha)) \\ &= g\bar{h}_\beta(\gamma(x_\alpha)) = g\gamma\bar{h}_\alpha\gamma^{-1}(\gamma(x_\alpha)) = g\gamma\bar{h}_\alpha(x_\alpha) = g\gamma\bar{h}(x_\alpha) \\ &= f(g(\bar{h}(x_\alpha))) = f(hg(x_\alpha)) = fh(x_\sigma) = fh_\sigma(x_\sigma). \end{aligned}$$

Hence $h_\tau f = fh_\sigma$; that is, $h_\tau = fh_\sigma f^{-1}$ for any $h \in H$. Thus the representations r_σ and r_τ are equivalent. Hence it follows that for any $\lambda \in K$ and $g \in \Gamma$ we have

$$g(Y_\lambda) \subset Y_\mu \tag{21}$$

for some $\mu \in K$. It follows from (21) that (17) is a partition of X into systems of imprimitivity for Γ. ∎

Since Γ is transitive, we can write $g(Y_\lambda) = Y_\mu$ instead of (21). If $Y_\lambda = \bigcup X_\alpha$, $\alpha \in I_\lambda$, then $Y_\mu = \bigcup g(X_\alpha)$, $\alpha \in I_\lambda$. Thus we have

COROLLARY 1. *The cardinality of the set of orbits of H occurring in Y_λ is the same for all $\lambda \in K$. In other words, all the representations r_α have the same multiplicity.*

COROLLARY 2. *If* card $I = m < \infty$, *the number of inequivalent representations r_α divides m.*

THEOREM 7. *A primitive permutation group has at most one nontrivial abelian normal subgroup.*

PROOF. Let H and F be nontrivial abelian normal subgroups of a primitive permutation group. By Corollary 4.1, H and F are transitive groups. Consider the intersection $D = H \cap F$. There are two possibilities: $D = (e)$ and $D \neq (e)$. If $D = (e)$, then HF is an abelian normal subgroup containing H. By Corollary 2.3.4, $HF = H$, and so $F \subset H$, contradicting the assumption $D = (e)$. But if $D \neq (e)$, then D is transitive by Corollary 4.1. By Corollary 2.3.4 we have $D = H$, and $D = F$, and so $H = F$. ∎

§4. Permutation groups having a regular normal subgroup. Primitive solvable groups

1. *Regular groups.* Let F be a regular subgroup of $S(X)$ and 0 some fixed element of X. We define a bijection

$$\lambda\colon X \to F, \quad a \mapsto f_a, \quad f_a(0) = a. \tag{1}$$

By means of this bijection we can define an operation + on the set X, setting

$$a+b=f_a f_b(0) \tag{2}$$

for $a, b \in X$. The set then becomes an additive group $\langle X, +\rangle$ isomorphic to F. Setting $b = x$ in (2), we see that $f_a(x) = a + x$ for any $x \in X$. Hence and from (1) it follows that λ is the left regular representation of the group $\langle X, +\rangle$. We have thus proved

LEMMA 1. *Every regular permutation group is an image of its left regular representation.*

LEMMA 2. (i) *Isomorphic regular permutation groups on a set X are conjugate in $S(X)$.*

(ii) *Let + and $+_1$ be group operations on a set X, γ: $\langle X, +\rangle \longrightarrow \langle X, +_1\rangle$ an isomorphism, λ: $\langle X, +\rangle \longrightarrow S(X)$ and λ_1: $\langle X, +_1\rangle \longrightarrow S(X)$ the left regular representations of the groups $\langle X, +\rangle$ and $\langle X, +_1\rangle$, $G = \operatorname{Im} \lambda$ and $\overline{G} = \operatorname{Im} \lambda_1$. Then $\overline{G} = \gamma G \gamma^{-1}$.*

PROOF. By Lemma 1, (i) follows from (ii). Let us prove (ii). We set $\lambda(a) = g_a$ and $\lambda_1(a) = \bar{g}_a$. For $\bar{g}_a \in \overline{G}$ and $x \in X$ we have

$$\bar{g}_a \gamma(x) = a +_1 \gamma(x) = \gamma(\gamma^{-1}(a) + x) = \gamma(g_{\gamma^{-1}(a)}(x)).$$

Consequently

$$\bar{g}_a \gamma = \gamma g_{\gamma^{-1}(a)}, \quad \bar{g}_a = \gamma g_{\gamma^{-1}(a)} \gamma^{-1}, \quad \overline{G} = \gamma G \gamma^{-1}. \ ■$$

An image of the regular representation of any group is obviously a regular permutation group. Therefore, *if A is an abstract group such that* card A = card X, *then by Lemma* 2 *$S(X)$ contains one and, to within conjugacy, only one regular subgroup F isomorphic to A. By means of* (2) *the group F in turn determines a group $\langle X, +\rangle$ isomorphic to A.*

2. *Theorem on the centralizer of a transitive group.*

THEOREM 1. *Let B be a transitive subgroup of $S(X)$ and C the centralizer of B in $S(X)$.*

(i) *If C is transitive, then $B \cong C$ and one can define on X a group operation + such that B is an image of the left regular representation of the group $\langle X, +\rangle$ and C an image of its right regular representation.*

(ii) *If C is intransitive and*

$$X = \bigcup X_\alpha, \quad \alpha \in I, \tag{3}$$

is the partition of X into orbits of C, then for any $\alpha \in I$ the representation r_α: $C \longrightarrow S(X_\alpha)$, $f \longmapsto f|X_\alpha$, is faithful, and for any $\alpha, \beta \in I$ the representations r_α and r_β are equivalent.

PROOF. Let C be transitive, $a \in X$ and B_a its stabilizer in B. For any $x \in X$, the group C contains a permutation f such that $f(a) = x$. If $g \in B_a$, then $gf = fg$, $gf(a) = fg(a)$, $g(x) = f(a) = x$ and $B_a = (e)$. Thus B is a regular group. By Lemma 1 we can define on X a group operation $+$ such that B is an image of the left regular representation of the group $\langle X, +\rangle$. Then, by Theorem 1.3, C is an image of the right regular representation of $\langle X, +\rangle$. Hence it follows that B and C are anti-isomorphic. This proves (i).

Now let C be intransitive. Then C is an intransitive normal subgroup of the group $\Gamma = BC$. If $C = (e)$, then (ii) is trivial. Let $C \neq (e)$. Then, by Theorem 3.4, (3) is a partition of X into systems of imprimitivity for Γ. Let ϵ be a fixed index in I. Since B is transitive, we can choose for any $\alpha \in I$ a permutation $g_\alpha \in B$ such that $g_\alpha(X_\epsilon) = X_\alpha$, $g_\epsilon = e$. Now let $f \in C$. Then $f(X_\alpha) = X_\alpha$. We set $f_\alpha = f|X_\alpha = r_\alpha(f)$ and $\gamma_\alpha = g_\alpha|X_\epsilon$. Obviously $\gamma_\alpha\colon X_\epsilon \longrightarrow X_\alpha$ is a bijection. For $x_\alpha \in X_\alpha$ we have $x_\alpha = g_\alpha(x_\epsilon) = \gamma_\alpha(x_\epsilon)$. Now,

$$r_\alpha(f)\,\gamma_\alpha(x_\epsilon) = f_\alpha g_\alpha(x_\epsilon) = fg_\alpha(x_\epsilon) = g_\alpha f(x_\epsilon) = \gamma_\alpha(r_\epsilon(f)(x_\epsilon)).$$

Hence $r_\alpha(f)\gamma_\alpha(x_\epsilon) = \gamma_\alpha(r_\epsilon(f)(x_\epsilon))$ for any $x_\epsilon \in X_\epsilon$. Thus the representations r_α and r_ϵ are equivalent.

It follows that the representations r_α, $\alpha \in I$, are pairwise equivalent. In particular, if $r_\alpha(f) = f_\alpha$ coincides for some $\alpha \in I$ and $f \in C$ with the identity permutation $e_\alpha \in S(X_\alpha)$, then also $r_\beta(f_\beta) = e_\beta$, so that $f = e$. Thus each representation r_α is faithful. ∎

COROLLARY 1. *The centralizer C of a primitive subgroup B of $S(X)$ in $S(X)$ is trivial, except when* card $X = p$ *(a prime) and B is a group of order p.*

PROOF. Let $C \neq (e)$. C is a normal subgroup of the primitive group BC, and so C is transitive by Corollary 3.4.1. By Theorem 1 B is a regular group. By Corollary 3.2.1, B is a group of prime order and prime degree p. ∎

COROLLARY 2. *Let C be the centralizer in $S(X)$ of a transitive subgroup B of $S(X)$ and H a subgroup of C.*

(i) *If H is transitive, then $H = C$.*

(ii) *If H is intransitive and $X = \bigcup X_\alpha$, $\alpha \in I$, is the partition of X into orbits of H, then the representations $r_\alpha\colon H \longrightarrow S(X_\alpha)$, $h \longmapsto h|X_\alpha$, $\alpha \in I$, are pairwise equivalent and faithful.*

PROOF. (i) follows directly from Theorem 1; (ii) is proved in the same way as in Theorem 1. ∎

COROLLARY 3. *If h is a permutation in $S(X)$ that commutes with every permutation in a transitive group B, then its cycles are of the same length. In particular, if* card $X < \infty$, *then the order of h divides* card X.

LEMMA 3. *Let Γ be a transitive subgroup of $S(X)$, expressible in the form of a product $\Gamma = HF$, where H and F are subgroups, H is intransitive and H and F commute elementwise. Let*

$$X = \bigcup X_\alpha, \quad \alpha \in I, \tag{4}$$

be the partition of X into orbits of H, and let $r_\alpha\colon H \longrightarrow S(X_\alpha)$, $r_\alpha(h) = h|X_\alpha$. Then the representations r_α are pairwise equivalent.

PROOF. Since H is normal in Γ, (4) is a partition of X into systems of imprimitivity for Γ. Let α and β be arbitrary indices in I. Then F contains a permutation f_β such that $f_\beta(X_\alpha) = X_\beta$. We set $\varphi = f_\beta|X_\alpha$. Obviously $\varphi\colon X_\alpha \longrightarrow X_\beta$ is a bijection. For $x_\alpha \in X_\alpha$ and $h \in H$ we can write

$$\begin{aligned} r_\beta(h)(\varphi(x_\alpha)) &= r_\beta(h)(f_\beta(x_\alpha)) = h_\beta f_\beta(x_\alpha) = hf_\beta(x_\alpha) \\ &= f_\beta h(x_\alpha) = f_\beta h_\alpha(x_\alpha) = \varphi(r_\alpha(h))(x_\alpha). \end{aligned}$$

Hence $r_\beta(h)\varphi = \varphi r_\alpha(h)$, which proves the equivalence of r_α and r_β. ∎

3. *Groups having a regular normal subgroup.* Let F be a regular subgroup of $S(X)$, N the normalizer of F in $S(X)$, 0 the zero element of the group $\langle X, +\rangle$ (see (1), (2)) and N_0 the stabilizer of 0 in N. Then we have

THEOREM 2 (THEOREM ON THE HOLOMORPH). (i) $N = FN_0$.

(ii) $N_0 = \mathrm{Aut}\langle X, +\rangle$ *is the group of all automorphisms of* $\langle X, +\rangle$.

(iii) *For any automorphism ψ of F, the group N_0 contains an element g_0 such that $\psi(f) = g_0 f g_0^{-1}$ for every $f \in F$.*

PROOF. (i) follows from Theorem 2.2. We prove (ii). Let $g_0 \in N_0$. Then for $a, b \in \langle X, +\rangle$

$$\begin{aligned} g_0(a) &= g_0 f_a(0) = g_0 f_a g_0^{-1}(0) = f_c(0) = c, \\ g_0(b) &= g_0 f_b(0) = f_d(0) = d, \end{aligned}$$

where $f_c = g_0 f_a g_0^{-1}$ and $f_d = g_0 f_b g_0^{-1}$. Further,

$$\begin{aligned} g_0(a+b) &= g_0(f_a f_b(0)) = g_0 f_a f_b g_0^{-1}(0) \\ &= f_c f_d(0) = c + d = g_0(a) + g_0(b). \end{aligned}$$

Consequently, $g_0 \in \mathrm{Aut}\langle X, +\rangle$, and so $N_0 \subset \mathrm{Aut}\langle X, +\rangle$. Now let $\varphi \in \mathrm{Aut}\langle X, +\rangle$. Then

$$\varphi(0) = 0, \qquad \varphi(a) = \varphi f_a \varphi^{-1}(0) = b = f_b(0),$$

where $a, b \in X$ and $f_a, f_b \in F$. We claim that for any $x \in X$

$$\varphi f_a \varphi^{-1}(x) = f_b(x). \tag{5}$$

We set $x = \varphi(y)$, $y \in X$. Then

$$\varphi f_a \varphi^{-1}(x) = \varphi f_a(y) = \varphi f_a f_y(0)$$
$$= \varphi(a+y) = \varphi(a) + \varphi(y) = b + x = f_b(x).$$

This proves (5). Therefore $\varphi f_a \varphi^{-1} = f_b \in F$ for any φ in $\mathrm{Aut}\langle X, +\rangle$; that is, $\varphi \in N$. Since $\varphi(0) = (0)$, it follows that $\varphi \in N_0$, and so $\mathrm{Aut}\langle X, +\rangle \subset N_0$. Hence and by the preceding inclusion, $N_0 = \mathrm{Aut}\langle X, +\rangle$.

It remains to prove (iii). Let ψ be an automorphism of F and let x range over the elements of X. If $\psi(f_x) = f_{\bar{x}}$, then the mapping $\varphi: x \longmapsto \bar{x}$ is an automorphism of $\langle X, +\rangle$, because (1) is an isomorphism of the groups $\langle X, +\rangle$ and F. By (ii) we have $\varphi \in N_0$. Consequently

$$\psi(f_x) = f_{\varphi(x)}, \tag{6}$$

where $\varphi \in N_0$. Further

$$\varphi f_x \varphi^{-1}(0) = \varphi(x) = f_{\varphi(x)}(0) \tag{7}$$

Since $f_{\varphi(x)}$ and $\varphi f_x \varphi^{-1}$ are elements of the regular group F, it follows from (7) that $\varphi f_x \varphi^{-1} = f_{\varphi(x)}$. Hence, by (6), $\psi(f_x) = \varphi f_x \varphi^{-1}$, $\varphi \in N_0$. ∎

Theorem 2 implies

THEOREM 3. *Let Γ be a subgroup of $S(X)$ containing a regular normal subgroup F, and let $\langle X, +\rangle$ be the group defined by* (2).

(i) $\Gamma = \Gamma_0 F$, $F \cap \Gamma_0 = (e)$, *where Γ_0 is the stabilizer of the zero element* 0 *of $\langle X, +\rangle$.*

(ii) Γ_0 *is a subgroup of* $\mathrm{Aut}\langle X, +\rangle$.

We need the following definition. Let A be an arbitrary group, and H a subgroup of Aut A. A subgroup B of A is said to be *H-admissible* if $h(B) = B$ for every h in H. A subgroup H of Aut A is said to be *irreducible* if the only H-admissible subgroups of A are A and (e).

THEOREM 4. *A subgroup Γ of $S(X)$ containing a regular normal subgroup F is primitive if and only if Γ_0 (see Theorem* 3) *is an irreducible subgroup of* $\mathrm{Aut}\langle X, +\rangle$. *In particular, the normalizer N of a regular subgroup F of $S(X)$ is primitive if and only if F is characteristically simple.*

PROOF. Let $\Gamma = F\Gamma_0$ be imprimitive. Then by our primitivity criterion Γ contains an intermediate subgroup U, $\Gamma \supset U \supset \Gamma_0$, $\Gamma \neq U \neq \Gamma_0$. Let $F_1 = U \cap F$. Let $F_1 \neq (e)$, $F_1 \neq F$, and $gUg^{-1} = U$ for $g \in \Gamma_0$. Thus, for any $g \in \Gamma_0$,

$$gF_1 g^{-1} = F_1. \tag{8}$$

We now set $X_1 = \lambda^{-1}(F_1)$ (see (1)); we claim that X_1 is a Γ_0-admissible subgroup of $\langle X, +\rangle$. For $x_1 \in X_1$ and $g \in \Gamma_0$ we can write

$$g(x_1) = gf_{x_1}(0) = gf_{x_1}g^{-1}(0) = f_{x_2}(0) = x_2,$$

where f_{x_2} is an element of F_1 by (8). Consequently $x_2 \in X_1$ and X_1 is indeed a Γ_0-admissible subgroup of $\langle X, +\rangle$.

Conversely, suppose that $\langle X, +\rangle$ contains a nontrivial Γ_0-admissible subgroup X_1. We set $F_1 = \lambda(X_1)$. It is clear that F_1 lies between F and (e). For $y \in X_1$ and $g \in \Gamma_0$ we have

$$gf_y g^{-1}(0) = gf_y(0) = g(y) = y_1 \in X_1.$$

Consequently $gf_y g^{-1} = f_{y_1} \in \lambda(X_1) = F_1$; that is, $gF_1g^{-1} = F_1$. Thus $F_1\Gamma_0$ is a subgroup of Γ. Obviously $F_1\Gamma_0$ is in between Γ and Γ_0. By the primitivity criterion 3.2, Γ is imprimitive. ∎

The theorem just proved can be stated without appeal to the group $\langle X, +\rangle$:

Let Γ be a permutation group having a regular normal subgroup F, Γ_1 some stabilizer in Γ, and H the group of all automorphisms h of F of the form $h(f) = g_1 f g_1^{-1}$, $g_1 \in \Gamma_1$, $f \in F$. Then Γ is primitive if and only if H is an irreducible subgroup of Aut F.

4. *Primitive groups having an abelian normal subgroup.*

THEOREM 5. *Let F be a nontrivial abelian subgroup of $S(X)$, N its normalizer in $S(X)$, A and B primitive subgroups of N containing F, and A_0, B_0, N_0 the stabilizers of a point $0 \in X$ in A, B, N, respectively. Then A and B are conjugate in $S(X)$ if and only if A_0 and B_0 are conjugate in N_0.*

PROOF. Since A is primitive, it follows from Corollary 3.4.1 that F is transitive. Since F is commutative, it is also regular (see Corollary 2.3.3). Thus, by Theorems 2 and 3,

$$N = FN_0, \qquad A = FA_0, \qquad B = FB_0. \tag{9}$$

If N_0 contains an element v such that $vA_0v^{-1} = B_0$, then, using (9), we obtain

$$vAv^{-1} = vFA_0v^{-1} = vFv^{-1}vA_0v^{-1} = FB_0 = B.$$

Conversely, suppose that there is a permutation r in $S(X)$ such that $rAr^{-1} = B$. Then rFr^{-1} is an abelian normal subgroup of B. By Theorem 3.5, $rFr^{-1} = F$, and so $r \in N$. By (9), $r = vf$, where $v \in N_0$ and $f \in F$. Then $rAr^{-1} = vAv^{-1} = FB_0$. Obviously the stabilizer of 0 in vAv^{-1} is precisely the group vA_0v^{-1}. Thus $B_0 = vA_0v^{-1}$. ∎

THEOREM 6. *Let Γ and G be two primitive subgroups of $S(X)$, each having a nontrivial abelian normal subgroup. If Γ and G are isomorphic, they are conjugate in $S(X)$.*

PROOF. Let $\Phi \neq (e)$ be an abelian normal subgroup of Γ, and $F \neq (e)$ an abelian normal subgroup of G. Since Γ and G are primitive groups, while Φ and F are abelian, by Theorem 3.4 Φ and F are regular subgroups of $S(X)$. By Theorem 3, $\Gamma = \Gamma_0\Phi$ and $G = G_0F$, where Γ_0 and G_0 are stabilizers in Γ and G. Now let $\psi\colon \Gamma \longrightarrow G$ be an isomorphism. Then $\psi(\Phi)$ is an abelian normal subgroup of G, which by Theorem 3.5 must coincide with F. Thus $\Phi \cong F$. By Lemma 2, $S(X)$ contains a permutation t such that $t\Phi t^{-1} = F$. We set $A = t\Gamma t^{-1}$; A is a primitive group isomorphic to G. By Theorem 3, $A = A_0F$, where A_0 is a stabilizer in A. If $\varphi\colon A \longrightarrow G$ is an isomorphism, then by Theorem 3.5 we have $\varphi(F) = F$. Consequently $\varphi|F$ is an automorphism of F. Now let N be the normalizer of F in $S(X)$. Then $N = N_0F$, where N_0 is a stabilizer in N. By Theorem 2, N_0 contains a permutation d such that $\varphi(f) = dfd^{-1}$ for any $f \in F$. Now consider the isomorphism $\gamma\colon A \longrightarrow \gamma(A)$ defined by $x \longmapsto d^{-1}\varphi(x)d$. By the choice of d, the restriction $\gamma|F$ is the unit element of Aut F. For any $f \in F$ and $a \in A$,

$$afa^{-1} \in F. \tag{10}$$

Hence

$$\gamma(a)f\gamma(a)^{-1} = afa^{-1}. \tag{11}$$

Consequently $a^{-1}\gamma(a)f = fa^{-1}\gamma(a)$ for any $f \in F$. By Corollary 2.3.4, $a^{-1}\gamma(a) \in F$. Now, $\gamma(a) = af_1$, $f_1 \in F$, and $\gamma(A) = \gamma(A_0F) = A_0F = A$. On the other hand, $\gamma(A) = d^{-1}\varphi(A)d = d^{-1}Gd$. Thus $t\Gamma t^{-1} = A = d^{-1}Gd$. ∎

A nontrivial abelian group F that is a normal subgroup of a primitive group is characteristically simple. The characteristically simple abelian groups are easily determined:

LEMMA 4. *The only characteristically simple abelian groups are the additive groups of linear spaces over prime fields.*(2)

PROOF. If A is a characteristically simple (additive) abelian group, clearly all its nonzero elements are of the same prime order p, unless A is torsion free. In the former case, A is the additive group of a linear space over the field $GF(p)$. Consider the second case. The equation $mx = na$, where m and n are integers, $m > 0$, obviously has a unique solution in A, say $(n/m)a = x$. Direct computations show that if multiplication of elements of A by rationals is defined in this way, A becomes a linear space over the field of rationals.

On the other hand, the *general linear group* $GL(V)$ of all invertible linear transformations of a space V over any field is transitive (see Chapter II, §7), so that the additive group of a linear space is characteristically simple. ∎

---

(2) It is obvious that the prime field may be replaced by an arbitrary field, but the present formulation is convenient for our purposes.

It is also known that an abelian group is characteristically simple if and only if it is isomorphic to the additive group of some field.

LEMMA 5. *Let V be a linear space over a prime field Δ, and A the additive group of V. Then* Aut A *is the general linear group* $GL(V)$.

PROOF. $GL(V)$ consists of all bijections $g: V \longrightarrow V$ such that $g(x+y) = g(x) + g(y)$ and $g(\lambda x) = \lambda g(x)$ for any $x, y \in V$ and $\lambda \in \Delta$. Thus $GL(V) \subset$ Aut A.

The reverse inclusion Aut $A \subset GL(V)$ also holds. For let $x \in A$, $\lambda \in \Delta$ and $\varphi \in$ Aut A. If Δ is a prime field of prime characteristic, then $\lambda x = lx$, where l is an integer depending only on λ. Consequently $\varphi(\lambda x) = \varphi(lx) = l\varphi(x) = \lambda\varphi(x)$, $\varphi \in GL(V)$. Now let Δ be a prime field of characteristic zero; that is, the field of rationals, and $\lambda \neq 0$. Then $\lambda = k/l$, where k and l are integers. We set $\lambda x = y$. It is clear that $kx = ly$, $\varphi(kx) = \varphi(ly)$, $k\varphi(x) = l\varphi(y)$, $\varphi(y) = \lambda\varphi(x)$, $\varphi(\lambda x) = \lambda\varphi(x)$, and so $\varphi \in GL(V)$. ∎

Obviously a linear space $V \neq (0)$ over a field Δ has no subgroups that are not subspaces if and only if Δ is a prime field of prime characteristic. Consequently, *if A is the additive group of a linear space V over Δ, then for $\Delta = GF(p)$ the class of all irreducible subgroups of* Aut A *coincides with the class of all irreducible subgroups of $GL(V)$*,(3) *and for $\Delta = Q$ (the field of rationals) the class of irreducible subgroups of* Aut A *is a proper subclass of the class of irreducible subgroups of $GL(V)$.*

THEOREM 7. *Let Γ be a primitive subgroup of $S(X)$ having an abelian normal subgroup $F \neq (e)$. Then the following assertions are true:*

(i) *F is isomorphic to the additive group of some linear space V over a prime field Δ and consists of all parallel translations f_a of V,*

$$f_a(x) = x + a, \quad a,\ x \in V = X, \quad f_a \in F.$$

(ii) *$\Gamma = \Gamma_0 F$ and $\Gamma_0 \cap F = (e)$, where 0 is the zero of the additive group $\langle X, +\rangle$ of V and Γ_0 the stabilizer of 0 in Γ.*

(iii) *Γ_0 is a subgroup of $GL(V)$ and is irreducible as a subgroup of* Aut$\langle X, +\rangle$.(4)

PROOF. Since Γ is primitive, F is transitive. Since F is commutative, it is also regular. By Theorem 4, F is characteristically simple. Together with Lemma 4, this implies (i). Now (ii) follows from Theorem 3 and (iii) from Theorem 3, Theorem 4, and Lemma 5. ∎

(3) Concerning irreducible linear groups, see §14.

(4) As mentioned above, when $\Delta = Q$ the irreducibility of Γ_0 as a subgroup of $GL(V)$ is not enough.

5. *Primitive solvable groups.* We now apply the results obtained so far to solvable primitive permutation groups. Let Γ be a solvable primitive subgroup of $S(X)$. Then, being solvable, Γ has a nontrivial abelian normal subgroup F (see Kuroš [1], p. 361 (English vol. II, p. 179)). Consequently Theorem 7 is applicable to Γ, so that

$$\Gamma = \Gamma_0 F, \tag{12}$$

where F is the group of parallel translations of some linear space V over a prime field Δ and Γ_0 the stabilizer of the zero of the additive group $\langle X, +\rangle$ of V in Γ. By Theorem 7, Γ_0 is a subgroup of $GL(V)$ and is irreducible as a subgroup of $\mathrm{Aut}\langle X, +\rangle$. We now present a method of specifying all primitive solvable subgroups of $S(X)$.

Let $\mathfrak{A} = \{A_\alpha\}$, $\alpha \in I$, denote the class of all characteristically simple abelian groups A_α such that $\mathrm{card}\, A_\alpha = \mathrm{card}\, X$. For each group $A_\alpha \in \mathfrak{A}$ we construct on X the additive group $\langle X, +_\alpha\rangle \cong A_\alpha$ and the regular representation $\lambda_\alpha \colon \langle X, +_\alpha\rangle \longrightarrow S(X)$. We set $F_\alpha = \mathrm{Im}\, \lambda_\alpha$. Now let $\mathfrak{G}_\alpha$ be the class of irreducible solvable subgroups of $\mathrm{Aut}\langle X, +_\alpha\rangle$ such that every irreducible solvable subgroup of $\mathrm{Aut}\langle X, +_\alpha\rangle$ is conjugate in $\mathrm{Aut}\langle X, +_\alpha\rangle$ to one and only one subgroup in $\mathfrak{G}_\alpha$. Since, by Theorem 2, $\mathrm{Aut}\langle X, +_\alpha\rangle$ is contained in the normalizer of F_α in $S(X)$, it follows that for $\Gamma_{0\alpha} \in \mathfrak{G}_\alpha$

$$\Gamma = \Gamma_{0\alpha} F_\alpha \tag{13}$$

is a group. Let $\mathfrak{M}_\alpha$ be the class of all groups of the form (13). We set

$$\mathfrak{M} = \bigcup \mathfrak{M}_\alpha, \qquad \alpha \in I. \tag{14}$$

THEOREM 8. (i) *Every group in $\mathfrak{M}$ is a primitive solvable subgroup of $S(X)$.*

(ii) *Any primitive solvable subgroup of $S(X)$ is conjugate in $S(X)$ to exactly one group in $\mathfrak{M}$.*

PROOF. Let $\Gamma \in \mathfrak{M}$. Then by (13) we have $\Gamma = \Gamma_{0\alpha} F_\alpha$ and $\Gamma_{0\alpha} \in \mathfrak{G}_\alpha$. Now $\Gamma_{0\alpha}$ is a solvable subgroup of Γ and F_α an abelian normal subgroup; therefore Γ is solvable. By Theorem 4, Γ is primitive, because $\Gamma_{0\alpha}$ is an irreducible subgroup of $\mathrm{Aut}\langle X, +_\alpha\rangle$. This proves (i).

We now consider (ii). We first show that no two distinct groups B and C in $\mathfrak{M}$ can be conjugate in $S(X)$. By (13) and (14), $B = H_\alpha F_\alpha$ and $C = G_\beta F_\beta$, where $H_\alpha \in \mathfrak{G}_\alpha$, $G_\beta \in \mathfrak{G}_\beta$ and $F_\alpha \not\cong F_\beta$ for $\alpha \neq \beta$. By Theorem 3.5, if B and C are conjugate in $S(X)$, so are F_α and F_β. Thus the groups B and C cannot be conjugate in $S(X)$ for $\alpha \neq \beta$. Now let $\alpha = \beta$. By the construction of $\mathfrak{G}_\alpha$, the groups H_α and G_α are not conjugate in $\mathrm{Aut}\langle X, +_\alpha\rangle$. Hence, by Theorem 5, B and C are not conjugate in $S(X)$.

It now remains to show that every primitive solvable subgroup Γ of $S(X)$ is conjugate in $S(X)$ to some group in $\mathfrak{M}$. By Theorem 7, $\Gamma = \Gamma_0 F$, where F is a regular abelian normal subgroup of $S(X)$ and Γ_0 an irreducible solvable subgroup of $\operatorname{Aut}\langle X, +\rangle$, where $\langle X, +\rangle$ is the group determined by F by means of (2). By construction, the class $\mathfrak{A}$ contains a group $A_\alpha \cong F \cong \langle X, +\rangle$. Consequently $\langle X, +\rangle \cong \langle X, +_\alpha\rangle$ for some $\alpha \in I$. By Lemma 2, the group $S(X)$ contains a permutation u such that $u\colon \langle X, +\rangle \longrightarrow \langle X, +_\alpha\rangle$ is an isomorphism, and $F_\alpha = uFu^{-1}$. Now if $N_0 = \operatorname{Aut}\langle X, +\rangle$, then $uN_0u^{-1} = \operatorname{Aut}\langle X, +_\alpha\rangle$. Then

$$u\Gamma u^{-1} = u\Gamma_0 u^{-1} F_\alpha, \qquad u\Gamma_0 u^{-1} \subset \operatorname{Aut}\langle X, +_\alpha\rangle.$$

$u\Gamma_0 u^{-1}$ is a solvable irreducible subgroup of $\operatorname{Aut}\langle X, +_\alpha\rangle$. Thus $u\Gamma_0 u^{-1}$ is conjugate in $\operatorname{Aut}\langle X, +_\alpha\rangle$ to some subgroup in $\mathfrak{G}_\alpha$. Then, by Theorem 5, $u\Gamma u^{-1}$ is conjugate in $S(X)$ to some group in $\mathfrak{M}_\alpha$. ∎

In connection with this theorem we note that the subclass $\mathfrak{M}_\alpha$ is not empty for any $\alpha \in I$ (see (14)). For the characteristically simple group A_α is isomorphic to the additive group of some field Ω (see the remark following Lemma 4). Consequently $\mathfrak{M}_\alpha$ contains the group Γ of all mappings of the form

$$g_{\lambda,\mu}\colon\ \Omega \to \Omega, \quad g_{\lambda,\mu}(x) = \lambda x + \mu, \quad x, \lambda, \mu \in \Omega, \quad \lambda \neq 0. \tag{15}$$

Finally we add that in the case of solvable groups Theorem 6 takes the following form:

THEOREM 6a. *If two solvable primitive subgroups of $S(X)$ are isomorphic, they are also conjugate in $S(X)$.*

6. *Primitive solvable groups of finite degree.*

THEOREM 9. *Let X be a finite set. Then $S(X)$ contains primitive solvable subgroups if and only if* card X *is a prime power.*

PROOF. Let card $X = m$, and let Γ be a primitive solvable subgroup of S_m. By Theorems 7 and 8, Γ has an abelian normal subgroup F of order m, which is the additive group of a linear space over a prime field Σ. Thus $\Sigma = GF(p)$, where p is a prime, and $m = p^n$.

Conversely, let card $X = p^n$. For X we choose the field $\Delta = GF(p^n)$. Then the group of all permutations of the form (15) is a primitive solvable subgroup of $S(X)$. ∎

LEMMA 6. *A primitive solvable group Γ is maximal among the solvable subgroups of $S(X)$ if and only if Γ_0 (see (12)) is maximal among the solvable subgroups of $GL(V)$. If Γ and G are primitive solvable subgroups of $S(X)$ and $G \supset \Gamma$, then an abelian normal subgroup of Γ is a normal subgroup of G.*

PROOF. We first prove the second assertion. We write the groups Γ and G as $\Gamma = F\Gamma_0$ and $G = HG_0$, where F and H are regular abelian normal subgroups, and Γ_0 and G_0 the stabilizers of 0 in Γ and G, respectively. Clearly $\Gamma \cap H$ is a normal subgroup of Γ. Hence, by Theorem 3.7, either $\Gamma \cap H = F$ or $\Gamma \cap H = 1$. In the first case $H \supset F$, and so, by the regularity of H and F, we have $H = F$. We now show that the second case is impossible. Suppose that $\Gamma \cap H = 1$. We construct the group $U = H\Gamma$ and write it as $U = HU_0$, where U_0 is the stabilizer of 0 in U. Since $\Gamma \cap H = 1$, it follows that U_0 is isomorphic to Γ. Consequently U_0 has an abelian normal subgroup of order $p^n = \operatorname{card} X$. Since U is primitive, U_0 is an irreducible subgroup of $GL(n, p) \equiv GL(V)$, where V is an n-dimensional linear space over $GF(p)$. By Clifford's first theorem (Theorem 16.1), U_0 cannot have a normal subgroup of order p^n. Thus the second case is indeed impossible.

Now the proof of the first assertion is easy. If Γ is a maximal primitive solvable subgroup of $S(X)$, then obviously Γ_0 is a maximal solvable subgroup of $GL(V)$. Conversely, let $\Gamma = F\Gamma_0$, where Γ_0 is a maximal solvable subgroup of $GL(V)$, and $F \subset G$, where G is a solvable subgroup of $S(X)$. Then, by the second assertion of our theorem, $G = FG_0$ and F is a normal subgroup of G. Consequently $GL(V) \supset G_0 \supset \Gamma_0$. The maximality of Γ_0 now implies that $G_0 = \Gamma_0$ and $G = \Gamma$. ∎

If Γ is a primitive solvable subgroup of $S(X)$, $\operatorname{card} X = p^n$, and $F \neq (e)$ is an abelian normal subgroup of Γ, then, as remarked above, the group $\langle X, +\rangle$ may be regarded as the additive group of the space Σ^n over the field $\Sigma = GF(p)$. The group F then becomes the group of all parallel translations $f_a = a + x$, $a, x \in \Sigma^n$, of Σ^n. By Theorem 8, $\Gamma = \Gamma_0 F$, where Γ_0 is a solvable irreducible subgroup of $GL(\Sigma^n)$. Hence and from Theorems 7 and 5 we derive

THEOREM 10. (i) *The group*

$$\Gamma = AF, \tag{16}$$

where F is the group of parallel translations of Σ^n, $\Sigma = GF(p)$, and A an irreducible solvable subgroup of $GL(\Sigma^n)$, is a primitive solvable subgroup of the symmetric group S_{p^n}.

(ii) *Any primitive solvable subgroup of S_{p^n} is conjugate in S_{p^n} to a group of the form* (16).

(iii) *Two primitive solvable subgroups $\Gamma = AF$ and $\Phi = BF$ of the form* (16) *are conjugate in S_{p^n} if and only if A and B are conjugate in $GL(\Sigma^n)$.*

(iv) *The group* (16) *is a maximal solvable subgroup of S_{p^n} if and only if A is a maximal solvable subgroup of $GL(\Sigma^n)$.*

Theorem 10 reduces the construction of primitive solvable subgroups of S_{p^n}

to the construction of irreducible solvable subgroups of $GL(\Sigma^n)$.

For groups of prime degree, that is, for $n = 1$, the question is fully answered by

THEOREM 11. (i) *The group S_p contains one and, to within conjugacy, only one maximal solvable transitive subgroup Γ_p. The order of Γ_p is $p(p-1)$; it consists of all permutations $h_{\alpha\beta}$ of the form $h_{\alpha\beta}(x) = \beta x + \alpha$, $\alpha, \beta, x \in GF(p)$, $\beta \neq 0$.*

(ii) *Let $\tau = \tau(p-1)$ be the number of positive divisors of $p-1$. Then the system of all solvable transitive subgroups of S_p can be partitioned into τ conjugacy classes of subgroups of S_p. As representatives of these classes we can take the groups $\Gamma_{p,s}$, $s|(p-1)$,*[5] *$s > 0$; the order of $\Gamma_{p,s}$ is ps, and it consists of all permutations h_s of the form $h_s(x) = \beta^{(p-1)/s}x + \alpha$, $\alpha, \beta, x \in GF(p)$, $\beta \neq 0$.*

PROOF. The theorem follows from Theorem 10, because for $n = 1$ the space Σ^n reduces to $\Sigma = GF(p)$, and the group $GL(\Sigma^n)$ to the group of permutations $g(x) = \beta x$, $\beta, x \in GF(p)$, $\beta \neq 0$. Moreover, by the corollary to Lemma 3.1, a transitive permutation group of prime degree is primitive. ■

7. *Imprimitive solvable groups of finite order.* Let θ be any abstract group-theoretical property satisfying the following conditions:

1. If a group has the property θ, its subgroups and homomorphic images also have the property.

2. If a normal subgroup N of a group Γ and its factor group Γ/N have the property θ, then Γ itself also has the property.*

Let us use the term *θ-groups* for groups possessing the property θ. Examples are solvable groups and Π-groups.

THEOREM 12. *Let Γ be a transitive subgroup of S_n and a maximal θ-subgroup of S_n (the property θ is fixed once and for all). Then either Γ is primitive or $\Gamma = F \operatorname{Wr} G$, where F is a maximal primitive θ-subgroup of S_m and G a maximal transitive θ-subgroup of S_l, $lm = n$, $1 < l < n$.*

This follows from the definition of θ-groups and from Theorem 3.3.

If θ is the property of solvability, Theorems 12 and 9 together give the following result.

COROLLARY 1. *A maximal solvable transitive subgroup Γ of S_n is either primitive, or $\Gamma = F \operatorname{Wr} G$, where F is a maximal primitive solvable subgroup of S_{p^α}, G a maximal transitive solvable subgroup of S_l, $p^\alpha l = n$, p a prime and α a natural number.*

---

(5) Here and in what follows we use the standard notation: $m|n$ means that m divides n; $m\nmid n$ means that m does not divide n.

*Editor's note. In the current terminology of closure operations: θ is S-, Q-, and P-closed.

If θ is the property of being a Π-group, we obtain

COROLLARY 2. *A transitive Sylow Π-subgroup Γ of S_n is either primitive or $\Gamma = F \operatorname{Wr} G$, where F is a primitive Sylow Π-subgroup of S_m and G a transitive Sylow Π-subgroup of S_l, $ml = n$, $1 < l < n$.*

If p is a prime, a finite p-group has a nontrivial center; hence Theorems 12 and 3.6 imply

COROLLARY 3. *A Sylow p-subgroup $\mathfrak{P}_{p^\alpha}$ of S_{p^α} can be expressed in the form*

$$\mathfrak{P}_{p^\alpha} = \mathfrak{P}_p \operatorname{Wr} \mathfrak{P}_{p^{\alpha-1}} = \mathfrak{P}_p \operatorname{Wr} (\mathfrak{P}_p \operatorname{Wr} \mathfrak{P}_p \operatorname{Wr} \dots \operatorname{Wr} \mathfrak{P}_p).$$

The order of $\mathfrak{P}_{p^\alpha}$ is p^ω, where $\omega = (p^\alpha - 1)(p-1)^{-1}$.

§5. Nilpotent and locally nilpotent permutation groups

1. *Preliminary lemmas.* A group G is called an *N-group*, or a *group satisfying the normalizer condition*, if each of its proper subgroups is different from its normalizer in G.

LEMMA 1. *Let Γ be a transitive subgroup of $S(X)$ satisfying the normalizer condition. If* card $X > 1$ *is not a prime number, then Γ is imprimitive.*

PROOF. Let card $X > 1$ and suppose that Γ is primitive. Consider the stabilizer Γ_a in Γ. Since Γ is an N-group, it has a subgroup U containing Γ_a as a nontrivial normal subgroup. By the primitivity criterion 3.2, $U = \Gamma$. Since Γ is transitive, it now follows by Corollary 3.4.1 that $\Gamma_a = (e)$, so that Γ is a regular group. By Corollary 3.2.1, card X must be a prime number. ∎

LEMMA 2. *Let* card $X > 1$ *and let Γ be a transitive N-subgroup of $S(X)$. If $a \in X$, there is a point $b \neq a$ in X such that $\Gamma_b = \Gamma_a$.*

PROOF. Let U be the normalizer of Γ_a in Γ. Then for $h \in U \backslash \Gamma_a$ we have $h(a) = b \neq a$ and $\Gamma_a = h\Gamma_a h^{-1} = \Gamma_b$. ∎

COROLLARY. *A doubly transitive subgroup of $S(X)$ is not an N-group.*

LEMMA 3. *If n is a natural number and Γ a transitive nilpotent subgroup of S_n, then $\Pi(\Gamma) = \Pi(n)$.*[6]

PROOF. First let $n = p$ be a prime number. Then Γ is primitive, and so by Corollary 3.4.1 its center $Z(\Gamma)$ is transitive. Since an abelian transitive subgroup is maximal among the abelian subgroups of S_n, it follows that $\Gamma = Z(\Gamma)$. Consequently

---

[6] Here $\Pi(n)$ is the set of all prime divisors of n and $\Pi(\Gamma)$ the set of all primes dividing the orders of the elements of a (periodic) group Γ. This notation will be used henceforth.

Γ is a cyclic group of order p. Thus for $n = p$ the proof is complete.

We proceed by induction on the number $\tau(n)$ of divisors of n. For $\tau(n) = 2$ the lemma is true. Let $\tau(n) > 2$ and suppose the lemma is true for all n_1 such that $\tau(n_1) < \tau(n)$. By Lemma 1, the group Γ is imprimitive. A partition of the set X on which Γ acts into systems of imprimitivity defines a homomorphism $\varphi: \Gamma \longrightarrow S_m$, where m is the number of systems. Now $\overline{\Gamma} = \operatorname{Im} \varphi$ is a nilpotent transitive group of degree m, where $m|n$ and $\tau(m) < \tau(n)$. By the inductive hypothesis, every prime divisor of the order of $\overline{\Gamma}$ divides m. The group $K = \operatorname{Ker} \varphi$ is contained in the direct product of m nilpotent transitive groups of degree nm^{-1}, and so by the inductive hypothesis every prime divisor of card K divides nm^{-1}. Since card Γ = (card K)(card $\overline{\Gamma}$), every prime divisor of card Γ divides n. Since Γ is transitive, the number card Γ is divisible by n. ■

COROLLARY 1. *A transitive nilpotent subgroup of* S_{p^ν}, *where* p *is a prime, is a* p*-subgroup.*

COROLLARY 2. *A maximal transitive nilpotent subgroup of* S_{p^ν}, *where* p *is a prime, is a Sylow* p*-subgroup.*

2. *Maximal transitive nilpotent subgroups of* S_n. In this subsection we present a complete characterization of the structure of maximal nilpotent transitive subgroups of the symmetric group S_n; in particular, we prove that they are conjugate in S_n.

Let

$$n = p_1^{\nu_1} \ldots, p_k^{\nu_k} \tag{1}$$

be the canonical representation of n. When $k = 1$ $(n = p^\nu)$, Corollary 2 to Lemma 3 tells us that a maximal transitive nilpotent subgroup of S_{p^ν} is a Sylow p-subgroup. Consequently, by Sylow's theorem, *all maximal transitive nilpotent subgroups of* S_{p^ν} *are conjugate in* S_{p^ν}.

Now let $k > 1$ and $q_j = p_j^{\nu_j}$, and let X_j be a set of q_j points, $j = 1, \ldots, k$. We shall now construct a certain transitive nilpotent subgroup N_n of S_n and show that every transitive nilpotent subgroup of S_n is conjugate in S_n to a subgroup of N_n. Clearly the number of elements in the cartesian product $X_1 \times \cdots \times X_k = X$ is n. We may thus assume that $S_n = S(X)$. Let N_{q_j} be a Sylow p_j-subgroup of $S(X_j)$, and g_j an arbitrary permutation in N_{q_j}, $j = 1, \ldots, k$. Let N_n be the set of all permutations $g \in S(X)$ of the form

$$g(x) = g(x_1, \ldots, x_k) = (g_1(x_1), \ldots, g_k(x_k)), \tag{2}$$

where $x = (x_1, \ldots, x_k) \in X$ and $g_j \in N_{q_j}$, $j = 1, \ldots, k$. The group N_n is clearly isomorphic to the direct product of the groups $N_{q_1}, \ldots, N_{q_k}$. Since the groups N_{q_j} are transitive and nilpotent, so is N_n, and card $N_n = p_1^{\omega_1} \cdots p_k^{\omega_k}$, where $\omega_j = (p_j^{\nu_j} - 1)(p_j - 1)^{-1}$, $j = 1, \ldots, k$.

Now let N be any transitive nilpotent subgroup of $S(X)$. Then, by Lemma 3,

$$\operatorname{card} N = p_1^{\alpha_1} \ldots p_k^{\alpha_k}, \quad \alpha_j \geqslant \nu_j.$$

Thus N is a direct product $N = H \times F$, where H is of order $p_1^{\alpha_1}$ and F of order $p_2^{\alpha_2} \cdots p_k^{\alpha_k}$.

LEMMA 4. *The set X can be expressed as a cartesian product $X = Y \times Z$ of two sets Y and Z with the following properties:*

(i) card $Y = p_1^{\nu_1}$ *and* card $Z = np_1^{-\nu_1}$.

(ii) *For $h \in H$ and $f \in F$,*

$$h(y, z) = (\bar{h}(y), z), \qquad f(y, z) = (y, \bar{f}(z)),$$
$$\bar{h} \in S(Y), \qquad \bar{f} \in S(Z), \qquad (y, z) \in Y \times Z.$$

(iii) *$h \mapsto \bar{h}$ is a faithful transitive representation of H and $f \mapsto \bar{f}$ a faithful transitive representation of F.*

PROOF. Since $n \nmid \operatorname{card} F$, it follows that F is intransitive. Let $X = X_1 \cup \cdots \cup X_s$ be the partition of X into orbits of F. By Lemma 4.3, the representations r_i: $f \mapsto f_i = f | X_i$, $i = 1, \ldots, s$, are pairwise equivalent. It is obvious that Im r_i is a transitive nilpotent subgroup of $S(X_i)$. Consequently, by Lemma 3, $\Pi(\operatorname{card} X_i) = \Pi(\operatorname{card} F)$. Since $N = H \times F$ is transitive, H permutes the orbits $X_1, \ldots, X_s$ of F transitively. Thus $s | \operatorname{card} H$. Hence

$$s = p_1^{\beta_1}, \qquad \operatorname{card} X_i = p_2^{\beta_2} \ldots p_k^{\beta_k}.$$

Since $(\operatorname{card} X_i)s = n$, we have

$$s = p_1^{\nu_1}, \qquad \operatorname{card} X_i = p_2^{\nu_2} \ldots p_k^{\nu_k}.$$

Now we set $t = \operatorname{card} X_i$ and denote the points of the orbit X_1 by the number pairs $(1, 1), (1, 2), \ldots, (1, t)$. Let $h_1 = e, h_2, \ldots, h_s$ be permutations in H such that $h_y(X_1) = X_y$. We write the points of the orbit X_y as $(y, 1), (y, 2), \ldots, (y, t)$ setting $(y, z) = h_y(1, z)$, $z = 1, \ldots, t$. Obviously $X = Y \times Z$, where $Y = \{1, \ldots, s\}$ and $Z = \{1, \ldots, t\}$. For $f \in F$ we have

$$f(y, z) = fh_y(1, z) = h_y f(1, z) = h_y(1, \bar{f}(z)) = (y, \bar{f}(z)),$$

where $\bar{f} \in S(Z)$ depends only on f. Now consider $h(y, z)$, $h \in H$. We have $h(1, z) = (y_1, z_1)$, $1 \leqslant y_1 \leqslant s$. If $z_1 \neq z$, then

$$h_{y_1}^{-1} h(1, z) = h_{y_1}^{-1}(y_1, z_1) = (1, z_1) \neq (1, z),$$

that is, $h_{y_1}^{-1}h|X_1 = h_0$ is a permutation of $S(X_1)$ different from e. Now we set $F_0 = F|X_1$. Then h_0 commutes with every permutation in F_0. Hence $(h_0)F_0$ is a transitive nilpotent subgroup of $S(X_1)$. By Lemma 3, the prime divisors of the order of h_0 divide t. But this is impossible. Consequently $z_1 = z$ and

$$h(1, z) = (y_1, z). \tag{3}$$

Now consider $h(y, z)$. By means of (3) we obtain $h(y, z) = hh_y(1, z) = (y', z) = (\bar{h}(y), z)$, where $\bar{h} \in S(Y)$ depends only on h. ■

Lemma 4 implies

THEOREM 1. *Any transitive nilpotent subgroup of the symmetric group S_n is conjugate in S_n to some subgroup of the group N_n defined by* (2).

This theorem in turn implies

THEOREM 2. *Any two maximal transitive nilpotent subgroups of S_n are conjugate in S_n.*

If (1) *is the canonical representation of n, then a maximal transitive nilpotent subgroup of S_n is isomorphic to a direct product of Sylow p_j-subgroups of the symmetric groups $S_{p_j^{\nu_j}}$, $j = 1, \ldots, k$.*

We prove one more theorem on transitive nilpotent subgroups of S_n.

THEOREM 3. *Let m be the product of all distinct prime divisors of n, Γ a maximal transitive nilpotent subgroup of $S_n = S(X)$, Γ_a a stabilizer in Γ and U its normalizer in Γ. Then the index of Γ_a in U is m; symbolically, $U\colon \Gamma_a = m$.*

Before proving this theorem we note that it is equivalent to the following proposition:

If $Y = \{y \mid y \in X, \Gamma_a(y) = y\}$, then card $Y = m$.

Yet another formulation of Theorem 3 is as follows:

The number of distinct stabilizers in the group Γ is nm^{-1}.

PROOF OF THEOREM 3. For $n = 1$ the theorem is trivial. Let $n > 1$ and set $U\colon \Gamma_a = t$. Since Γ is an N-group, it follows that $t > 1$. Obviously $t|n$.

We first prove that if $n = p^\nu$, p a prime, then $t = p$. Let $l = p^{\nu-1}$. Now Γ is a Sylow p-subgroup of S_{p^ν} and contains, in particular, permutations of the form $d_1^{s_1} \cdots d_l^{s_l}$, where the d_j are pairwise disjoint p-cycles and $0 \leqslant s_j \leqslant p$. If $d_1(a) \neq a$, the permutation $d_2 \cdots d_l \in \Gamma_a$ leaves exactly p points invariant. Thus $t \leqslant p$. Since $t > 1$ and $t|p^\nu$, it follows that $t = p$.

Now let (1) be the canonical representation of n, and let $k > 1$. We may assume that $\Gamma = N_n$, where N_n is defined by (2). We consider the stabilizer Γ_a in N_n of the element $a = (a_1, \ldots, a_k) \in X = X_1 \times \cdots \times X_k$. If $g \in \Gamma_a$, the components g_j of g are such that $g_j(a_j) = a_j$. When g ranges over the group Γ_a, the g_j range independently over the stabilizers $(N_{q_j})_{a_j}$ in the groups N_{q_j}. By what we have already proved, the group $(N_{q_j})_{a_j}$ leaves exactly p_j points invariant. Consequently Γ_a leaves exactly $p_1 \cdots p_k = m$ points invariant. ■

3. *Maximal nilpotent intransitive subgroups of S_n.* Let Γ be a maximal nilpotent intransitive subgroup of $S_n = S(X)$ and $X = Y_1 \cup \cdots \cup Y_k$ the partition of X into orbits of Γ. Then $\Gamma_j = \Gamma | Y_j$, $j = 1, \ldots, k$, is a maximal nilpotent transitive subgroup of $S(Y_j)$, and Γ is the direct product of its restrictions Γ_j. Hence and from Theorem 2 it follows that *any maximal nilpotent intransitive subgroup of S_n is determined to within conjugacy in S_n by an additive representation of n*:

$$n = n_1 + \ldots + n_k, \tag{4}$$

where $k > 1$, $n_j = $ card Y_j, $j = 1, \ldots, k$, and Y_j is an orbit of Γ. In other words, *a maximal nilpotent subgroup of S_n is determined to within conjugacy in S_n by its orbital type.*

It is not true, however, that for every representation (4) we can find a maximal nilpotent intransitive subgroup of S_n whose orbits have the cardinalities

$$n_1, \ldots, n_k. \tag{5}$$

Let (4) be a representation of n (the n_j are natural numbers) and

$$X = Y_1 \cup \ldots \cup Y_k \tag{6}$$

a partition of X such that card $Y_j = n_j$, $j = 1, \ldots, k$. We construct a group $H = H(n_1, \ldots, n_k)$ such that (6) is a partition of X into orbits of H, $H_j = H|Y_j$ is a transitive maximal nilpotent subgroup of $S(Y_j)$, $j = 1, \ldots, k$, and H is the direct product of its restrictions to the orbits. Let us clarify for what decompositions (4) $H(n_1, \ldots, n_k)$ is a maximal nilpotent subgroup of S_n.

LEMMA 5. *If the numbers* (5) *are all distinct, then $H(n_1, \ldots, n_k)$ is a maximal nilpotent subgroup of S_n.*

PROOF. Suppose that the numbers (5) are all distinct, but that $H = H(n_1, \ldots, n_k)$ is contained in some other nilpotent subgroup of S_n. Then H is a proper normal subgroup of some nilpotent group U in S_n. It is obvious that an orbit Y of U is the union of several orbits of H. The transitive group $U|Y$ has an intransitive normal subgroup $H|Y$. By Theorem 3.4, the orbits of $H|Y$ are systems of imprimitivity for $U|Y$. Thus the orbits of H occurring in Y all have the same cardinality, and this contradicts the assumption of the lemma. ■

LEMMA 6. *Let $n_1 = \cdots = n_k = m > 1$, where $k > 1$. The group*

$H = H(n_1, \ldots, n_k)$ is contained in a transitive nilpotent subgroup of S_n if and only if $m = p^\alpha$ and $k = p^\beta$, where p is a prime.

PROOF. Let $H \subset G$, where G is a transitive nilpotent subgroup of S_n (which we may assume to be maximal). We set $m = p^\alpha t$, $(t, p) = 1$, $\alpha > 0$, p a prime. Let us find the exponent u to which p occurs in $d = \text{card } H$. We first consider the order d_1 of the group $H|Y_1$. By Theorem 2, p occurs in d_1 to the exponent $\omega = (p^\alpha - 1)(p - 1)^{-1}$. Thus $u = k\omega = k(p^\alpha - 1)(p - 1)^{-1}$. If now $k = p^\beta r$, $(p, r) = 1$, then p occurs in the order of G to the exponent $v = (p^{\alpha+\beta} - 1)(p - 1)^{-1}$, because $n = km = p^{\alpha+\beta} rt$, $(p, rt) = 1$. Since H is a subgroup of G, we have $u \leqslant v$, and so

$$u^{-1}v = \frac{p^{\alpha+\beta} - 1}{(p^\alpha - 1)\, p^\beta r} = f(r) \geqslant 1. \tag{7}$$

For fixed p, α and β, $f(r)$ depends only on r. It is clear that $f(r)$ is a decreasing function of r. On the other hand, $f(2) < 1$. Hence and from (7) it follows that $r = 1$; that is,

$$k = p^\beta. \tag{8}$$

It is easy to show that $m = p^\alpha$. For let $q \neq p$ be a prime such that $q|m$. Then just as in deriving (8) we find that $k = q^\gamma$, which contradicts (8). This proves the necessity.

Now let $m = p^\alpha$ and $k = p^\beta$. Then $H = H(n_1, \ldots, n_k)$, $n_1 = \cdots = n_k = p^\alpha$, is a p-subgroup of $S_n = S_{p^{\alpha+\beta}}$. Consequently H is contained in a Sylow p-subgroup $N_{p^{\alpha+\beta}}$ of $S_{p^{\alpha+\beta}}$. As we have seen above, $N_{p^{\alpha+\beta}}$ is a transitive subgroup of $S_{p^{\alpha+\beta}}$. ■

We also note that if at least two of the numbers (5) are equal to 1, then $H(n_1, \ldots, n_k)$ is contained in another nilpotent subgroup of S_n. For example, if $n_1 = n_2 = 1$, then $H(n_1, \ldots, n_k)$ is a proper subgroup of $H(2, n_3, \ldots, n_k)$.

From Lemmas 5, 6 and the last remark we infer the next result.

THEOREM 4. *Every intransitive maximal nilpotent subgroup of S_n is conjugate in S_n to a group $H(n_1, \ldots, n_k)$.*

$H(n_1, \ldots, n_k)$ is a maximal nilpotent subgroup of S_n if and only if the following two conditions hold:

(i) *The polynomial $(x - n_1) \cdots (x - n_k) = \varphi(x)$ is not divisible by $(x - 1)^2$.*

(ii) *If p^α ($\alpha > 0$, p a prime) is a root of the polynomial $\varphi(x)$, its multiplicity is less than p.*

Theorem 4 together with Theorems 1 and 2 provides a complete description of the maximal nilpotent subgroups of the symmetric group S_n.

4. *On nilpotent and locally nilpotent permutation groups of infinite degree.* Let X be an infinite set. We define $SF(X)$ to be the group of all $f \in S(X)$ such that $f(x) \neq x$ for only finitely many points $x \in X$.

It is easy to see that *$SF(X)$ is a locally finite normal subgroup of $S(X)$.*

Since in a regular subgroup of $S(X)$ every permutation other than the identity actually permutes all points of X, the intersection of any regular subgroup of $S(X)$ with $SF(X)$ is trivial. In particular, *in $SF(X)$ there are no transitive abelian subgroups.* Consequently *the orbits of its abelian subgroups are finite.*

LEMMA 7. *The center of a transitive subgroup of $SF(X)$ is trivial.*

PROOF. Let H be a transitive subgroup of $SF(X)$ and Z its center. All the orbits of Z are finite. Consequently, if

$$X = \bigcup X_\alpha, \quad \alpha \subset I, \tag{9}$$

is the partition of X into orbits of Z, then I is infinite. Now let $Z \neq (e)$, $c \in Z$ and $c \neq e$. Then by Theorem 4.1 we have $c|X_\alpha \neq e$ for any $\alpha \in I$. It follows that c permutes all the points of an infinite set and is therefore not an element of $SF(X)$. ∎

This lemma implies

THEOREM 5. *The orbits of a nilpotent subgroup of $SF(X)$ are finite.*

PROOF. Let Γ be a nilpotent subgroup of $SF(X)$ and X_α one of its orbits. Then $\Gamma_\alpha = \Gamma|X_\alpha$ is a nilpotent group. If card $X_\alpha > 1$, the center of Γ_α is nontrivial. By Lemma 7 the set X_α is finite. ∎

THEOREM 6. *A transitive subgroup Γ of $SF(X)$ cannot be expressed as the product of two nontrivial elementwise permutable normal subgroups. In particular, Γ is not expressible as a direct product.*

PROOF. Suppose that a transitive subgroup Γ of $SF(X)$ is expressible as a product $\Gamma = HF$, where H and F are nontrivial subgroups such that $(H, F) = (e)$. There are two possibilities: (1) H and F are transitive; (2) at least one of them is intransitive. We claim that both possibilities lead to a contradiction. In the first case, by Theorem 4.1, H and F are regular groups, which contradicts the fact that $H \subset SF(X)$. Now consider the second case. Suppose, say, that H is intransitive, and let (9) be the partition of X into orbits of H. Then, by Lemma 4.3, the representations $r_\alpha: H \longrightarrow S(X_\alpha)$, $h \longmapsto h|X_\alpha$, $\alpha \in I$, are pairwise equivalent. Hence and from $H \subset SF(X)$ it follows that I is finite. Thus the orbits of H are infinite. On the other hand, since H is normal in Γ, it follows from Theorem 3.4 that (9) is a partition of X into systems of imprimitivity for Γ. But it is clear that systems of imprimitivity for a transitive subgroup of $SF(X)$ cannot be infinite. ∎

Theorem 6 implies

THEOREM 7. *A locally nilpotent transitive subgroup of $SF(X)$ is a p-group.*

PROOF. Let Γ be a transitive locally nilpotent subgroup of $SF(X)$. Then Γ, as a periodic locally nilpotent group, is the direct product of its Sylow subgroups. But by Theorem 6, Γ is indecomposable, and therefore is a p-group. ∎

Theorem 7 in turn implies

THEOREM 8. *A maximal locally nilpotent transitive subgroup of $SF(X)$ is a Sylow p-subgroup.*

The structure of the Sylow p-subgroups of $SF(X)$ has been studied by Ivanjuta [1], [2]. He has shown, in particular, that *if* card $X > \aleph_0$, *then $SF(X)$ contains no transitive p-subgroups, and if X is countable, then any two transitive Sylow p-subgroups of $SF(X)$ are conjugate in $S(X)$.* Hence and from Theorem 7 it follows that *the orbits of a locally nilpotent subgroup of $SF(X)$ are either finite or countable.*

In connection with Theorem 6, we state the following problem: *Under what conditions is a locally finite group isomorphic to some transitive subgroup of $SF(X)$? What locally finite groups can be isomorphically embedded in $SF(X)$?*

SECOND CHAPTER

THE GENERAL LINEAR GROUP

§6. Definitions. Preliminary propositions

1. *The multiplicative group of a ring. Properties of division rings.* Throughout this chapter the term *ring* denotes an associative ring with unit element containing at least two elements.

Let Δ be a ring with unit element 1. An element α of Δ is said to be *invertible* in Δ if there is an element β in Δ such that $\alpha\beta = \beta\alpha = 1$. It is evident that the set of all invertible elements in Δ is a group under multiplication. This is called the *multiplicative group of* Δ and is denoted by Δ^*. If $\gamma \in \Delta \backslash \Delta^*$, we say that γ is *noninvertible* or *singular* in Δ.

An element η of a ring Δ is said to be *unipotent* if $\eta - 1$ is nilpotent; that is, $(\eta - 1)^k = 0$ for some natural number k. Let U denote the set of all unipotent elements of Δ. *A unipotent element η is invertible in Δ.* For, $\eta = 1 + \theta$, $\theta^k = 0$ and

$$(1+\theta)[1-\theta+\theta^2- \dots +(-1)^{k-1}\theta^{k-1}]=1-\theta^k=1.$$

The element $\theta_1 = \sum_{\nu=1}^{k-1} (-1)^\nu\theta^\nu$ is nilpotent, because

$$\theta_1^k = \theta^k\left[\sum_{\nu=1}^{k-1}(-1)^\nu\theta^{\nu-1}\right]^k = 0.$$

Consequently $\eta^{-1} = 1 + \theta_1$ is also unipotent.

It is obvious that *the product of two permutable unipotent elements is also unipotent,* so that *if Δ is a commutative ring, then U is a subgroup of Δ^*.*

Now let $\Delta = T$ be a division ring.

PROPOSITION 1. *A periodic abelian subgroup G of the multiplicative group of a division ring T is locally cyclic. In particular, the multiplicative group of a finite field $GF(q)$ is cyclic of order $q - 1$.*

PROOF. As a periodic abelian group, G is locally finite. It therefore suffices to prove that every finite subgroup H of G is cyclic. The orders of the elements of a finite abelian group divide their maximal order; therefore all the elements of H are roots of the equation

$$x^d = 1, \tag{1}$$

where d is the order of some element h of H. Obviously H is contained in some field $\Sigma \subset T$. In Σ equation (1) has at most d roots. On the other hand, $1, h, h^2, \ldots, h^{d-1}$ are roots of the equation and lie in H. Thus $H = (h)$ is a cyclic group. ∎

PROPOSITION 2. *Suppose that T is either not a field of characteristic 2, or a field of characteristic 2 containing no elements that are transcendental over its prime field. Let S be the subgroup of the additive group of T generated by all the products of squares in T. Then $S = T$.*

PROOF. S is the set of all elements of T of the form

$$\sum \pm \alpha_1^2 \ldots \alpha_k^2, \qquad \alpha_i \in T.$$

If the characteristic of T is not 2, the fact that S and T coincide follows from

$$\alpha = \left(\frac{\alpha+1}{2}\right)^2 - \left(\frac{\alpha-1}{2}\right)^2, \qquad \alpha \in T.$$

Now let char $T = 2$, T a noncommutative division ring, and $Z(T)$ its center. If $\alpha \in T \backslash Z(T)$, there is an element $\beta \in T$ such that $\gamma = \alpha\beta - \beta\alpha \neq 0$. Now the element

$$\gamma = \alpha\beta + \beta\alpha \tag{2}$$

lies in S, because $\gamma = (\alpha + \beta)^2 - \alpha^2 - \beta^2$. We have also $\gamma\alpha = \alpha(\beta\alpha) + (\beta\alpha)\alpha \in S$, since this element has the form (2). It is clear that S is closed under multiplication, and so $\gamma^{-1} = \gamma(\gamma^{-1})^2 \in S$ and $\alpha = \gamma^{-1}(\gamma\alpha) \in S$. Thus $T\backslash Z(T) \subset S$.

If $\delta \in Z(T)$, then $\delta\gamma = \alpha(\delta\beta) + (\delta\beta)\alpha \in S$, $\delta \in S$. Hence $Z(T) \subset S$, and so $S = T$.

It remains to consider the case of a field T of characteristic 2 whose elements are all algebraic over its prime field. Every nonzero element ϵ of T is of odd order, since it lies in the multiplicative group of some finite field $GF(2^k)$. Therefore $\epsilon = (\epsilon^\nu)^2 \in S$. ∎

Proposition 2 may fail to hold for a field T of characteristic 2 containing transcendental elements. This can be seen from the simple example of the field of fractions $\Delta(x)$, where $\Delta = GF(2)$.

2. *Modules.* Let Δ be a ring. A set M is called a *right Δ-module* if it satisfies the following conditions:

(1) M is an additive abelian group.

(2) For any $\alpha \in \Delta$ and $a \in M$ an element $a\alpha \in M$ is defined, and if $b \in M$ and $\beta \in \Delta$, then

$$(a+b)\alpha = a\alpha + b\alpha, \quad a(\alpha+\beta) = a\alpha + a\beta,$$
$$a(\alpha\beta) = (a\alpha)\beta, \quad a \cdot 1 = a.$$

A *left* Δ-*module* M is defined similarly:

(1) M is an additive abelian group.

(2) For any $\alpha \in \Delta$ and $a \in M$ an element $\alpha a \in M$ is defined, and if $b \in M$ and $\beta \in \Delta$, then

$$\alpha(a+b) = \alpha a + \alpha b, \quad (\alpha+\beta)a = \alpha a + \beta a,$$
$$(\alpha\beta)a = \alpha(\beta a), \quad 1 \cdot a = a.$$

A subset $N \subset M$ is called a *submodule* of M if it is a subgroup of the additive group M and $c\alpha \in N$ for any $\alpha \in \Delta$ and $c \in N$.

For example, for any $a \in M$ the set $a\Delta = \{a\alpha | \alpha \in \Delta\}$ is a submodule of M.

A submodule P of a Δ-module M is called a *hyperplane* of M if M is a *direct sum*: $M = P \dot{+} v\Delta$ for some $v \in M$, $v \neq 0$. A system B of elements of M is said to be *linearly independent over* Δ if for any finite subsystem $b_1, \ldots, b_k$ of B

$$b_1\lambda_1 + \ldots + b_k\lambda_k = 0, \ \lambda_j \in \Delta \Rightarrow \lambda_1 = \ldots = \lambda_k = 0.$$

An element u of a Δ-module M is said to be *free over* Δ if $u\lambda = 0$, $\lambda \in \Delta$, implies $\lambda = 0$. If M contains a linearly independent system of elements B such that every element of M is a linear combination of finitely many elements of B, then B is called a λ-*basis* of M, and M is called a *free* Δ-*module.* The zero module is also said to be free.

We often omit the reference to Δ and write simply "basis", "module", etc.

The symbol Δ^n denotes the right free Δ-module consisting of all elements $(\alpha_1, \ldots, \alpha_n)$, $\alpha_j \in \Delta$, with

$$(\alpha_1, \ldots, \alpha_n) + (\beta_1, \ldots, \beta_n) = (\alpha_1 + \beta_1, \ldots, \alpha_n + \beta_n),$$
$$(\alpha_1, \ldots, \alpha_n)\gamma = (\alpha_1\gamma, \ldots, \alpha_n\gamma), \quad \gamma \in \Delta.$$

The elements $(1, 0, \ldots, 0), (0, 1, \ldots, 0), \ldots, (0, 0, \ldots, 1)$ form a basis of this module.

The symbol Δ^n_L denotes the left free Δ-module defined in the same way as Δ^n, except that $(\alpha_1, \ldots, \alpha_n)\gamma$ is replaced by

$$\gamma(\alpha_1, \ldots, \alpha_n) = (\gamma\alpha_1, \ldots, \gamma\alpha_n).$$

If the cardinality of a Δ-basis of a free module M over a ring Δ is independent of the choice of the basis, this cardinality is known as the *dimension* of M. Let Δ be a principal ideal domain and M a free Δ-module. Then any two Δ-bases of M have the same cardinality, every submodule of M is free, and its dimension does not exceed that of M (Lang [1], p. 387).

In particular, if Δ is a division ring, a Δ-module is a *vector space over* Δ. Every vector space M over Δ is a free module; the cardinality of its basis is an invariant and is called the *dimension* of the vector space, denoted by $M : \Delta$. Steinitz' exchange theorem is valid. Submodules of a Δ-module M over Δ are called *subspaces* of the vector space M. Every subspace of a vector space M is a direct summand of M. Every linearly independent system of elements is contained in some basis of the space (see Kuroš [2], Chapter 5, §3).

§7. Endomorphisms

1. End M. Let M be a right Δ-module. A mapping $h\colon M \longrightarrow M$ is called an *endomorphism* of M if for $x, y \in M$ and $\alpha \in \Delta$ we have

$$h(x+y)=h(x)+h(y), \qquad h(x\alpha)=h(x)\alpha. \tag{1}$$

We denote by End M the set of all endomorphisms of M. The product gh of two endomorphisms g and h is defined to be their product as mappings. The sum $g + h$ is defined by setting for any $x \in M$

$$(g+h)(x)=g(x)+h(x). \tag{2}$$

LEMMA 1. *End M is a ring with unit element.*

PROOF. Since M is an additive abelian group, it follows from (1) and (2) that End M is also an additive abelian group. Associativity of multiplication in End M follows from associativity of multiplication of mappings. Now consider $f(g + h), f, g, h \in$ End M. For $x \in M$,

$$f(g+h)(x)=f(g(x)+h(x))=fg(x)+fh(x)=(fg+fh)(x).$$

Consequently $f(g + h) = fg + fh$. Similarly one proves that $(g + h)f = gf + hf$.

The unit element of End M is obviously the identity mapping. ∎

Instead of End M we sometimes write Hom(M, M).

An endomorphism of a left Δ-module M is defined as a mapping $h\colon M \longrightarrow M$ such that $(x + y)h = xh + yh$ and $(\alpha x)h = \alpha(xh)$ for $x, y \in M$ and $\alpha \in \Delta$. Multiplication of endomorphisms g and h is defined by $x(gh) = (xg)h$.

The following lemma is easy to verify.

LEMMA 2. *Let M be a free Δ-module and B a basis of M. Then any mapping $f\colon B \longrightarrow M$ can be extended uniquely to an endomorphism of M.*

2. *The general linear group.* A bijective endomorphism of a module M is called an *automorphism* of M. It is obvious that *the set* Aut M *of all automorphisms of a module M is a group under multiplication, namely, the multiplicative group of* End M.

The group of automorphisms of a free right Δ-module M is called the *general linear group* and is denoted by $GL(M)$.

LEMMA 3. *Let M be a free Δ-module, h an endomorphism and B a basis of M. Then h is an automorphism if and only if $h(B)$ is a basis of M and*

$$B \to h(B), \qquad b \mapsto h(b), \qquad b \in B, \tag{3}$$

is a bijection.

PROOF. If a mapping (3) satisfies the conditions, by Lemma 2 its inverse $h(b) \longmapsto b$ can be extended uniquely to an endomorphism g of M; $gh = hg = e$ is the identity automorphism. Thus $h \in \operatorname{Aut} M$.

Conversely, let h be an automorphism of M. Then $h(M) = M$. Consequently, every element of M is a linear combination of the finitely many elements of the form $d = h(b)$, $b \in B$. Every finite system $d_1 = h(b_1), \ldots, d_n = h(b_n)$, $b_i \in B$, is linearly independent:

$$d_1\lambda_1 + \ldots + d_n\lambda_n = 0 \Rightarrow h(a) = 0,$$

where $\lambda_i \in \Delta$, and $b_1\lambda_1 + \cdots + b_n\lambda_n = a$; but h is a bijection, and so $a = 0$ and $\lambda_i = 0$ for $i = 1, \ldots, n$. Thus $h(B)$ is a module basis. ■

The next result is an obvious consequence of the last two lemmas.

LEMMA 4. *For any pair A and B of equipotent bases of a free module M there exists an automorphism taking A onto B. Every bijection $A \longrightarrow B$ can be extended uniquely to an automorphism of M.*

Hence it follows, in particular, that if M is a vector space, then for any pair a and b of nonzero vectors there is an automorphism of M taking a to b; in other words, *$GL(M)$ is transitive on the set $M \setminus \{0\}$.*

EXAMPLE. Let M be an n-dimensional space over $GF(q)$. Then $GL(M)$ is a finite group of order

$$(q^n - 1)(q^n - q) \ldots (q^n - q^{n-1}).$$

3. *Homotheties.* Let M be a right Δ-module. We consider the mapping $\rho_\alpha: M \longrightarrow M$, $x \longmapsto x\alpha$, $x \in M$, $\alpha \in \Delta$. It is obvious that ρ_α is not always an endomorphism of M. But if $\alpha \in Z(\Delta)$, where $Z(\Delta)$ is the center of Δ, then $\rho_\alpha \in \operatorname{End} M$. In general:

LEMMA 5. *If a module M contains a free element, then*

$$\rho_\alpha \in \operatorname{End} M \tag{4}$$

if and only if $\alpha \in Z(\Delta)$.

PROOF. Suppose that (4) is true. Then for $v \in M$ and $\lambda \in \Delta$ we have $\rho_\alpha(v\lambda) = \rho_\alpha(v)\lambda = v(\alpha\lambda)$. On the other hand, $\rho_\alpha(v\lambda) = (v\lambda)\alpha = v(\lambda\alpha)$. Hence $v(\alpha\lambda - \lambda\alpha) = 0$. Now, if v is a free element, then $\alpha\lambda = \lambda\alpha$, and so $\alpha \in Z(\Delta)$. ■

If M is a free element, then obviously *the automorphisms ρ_α, $\alpha \in Z(\Delta)^*$, form a group $H(M)$ isomorphic to $Z(\Delta)^*$.* We call $H(M)$ the *homothety group of M*, and its elements *homotheties.*

A submodule N of M is said to be *invariant under an endomorphism f* if $f(N) \subset N$.

LEMMA 6. *Let Δ be a ring without divisors of zero, M a right free Δ-module containing at least two linearly independent elements, and σ an automorphism of M. If every submodule of M of the form $w\Delta$ is invariant under σ, then $\sigma \in H(M)$.*

PROOF. For any u and v in M,

$$\sigma(u) = u\alpha, \quad \sigma(v) = v\beta, \quad \sigma(u+v) = (u+v)\gamma, \quad \alpha, \beta, \gamma \in \Delta.$$

On the other hand, $\sigma(u + v) = \sigma(u) + \sigma(v) = u\alpha + v\beta$. Consequently $u\gamma + v\gamma = u\alpha + v\beta$. If now u and v are linearly independent, it follows that $\alpha = \gamma = \beta$.

Since Δ contains no divisors of zero, linear independence of u and v implies that of w and v, where $w = u\lambda$, $\lambda \neq 0$, $\lambda \in \Delta$. Therefore $\sigma(w) = w\alpha = u\lambda\alpha$. On the other hand, $\sigma(w) = \sigma(u\lambda) = u\alpha\lambda$. Consequently

$$u(\lambda\alpha - \alpha\lambda) = 0, \quad \lambda\alpha - \alpha\lambda = 0, \quad \alpha \in Z(\Delta),$$

where $Z(\Delta)$ is the center of Δ.

In particular, if B is a basis of M, then for all $b \in B$ we have $\sigma(b) = b\alpha$, $\alpha \in Z(\Delta)$. Thus $\sigma(a) = a\alpha$ for any $a \in M$, and $\sigma \in H(M)$. ■

4. *Transvections.* An endomorphism τ of a Δ-module M is called a *transvection* if there exists a hyperplane P in M each element of which is invariant under τ, and such that $\tau(a) - a \in P$ for any $a \in M$.

The set of all transvections associated with a fixed hyperplane P is denoted by $K(P)$.

By the definition of a hyperplane, M can be expressed as a direct sum

$$M = P \dotplus v\Delta, \quad v \in M, \quad v \neq 0. \tag{5}$$

It is obvious that *an endomorphism τ of M belongs to $K(P)$ if and only if*

(i) $\tau(a) = a$ *for any* $a \in P$;

(ii) $\tau(v) = v + p$, $p \in P$.

THEOREM 1. *Let $M = P + v\Delta$, where P is a hyperplane and v a free element. Then the set $K(P)$ is a group isomorphic to the additive group of P.*

PROOF. By (ii), for a fixed decomposition (5), any automorphism $\tau \in K(P)$ is determined by an element p of P. We set $\tau = \tau_p$. For $a \in P$ and $p, q \in P$ we have

$$\tau_p\tau_q(a) = a, \qquad \tau_p\tau_q(v) = \tau_p(v + q) = v + p + q.$$

Consequently $\tau_p\tau_q = \tau_{p+q}$. It is clear that the mapping $\gamma\colon K(P) \longrightarrow P$, $\tau_p \longmapsto p$, is bijective. Since

$$\gamma(\tau_p\tau_q) = \gamma(\tau_p) + \gamma(\tau_q),$$

γ is an isomorphism. ■

Note that *a transvection is a unipotent element of* End M. For,

$$(\tau_p - e)(v) = v + p - v = p,$$

and

$$(\tau_p - e)^2(v) = (\tau_p - e)(p) = p - p = 0,$$

whence $(\tau_p - e)^2 = 0$.

THEOREM 2. *Let M be a right Δ-module admitting a decomposition* (5), *where v is a free element and P contains a free element. Then the following assertions are true*:

(i) *The centralizer of the subgroup $K(P)$ in* Aut M *is $K(P)H(M)$.*

(ii) *$K(P)H(M)$ is a maximal abelian subgroup of* Aut M.

(iii) *The set U of all unipotent elements of $K(P)H(M)$ is $K(P)H_u$, where H_u is the group of all unipotent homotheties. In particular, if the center of Δ contains no unipotent elements other than* 1, *then $U = K(P)$.*

PROOF. Let σ be an automorphism that commutes with every transvection in $K(P)$, and let $\tau_p \in K(P)$. Since $\sigma(v) = a + v\lambda$, $a \in P$, $\lambda \in \Delta$, it follows from $\tau_p(a + v\lambda) = \sigma(v + p)$ that

$$a + (v + p)\lambda = a + v\lambda + \sigma(p),$$

that is, $\sigma(p) = p\lambda$ for any $p \in P$. Now for $\delta \in \Delta$ we have $\sigma(p\delta) = (p\delta)\lambda = \sigma(p)\delta = p\lambda\delta$. If p is a free element, this implies that $\delta\lambda = \lambda\delta$, so that λ lies in the center $Z(\Delta)$ of Δ. Thus

$$\sigma(p) = p\lambda, \quad p \in P, \quad \sigma(v) = a + v\lambda, \quad a \in P, \quad \lambda \in Z(\Delta). \tag{6}$$

The automorphism σ^{-1} also commutes with every transvection in $K(P)$, and so by (6)

$$\sigma^{-1}(p) = p\mu, \quad \sigma^{-1}(v) = v\mu + b, \quad \mu \in Z(\Delta), \quad b \in P.$$

Furthermore,

$$v = e(v) = \sigma^{-1}\sigma(v) = \sigma^{-1}(a + v\lambda) = a\mu + v\mu\lambda + b\lambda,$$

whence

$$v(1 - \mu\lambda) - (a\mu + b\lambda) = 0. \tag{7}$$

Since (5) is a direct sum, it follows from (7) that $v(1 - \mu\lambda) = 0$. But v is a free element, and so $1 = \mu\lambda$, $\lambda \in Z(\Delta)^*$.

Now consider the product $\tau^{-1}\sigma$, where $\tau = \tau_{a\lambda^{-1}}$. For $p \in P$

$$\tau^{-1}\sigma(p) = \tau^{-1}(p\lambda) = p\lambda.$$

Since $\tau^{-1} = \tau_{-a\lambda^{-1}}$, we have

$$\tau^{-1}\sigma(v) = \tau^{-1}(a + v\lambda) = a + (v - a\lambda^{-1})\lambda = v\lambda.$$

Moreover, $\tau^{-1}\sigma(v\delta) = (v\lambda)\delta = (v\delta)\lambda$, $\delta \in \Delta$. Therefore $\tau^{-1}\sigma(x) = x\lambda$ for any element x of M; that is, $\tau^{-1}\sigma \in H(M)$. Hence $\sigma \in K(P)H(M)$.

On the other hand, every element of $K(P)H(M)$ obviously commutes with every transvection in $K(P)$. This proves (i).

It is clear that $K(P)H(M)$ is an abelian group, so that (ii) follows from (i).

Now let $\sigma \in K(P)H(M)$. Then for $p \in P$ we have

$$\sigma(p) = p\lambda, \quad \sigma(v) = v\lambda + a_1, \quad a_1 \in P, \quad \lambda \in Z(\Delta)^*.$$

Next, $(\sigma - e)(v) = v(\lambda - 1) + a_1$, and

$$(\sigma - e)^k(v) = v(\lambda - 1)^k + a_k, \quad a_k \in P. \tag{8}$$

If σ is a unipotent element, then $(\sigma - e)^k = 0$ for some k. It now follows from (8) that $v(\lambda - 1)^k = 0$ and, since v is a free element, $(\lambda - 1)^k = 0$, so that λ is unipotent. On the other hand, the product of a transvection τ and a unipotent homothety ρ is again a unipotent element, because τ and ρ commute. ■

Theorem 2 is valid, in particular, for a free module having a basis of at least two elements.

COROLLARY 1. *If Δ is a ring without divisors of zero and a Δ-module M has two decompositions $M = P \dot{+} v\Delta$ and $M = Q \dot{+} w\Delta$, where v and w are free elements, the hyperplanes P and Q each contain at least one free element, and $P + Q = M$, then the centralizer C of the union $K(P) \cup K(Q)$ in* Aut M *is $H(M)$.*

PROOF. We have $C = K(P)H(M) \cap K(Q)H(M)$. If $\sigma \in C$, then $\sigma = \tau_p\rho$, $\rho \in H(M)$, $p \in P$. But $H(M) \subset C$; therefore $\tau_p \in C$, and hence

$$\tau_p = \bar{\tau}_q \rho_\alpha, \quad q \in Q, \quad \bar{\tau}_q \in K(Q), \quad \rho_\alpha \in H(M).$$

Since $P + Q = M$, the hyperplane P contains an element a such that $a = b + w\delta$, $b \in Q$, $\delta \in \Delta$, $\delta \neq 0$. Now,

$$a = \tau_p(a) = \bar{\tau}_q \rho_\alpha(a) = \rho_\alpha \bar{\tau}_q(b + w\delta) = (b + (w + q)\delta)\alpha.$$

Therefore $b + w\delta = (b + (w + q)\delta)\alpha$ and $b - b\alpha - q\delta\alpha = w\delta(\alpha - 1)$. Hence $w\delta(\alpha - 1) = 0$, and, since w is a free element, $\alpha = 1$. Thus

$$\tau_p = \bar{\tau}_q = e, \quad \sigma = \rho, \quad C = H(M). \ \blacksquare$$

COROLLARY 2. *Let Δ be an arbitrary ring, M a free Δ-module with a basis $B = \{a, b, \dots\}$ containing at least two elements, and P and Q the hyperplanes generated by the sets $B \backslash a$ and $B \backslash b$, respectively. Then the centralizer C of the union $K(P) \cup K(Q)$ is $H(M)$.*

PROOF. We have $M = P + a\Delta$, $b \in P$ and $M = Q + b\Delta$. Thus, by Theorem 2, $C = K(P)H(M) \cap K(Q)H(M)$. Let $\sigma \in C$. Then $\sigma = \tau_p \rho$, $p \in P$, $\rho \in H(M)$. Since $H(M) \subset C$, we have $\tau_p \in C$. Hence

$$\tau_p = \rho_\alpha \bar{\tau}_q, \quad \rho_\alpha \in H(M), \quad q \in Q, \quad \bar{\tau}_q \in K(Q).$$

Now $a + p = \tau_p(a) = \rho_\alpha \bar{\tau}_q(a) = a\alpha$. Consequently $p = 0$, $\alpha = 1$, $\tau_p = e$, $\sigma = \rho \in H(M)$ and $C = H(M)$. ■

Now suppose that either $Z(\Delta^*) = Z(\Delta)^*$ or M contains a basis B of at least two elements. Then we have

COROLLARY 3. *The center of $GL(M)$ is $H(M)$.*

PROOF. If $M = u\Delta$, where u is a free element, then the mapping $\gamma \longmapsto \sigma_\gamma$, where $\sigma_\gamma(u) = u\gamma$, is an isomorphism of Δ^* onto $GL(M)$. Consequently $H(M)$ is the center of $GL(M)$. If some basis of M contains more than one element, the assertion follows from Corollary 2. ■

Now let $\Delta = T$ be a division ring, and M a vector space over T of dimension >1.

LEMMA 7. *Any two nonidentity transvections of M are conjugate in $GL(M)$.*

PROOF. Suppose that

$$M = P \dotplus vT = Q \dotplus wT, \quad \tau_p \in K(P), \quad \bar{\tau}_q \in K(Q),$$
$$\tau_p, \ \bar{\tau}_q \neq e.$$

The module M has bases $\{A, v\}$ and $\{B, w\}$ such that A is a basis of P and B a basis of Q, with $p \in A$ and $q \in B$. By Lemma 4, $GL(M)$ contains an automorphism σ such that $\sigma(A) = B$, $\sigma(p) = q$ and $\sigma(v) = w$. It is easy to see that $\sigma\tau_p\sigma^{-1} = \bar{\tau}_q$. In fact,

$$\sigma\tau_p\sigma^{-1}(b) = \sigma\tau_p(a) = \sigma(a) = b, \qquad a = \sigma^{-1}(b), \qquad b \in B$$

and

$$\sigma\tau_p\sigma^{-1}(w) = \sigma\tau_p(v) = \sigma(v + p) = w + q. \ \blacksquare$$

§8. Matrix representations of an endomorphism

1. *The matrix of an endomorphism.* Let M be a right free Δ-module having a finite basis

$$u_1, \ldots, u_n. \tag{1}$$

If σ is an endomorphism of M, then

$$\sigma(u_\nu) = \sum_{\mu=1}^{n} u_\mu \alpha_{\mu\nu}, \quad \alpha_{\mu\nu} \in \Delta, \quad \nu = 1, \ldots, n. \tag{2}$$

We set $a = \|\alpha_{\mu\nu}\|$ and call *a the matrix of σ in the basis* (1).

We denote by Δ_n the set of all $n \times n$ matrices over Δ, and consider the mapping

$$\varphi\colon \operatorname{End} M \to \Delta_n, \quad \sigma \mapsto a. \tag{3}$$

By Lemma 7.2, φ is a bijection.

THEOREM 1. *The mapping φ is an isomorphism of the rings* End M *and* Δ_n.

PROOF. Let $\sigma, \tau \in \operatorname{End} M$ and $\varphi(\sigma) = a = \|\alpha_{\mu\nu}\|$, $\varphi(\tau) = b = \|\beta_{\mu\nu}\|$. It follows from (3) that $\varphi(\sigma + \tau) = a + b = \varphi(\sigma) + \varphi(\tau)$. For $\sigma\tau$ we can write

$$\begin{aligned}
\sigma\tau(u_\nu) &= \sigma\left(\sum_{\mu=1}^{n} u_\mu\beta_{\mu\nu}\right) \\
&= \sum_{\mu=1}^{n} \sigma(u_\mu)\beta_{\mu\nu} = \sum_{\mu=1}^{n}\left(\sum_{k=1}^{n} u_k\alpha_{k\mu}\right)\beta_{\mu\nu} \\
&= \sum_{k=1}^{n} u_k\left(\sum_{\mu=1}^{n}\alpha_{k\mu}\beta_{\mu\nu}\right) = \sum_{k=1}^{n} u_k\gamma_{k\nu},
\end{aligned}$$

where $\|\gamma_{k\nu}\| = ab$. Consequently $\varphi(\sigma\tau) = \varphi(\sigma)\varphi(\tau)$. ■

The multiplicative group of Δ_n is denoted by $GL(n, \Delta)$. Since the multiplicative groups of isomorphic rings are isomorphic, Theorem 1 implies

COROLLARY 1. $GL(M) \cong GL(n, \Delta)$.

$GL(n, \Delta)$ is called the *general linear group* of degree n over Δ. In what follows we make no essential distinction between the groups $GL(M)$ and $GL(n, \Delta)$.

LEMMA 1a. *A matrix of Δ_n is invertible in Δ_n if and only if its columns form a basis of the right free module Δ^n.*

PROOF. Let $a = \varphi(\sigma)$ (see (3)). Then a is invertible in Δ_n if and only if σ is an automorphism; that is,

$$\sigma(u_1), \ldots, \sigma(u_n) \tag{4}$$

is a basis of M. If

$$a_1, \ldots, a_n \tag{5}$$

are the columns of a, then by (2) the equalities

$$\sigma(u_1)\gamma_1 + \ldots + \sigma(u_n)\gamma_n$$
$$= u_1\lambda_1 + u_2\lambda_2 + \ldots + u_n\lambda_n, \qquad \gamma_i, \lambda_k \in \Delta$$

$$a_1\gamma_1 + \ldots + a_n\gamma_n = \begin{bmatrix} \lambda_1 \\ \cdot \\ \cdot \\ \cdot \\ \lambda_n \end{bmatrix}$$

are equivalent. Thus (4) is a basis of M if and only if (5) is a basis of Δ^n. ∎

Similar arguments for left modules prove the following result.

LEMMA 1b. *A matrix of Δ_n is invertible in Δ_n if and only if its rows form a basis of the left Δ-module Δ_L^n.*

THEOREM 2. *If a is the matrix of an endomorphism σ in the basis* (1), *and b the matrix of an automorphism τ in the same basis, then $b^{-1}ab$ is the matrix of σ in the basis*

$$\tau(u_1), \ldots, \tau(u_n). \tag{6}$$

PROOF. We set $\sigma_1 = \tau^{-1}\sigma\tau$. By Theorem 1, $b^{-1}ab$ is the matrix of σ_1 in the basis (1). On the other hand, the matrix of σ in the basis (6) is the same as that of σ_1 in the basis (1). Indeed, $\sigma = \tau\sigma_1\tau^{-1}$, and if

$$\sigma_1(u_\nu) = \sum_{\mu=1}^{n} u_\mu \lambda_{\mu\nu},$$

then

$$\tau\sigma_1\tau^{-1}(\tau(u_\nu)) = \tau\sigma_1(u_\nu) = \tau\left(\sum_{\mu=1}^{n} u_\mu\lambda_{\mu\nu}\right) = \sum_{\mu=1}^{n} \tau(u_\mu)\lambda_{\mu\nu}.$$

This implies the assertion. ∎

2. *Burnside's theorem.*

THEOREM 3 (BURNSIDE [3]). *Let Δ be a principal ideal domain, Σ the field of fractions of Δ, and G a subgroup of $GL(n, \Sigma)$ with the following property:*

Δ contains a nonzero element δ such that $g\delta \in \Delta_n$ for every $g \in G$. Then $GL(n, \Sigma)$ contains a matrix t such that $t^{-1}Gt \subseteq GL(n, \Delta)$.

PROOF. We select a basis

$$u_1, \ldots, u_n \tag{7}$$

in Σ^n and make no distinction between matrices $g \in G$ and the automorphisms of Σ^n that they define relative to the basis (7). For $g \in G$ we have

$$g(u_j) = \sum_{i=1}^{n} u_i\gamma_{ij}, \quad \gamma_{ij} \in \Sigma, \quad \gamma_{ij}\delta \in \Delta, \quad j = 1, \ldots, n. \tag{8}$$

Let M be a right free Δ-module with the basis (7). The space Σ^n can also be regarded as a Δ-module. By means of (8), each element g of G determines a homomorphism of Δ-modules

$$g\colon M \to \Sigma^n. \tag{9}$$

Now let N be the subset of all $v \in M$ such that $g(v) \in M$ for any $g \in G$. Obviously N is a submodule of M. The Δ-module N contains n linearly independent elements $u_1\delta, \ldots, u_n\delta$. Since Δ is a principal ideal domain, N is a free Δ-module of dimension n. Finally, N is invariant under G, because for $v \in N$ and $h, g \in G$ we have $g(v) \in M$ and $hg(v) \in M$; that is, $g(v) \in N$. If

$$v_1, \ldots, v_n \tag{10}$$

is a Δ-basis of N, then the elements (10) are linearly independent over Σ. For every element $\sigma \in \Sigma$ has the form $\delta\nu^{-1}$, where $\delta, \nu \in \Delta$. If

$$v_1\delta_1\nu_1^{-1} + \ldots + v_n\delta_n\nu_n^{-1} = 0 \tag{11}$$

and $\delta_n \neq 0$, then, multiplying (11) by $\nu_1 \cdots \nu_n$, we obtain

$$v_1x_1 + \ldots + v_nx_n = 0, \quad x_i \in \Delta, \quad x_n \neq 0. \tag{12}$$

This is impossible, and so (10) is a basis of Σ^n. Since

$$g(v_j) = \sum_{i=1}^{n} v_i\alpha_{ij}, \quad \alpha_{ij} \in \Delta, \quad j = 1, \ldots, n,$$

the theorem now follows from Theorem 2. ∎

COROLLARY 1. *If Δ is a principal ideal domain, Σ the field of fractions of Δ, and G a finite subgroup of $GL(n, \Sigma)$, then $GL(n, \Sigma)$ contains a matrix t such that $t^{-1}Gt \subset GL(n, \Delta)$.*

3. *Elementary matrices $t_{ij}(\lambda)$. The special linear group.* Let M be a right free Δ-module with the basis

$$u_1, \ldots, u_n, \quad n > 1. \tag{13}$$

We consider the transvection $\tau = \tau_{ij}(\lambda)$, $i \neq j$, defined by

$$\tau(u_\alpha) = u_\alpha, \qquad \alpha \neq j,$$
$$\tau(u_j) = u_j + u_i\lambda, \quad \lambda \in \Delta.$$

It is obvious that the matrix of τ relative to the basis (13) is

$$t_{ij}(\lambda) = E + e_{ij}\lambda, \tag{14}$$

where E is the identity matrix and $e_{ij} = \|a_{\mu\nu}\|$, $a_{ij} = 1$, and $a_{\mu\nu} = 0$ if $\mu \neq i$ or $\nu \neq j$. The matrix (14) is said to be *elementary*.

In particular, $E = t_{ij}(0)$.

The subgroup $SL(n, \Delta)$ of $GL(n, \Delta)$ generated by all elementary matrices $t_{ij}(\lambda)$ is called the *special linear group*. We can also regard the special linear group as the subgroup $SL(M)$ of $GL(M)$ generated by all the transvections $\tau_{ij}(\lambda)$.

If P_j is the submodule of M with the basis $u_1, \ldots, u_{j-1}, u_{j+1}, \ldots, u_n$, (see (13)), then $M = P_j \dotplus u_j\Delta$, and for $p_j \in P_j$ we have

$$p_j = \sum_{i \neq j} u_i\lambda_i, \quad \lambda_i \in \Delta.$$

Consequently, for any transvection τ_{p_j} in $K(P_j)$,

$$\tau_{p_j} = \prod_{i \neq j} \tau_{ij}(\lambda_i).$$

Together with Corollary 7.2.1, this implies

LEMMA 2. *The centralizer of the set of all the transvections of the form $\tau_{ij}(\lambda)$ in $GL(M)$ is $H(M)$.*

This lemma implies

THEOREM 4. *If $n > 1$, then the centralizer of $SL(n, \Delta)$ in $GL(n, \Delta)$ is $Z(\Delta)^*E_n$, where $Z(\Delta)$ is the center of Δ.*

LEMMA 3 (BASS [1]). *If $n > 2$ and i, j, k are distinct indices $\leqslant n$, then*

$$t_{ij}(\lambda\mu) = t_{ik}(\lambda)\, t_{kj}(\mu)\, t_{ik}^{-1}(\lambda)\, t_{kj}^{-1}(\mu).$$

In particular, $t_{ij}(\lambda) = t_{ik}(\lambda)t_{kj}(1)t_{ik}^{-1}(\lambda)t_{kj}^{-1}(1)$.

For if $i \neq j$, then

$$t_{ij}(\lambda)\, t_{ij}(\mu) = t_{ij}(\lambda + \mu), \quad t_{ij}^{-1}(\lambda) = t_{ij}(-\lambda).$$

Further,

$$\begin{aligned} t_{ik}(\lambda)\, t_{kj}(\mu) &= E + e_{ik}\lambda + e_{kj}\mu + e_{ij}\lambda\mu, \\ t_{ik}(-\lambda)\, t_{kj}(-\mu) &= E - e_{ik}\lambda - e_{kj}\mu + e_{ij}\lambda\mu. \end{aligned} \tag{15}$$

To complete the proof we need only multiply the two equalities of (15). ∎

An immediate consequence of Lemma 3 is

THEOREM 5. *If $n > 2$, then $SL(n, \Delta)$ is its own derived group and is therefore contained in the derived group of $GL(n, \Delta)$.*

We now turn to the case $n = 2$.

LEMMA 4. *If Δ^* contains an element α such that $\alpha - 1 \in \Delta^*$, then every elementary matrix in $GL(2, \Delta)$ is a commutator in $GL(2, \Delta)$.*

PROOF. If

$$a=\begin{bmatrix}\alpha & 0\\ 0 & 1\end{bmatrix},\quad b=\begin{bmatrix}1 & \mu\\ 0 & 1\end{bmatrix},\quad \mu=(\alpha-1)^{-1}\lambda,$$

then $t_{12}(\lambda) = aba^{-1}b^{-1}$. Now let

$$c=t_{21}(\gamma),\quad \gamma=-\lambda(\alpha-1)^{-1},\quad d=a^{-1};$$

then $t_{21}(\lambda) = cdc^{-1}d^{-1}$. ■

COROLLARY 1. *If $\Delta = T$ is a division ring other than $GF(2)$, then every elementary matrix in $GL(2, T)$ is a commutator.*

However, if $T = GF(2)$, no elementary matrix (other than the identity) in $GL(2, T)$ is a commutator.

COROLLARY 2. *If Δ^* contains an element α such that $\alpha - 1 \in \Delta^*$, then $SL(2, \Delta)$ is a subgroup of the derived group of $GL(2, \Delta)$.*

LEMMA 5. *If (13) is a basis of M, then for any pair α, β, $1 \leqslant \alpha, \beta \leqslant n$, $\alpha \neq \beta$, the group $SL(M)$ contains an element $s_{\alpha\beta}$ such that*

$$s_{\alpha\beta}(u_\alpha)=u_\beta,\quad s_{\alpha\beta}(u_\beta)=-u_\alpha,\quad s_{\alpha\beta}(u_i)=u_i,\qquad i\neq\alpha,\ \beta.$$

PROOF. We set $s_{\alpha\beta} = \tau_{\beta\alpha}(1)\tau_{\alpha\beta}(-1)\tau_{\beta\alpha}(1)$. It is obvious that this element has the required properties. ■

Let t be a permutation acting on the set $\{u_1, \ldots, u_n\}$ of basis elements of the module M. This permutation can be uniquely extended to an automorphism f of M. We can embed the symmetric group S_n in $GL(n, \Delta)$ by identifying t with the matrix of f in the basis (13). In particular, a cycle $(u_\alpha, u_\beta, u_\gamma)$ of length 3 can be represented as

$$(u_\alpha, u_\beta, u_\gamma)=s_{\alpha\beta}s_{\beta\gamma}\in SL(n, \Delta).$$

Thus from Lemma 5 we obtain

COROLLARY 1. *Every even permutation is an element of $SL(n, \Delta)$. If t is an odd permutation, then $t = f_\alpha s$, $f_\alpha(u_\alpha) = -u_\alpha$ and $f_\alpha(u_i) = u_i$ $(i \neq \alpha)$ for $s \in SL(n, \Delta)$.*

4. *Formal lemmas.* Let Δ^n be the right free Δ-module of column vectors. If $x \in \Delta^n$ and $c \in \Delta_n$, then $cx \in \Delta^n$. It is readily checked that the mapping $h\colon \Delta^n \longrightarrow \Delta^n$, $x \longmapsto cx$, is an endomorphism of Δ^n. Let $c = \|\gamma_{ij}\|$. The vectors

$$b_1=\begin{bmatrix}1\\0\\ \vdots\\ 0\end{bmatrix},\quad b_2=\begin{bmatrix}0\\1\\ \vdots\\ 0\end{bmatrix},\ \ldots,\ b_n=\begin{bmatrix}0\\0\\ \vdots\\ 1\end{bmatrix}\tag{16}$$

form a basis of Δ^n.

Now,

$$h(b_\nu) = cb_\nu = \begin{bmatrix} \gamma_{1\nu} \\ \gamma_{2\nu} \\ \cdot \\ \cdot \\ \cdot \\ \gamma_{n\nu} \end{bmatrix} = b_1\gamma_{1\nu} + b_2\gamma_{2\nu} + \dots + b_n\gamma_{n\nu},$$

$$\nu = 1, \dots, n.$$

Consequently the matrix of h in the basis (16) is c. Conversely, any endomorphism h of Δ^n reduces to multiplication of column vectors by the matrix of h in the basis (16). This justifies our making no distinction between an endomorphism and its matrix in the basis (16).

LEMMA 6 (BASS [1]). *For $n > 1$, the additive group generated by $SL(n, \Delta)$ is the same as the additive group of the matrix ring Δ_n.*

PROOF. Let D be the additive group generated by $SL(n, \Delta)$. For $i \neq j$ and arbitrary $\lambda \in \Delta$ we have $e_{ij}\lambda = t_{ij}(\lambda) - E_n \in D$. Furthermore, $e_{ii}\lambda = t_{ij}(\lambda)t_{ji}(1) - t_{ij}(\lambda) - t_{ji}(1) - E_n \in D$. ■

If Φ is a subgroup of the additive group of a ring Δ, we denote by Φ^n the subgroup of the additive group of Δ^n consisting of all column vectors

$$\begin{bmatrix} \varphi_1 \\ \cdot \\ \cdot \\ \cdot \\ \varphi_n \end{bmatrix}, \quad \varphi_\nu \in \Phi.$$

LEMMA 7 (BASS [1]). *Let $n > 1$. A subgroup L of the additive group of the module Δ^n is invariant under $SL(n, \Delta)$ if and only if $L = \Phi^n$, where Φ is a left ideal in Δ.*

PROOF. By virtue of Lemma 6, we may replace $SL(n, \Delta)$ in the statement of the lemma by the ring Δ_n. Let L be a subgroup of the additive group of Δ^n invariant under Δ_n. Consider the mapping

$$\pi_\nu\colon L \to \Delta, \qquad x = \begin{bmatrix} x_1 \\ \cdot \\ \cdot \\ \cdot \\ x_n \end{bmatrix} \mapsto x_\nu.$$

We set $\mathrm{Im}(\pi_\nu) = \Phi_\nu$. It is clear that Φ_ν is a subgroup of the additive group of Δ. But Φ_ν is also a left ideal in Δ. For let $a_\nu \in \Phi_\nu$. Then L contains the column vector

$$a = \begin{bmatrix} \alpha_1 \\ \cdot \\ \cdot \\ \alpha_\nu \\ \cdot \\ \cdot \\ \cdot \\ \alpha_n \end{bmatrix}.$$

If $\lambda \in \Delta$, then $(e_{\nu\nu}\lambda)a \in L$. But $(e_{\nu\nu}\lambda)a = b_\nu(\lambda\alpha_\nu)$. Therefore $\lambda\alpha_\nu \in \Phi_\nu$, and Φ_ν is a left ideal. Again, let $\alpha_\nu \in \Phi_\nu$ and

$$a = \begin{bmatrix} \alpha_1 \\ \cdot \\ \cdot \\ \cdot \\ \alpha_\nu \\ \cdot \\ \cdot \\ \cdot \\ \alpha_n \end{bmatrix} \in L.$$

Then for any $i = 1, \dots, n$,

$$e_{i\nu}a \in L, \qquad e_{i\nu}a = b_i\alpha_\nu.$$

Hence $\alpha_\nu \in \Phi_i$. Thus $\Phi_1 = \cdots = \Phi_n = \Phi$. It is obvious that $L \subset b_1\Phi + \cdots + b_n\Phi = \Phi^n$. On the other hand, if $\alpha_\nu \in \Phi$, then

$$a = \begin{bmatrix} \alpha_1 \\ \cdot \\ \cdot \\ \cdot \\ \alpha_\nu \\ \cdot \\ \cdot \\ \cdot \\ \alpha_n \end{bmatrix} \in L, \quad e_{\nu\nu}a = b_\nu\alpha_\nu \in L, \qquad L = \Phi^n.$$

This proves necessity.

Now for sufficiency. Let Φ be an arbitrary left ideal in Δ. We claim that the additive subgroup Φ^n of Δ^n is invariant under Δ_n. Let

$$b = \begin{bmatrix} \beta_1 \\ \cdot \\ \cdot \\ \cdot \\ \beta_n \end{bmatrix} \in \Phi^n.$$

Then $\beta_j \in \Phi$ and $(e_{ij}\lambda)b = b_i(\lambda\beta_j) \in \Phi^n$, since $\lambda\beta_j \in \Phi$. ∎

5. *The affine group.* Let $\mathrm{Aff}(n, \Delta)$ be the set of all pairs of the form

$$\langle g, a\rangle, \qquad g \in GL(n, \Delta), \qquad a \in \Delta^n. \tag{17}$$

A pair (17) will be interpreted as a mapping

$$\langle g, a\rangle\colon \Delta^n \to \Delta^n, \qquad x \mapsto gx + a.$$

It is clear that (17) is a bijection; that is, $\mathrm{Aff}(n, \Delta) \subset S(\Delta^n)$. If $\langle h, b\rangle$ is another pair in Aff(n, Δ), then

$$\langle g, a\rangle\langle h, b\rangle(x) = \langle g, a\rangle(hx+b) = ghx + gb + a$$
$$= \langle gh, gb+a\rangle(x);$$

that is,

$$\langle g, a\rangle\langle h, b\rangle = \langle gh, gb+a\rangle. \tag{18}$$

Thus Aff(n, Δ) is a group, which we call the *affine group*.

The unit element Aff(n, Δ) is obviously $\langle E_n, (0)\rangle$, and

$$\langle g, a\rangle^{-1} = \langle g^{-1}, -g^{-1}a\rangle. \tag{19}$$

It is also clear from (18) that

$$\gamma\colon \mathrm{Aff}(n, \Delta) \to GL(n, \Delta), \qquad \langle g, a\rangle \mapsto g, \tag{20}$$

is an epimorphism. Further,

$$\mathrm{Ker}\,\gamma = \{\langle E_n, a\rangle \mid a \in \Delta^n\} = \langle E_n, \Delta^n\rangle,\text{[1]}$$

so that

$$GL(n, \Delta) \cong \mathrm{Aff}(n, \Delta)/\langle E_n, \Delta^n\rangle.$$

A pair $\langle g, a\rangle \in$ Aff(n, Δ) can also be treated as an $(n+1)\times(n+1)$ matrix

$$\begin{bmatrix} \gamma_{11} & \cdots & \gamma_{1n} & \alpha_1 \\ \cdot & \cdot & \cdot & \cdot \\ \gamma_{n1} & \cdots & \gamma_{nn} & \alpha_n \\ 0 & \cdots & 0 & 1 \end{bmatrix}, \qquad \|\gamma_{ij}\| = g, \qquad \begin{bmatrix} \alpha_1 \\ \vdots \\ \alpha_n \end{bmatrix} = a,$$

which we write thus:

$$\begin{bmatrix} g & a \\ 0 & 1 \end{bmatrix}.$$

This representation of the elements of the affine group makes sense, because

$$\begin{bmatrix} g & a \\ 0 & 1 \end{bmatrix}\begin{bmatrix} h & b \\ 0 & 1 \end{bmatrix} = \begin{bmatrix} gh & gb+a \\ 0 & 1 \end{bmatrix}.$$

Thus Aff(n, Δ) $\subset GL(n+1, \Delta)$. The elements $\langle E_n, a\rangle \in \langle E_n, \Delta^n\rangle$ are represented here by matrices of the form

$$\begin{bmatrix} E_n & a \\ 0 & 1 \end{bmatrix},$$

so that $\langle E_n, \Delta^n\rangle \subset SL(n+1, \Delta)$.

From (18) and (19) it is easy to obtain the formula

$$\langle g, 0\rangle\langle E_n, a\rangle\langle g, 0\rangle^{-1} = \langle E_n, ga\rangle,$$

from which we derive

LEMMA 8. *Let L be a subgroup of the additive group of Δ^n and H a subgroup of $GL(n, \Delta)$. Then $\langle H, (0)\rangle$ is a subgroup of the normalizer of $\langle E_n, L\rangle$ in Aff(n, Δ) if and only if $HL = L$.*

[1]Henceforth we use notation of the kind $\langle G, L\rangle = \{\langle g, l\rangle \mid g \in G, l \in L\}$ without further explanation.

In what follows we often use expressions like "a subgroup A is *normalized* by a subgroup B of a group G" to mean that "B is contained in the normalizer of A in G."

LEMMA 9 (BASS [1]). *Let G be a subgroup of* $\mathrm{Aff}(n, \Delta)$, γ *the epimorphism* (20), *and* $H = \gamma(G)$. *Then the following assertions are true:*

(i) *The commutator subgroup K of G and $\langle E_n, \Delta^n\rangle$ satisfies the relation*

$$K = (G, \langle E_n, \Delta^n\rangle) = (\langle H, (0)\rangle, \langle E_n, \Delta^n\rangle) = \langle E_n, A\rangle,$$

where $A = \Sigma(h - E_n)\Delta^n$, $h \in H$, *is the additive group generated by all column vectors* $(h - E_n)a$, $h \in H$, $a \in \Delta^n$.

(ii) *If G is normalized by $\langle E_n, \Delta^n\rangle$ and*

$$G \cap \langle E_n, \Delta^n\rangle = \langle E_n, (0)\rangle, \tag{21}$$

then $G = \langle E_n, (0)\rangle$.

(iii) *If G is normalized by* $\langle SL(n, \Delta), (0)\rangle$, *then* $K = \langle E_n, \Phi^n\rangle$, *where Φ is a left ideal in Δ.*

PROOF. It follows from (18) and (19) that for $\langle h, b\rangle \in G$ and $\langle E_n, a\rangle \in \langle E_n, \Delta^n\rangle$ we have

$$\langle h, b\rangle \langle E_n, a\rangle \langle h, b\rangle^{-1} \langle E_n, a\rangle^{-1} = \langle E_n, (h - E_n)\, a\rangle,$$

and this gives (i).

Now (ii) follows from (i). For if A is normalized by $\langle E_n, \Delta^n\rangle$, then $K \subset A \cap \langle E_n, \Delta^n\rangle$. If, moreover, (21) is true, then by (i) $H = (E_n)$, $A \subset \langle E_n, \Delta^n\rangle$ and $A = \langle E_n, (0)\rangle$.

It remains to prove (iii). Suppose that A is normalized by $\langle SL(n, \Delta), (0)\rangle$. By Lemma 8 $\langle E_n, \Delta^n\rangle$ is also normalized by this group, and so is K. By (i) we have $K = \langle E_n, D\rangle$, and $SL(n, \Delta)D = D$ by Lemma 8. Consequently, by Lemma 7, $D = \Phi^n$, where Φ is a left ideal of Δ. ∎

§9. The Dieudonné determinant

1. *The matrices* $d(\mu)$. Let $\Delta = T$ be a division ring. We define matrices $d(\mu) \in GL(n, T)$, $n > 1$, by

$$d(\mu) = \mathrm{diag}\,[1, \ldots, 1, \mu], \quad \mu \in T^{*}, \tag{1}$$

so that $d(\mu)$ is the diagonal matrix with the elements $1, \ldots, 1, \mu$ along the principal diagonal.

LEMMA 1. *If* $a \in GL(n, T)$, *then* $a = sd(\mu)$, $s \in SL(n, T)$.

PROOF. Multiplication of a on the left by an elementary matrix $t_{ij}(\lambda)$ is equivalent to the following elementary operation: addition of the jth row of a, multiplied on the left by λ, to its ith row. The lemma will be proved if we can show that a matrix $a = \|\alpha_{ij}\|$ of $GL(n, T)$ can be brought to the form (1) by a sequence of such elementary operations.

By Lemma 8.1a, the columns of $a = \|\alpha_{ij}\|$ form a basis of T^n. Consequently the first column contains a nonzero element. We may assume that $\alpha_{21} \neq 0$. (If α_{11} is the only nonzero element in the first column, we begin by adding the first row to the second, unless $\alpha_{11} = 1$, in which case we proceed at once to the next step.*) Adding the second row of a, multiplied on the left by $(1 - \alpha_{11})\alpha_{21}^{-1}$, to its first row, we obtain a matrix with 1 in the top left corner. Similarly, adding the first row, multiplied on the left by a suitable constant, to each row in succession, we bring the matrix a to the form

$$\begin{bmatrix} 1 & \alpha_{12} \ \dots \ \alpha_{1n} \\ 0 & \\ \cdot & \\ \cdot & A_{n-1} \\ \cdot & \\ 0 & \end{bmatrix} = a_1.$$

But $a_1 \in GL(n, T)$, and therefore the columns of this matrix again form a basis for T^n. Taking the form of a_1 into account, we conclude that the first column of A_{n-1} contains an element of T^*. If $n - 1 > 1$, we can apply the previous reasoning to A_{n-1}. Thus, if $n > 2$ we may assume that the first column of A_{n-1} is

$$\begin{bmatrix} 1 \\ 0 \\ \cdot \\ \cdot \\ \cdot \\ 0 \end{bmatrix}.$$

Now α_{12} can be eliminated by an elementary operation. Proceeding in this way we arrive at a matrix of the form

$$a_2 = \begin{bmatrix} 1 & 0 & \dots & 0 & \alpha_{1n} \\ 0 & 1 & \dots & 0 & \alpha_{2n} \\ \cdot & \cdot & \cdot & \cdot & \cdot \\ 0 & 0 & \dots & 1 & \alpha_{n-1,\,n} \\ 0 & 0 & \dots & 0 & \alpha_{nn} \end{bmatrix}.$$

Since $a_2 \in GL(n, T)$, the columns of a_2 again form a basis for T^n, so that $\alpha_{nn} = \mu \neq 0$. Obviously a_2 can be brought to the form (1) by elementary operations. ∎

COROLLARY 1. *The matrices $t_{ij}(\lambda)$ and $d(\mu)$ together form a system of generators for $GL(n, T)$.*

LEMMA 2. *Let $n > 1$, $\alpha, \beta \in T^*$ and $\mu = \alpha\beta\alpha^{-1}\beta^{-1}$. Then $d(\mu) \in SL(n, T)$.*

PROOF. In the sequence

*Editor's insertion

$$\begin{bmatrix} 1 & 0 \\ 0 & 1 \end{bmatrix}, \begin{bmatrix} 1 & 0 \\ \alpha^{-1} & 1 \end{bmatrix}, \begin{bmatrix} 0 & -\alpha \\ \alpha^{-1} & 1 \end{bmatrix}, \begin{bmatrix} 0 & -\alpha \\ \alpha^{-1} & \beta^{-1} \end{bmatrix},$$

$$\begin{bmatrix} \alpha\beta\alpha^{-1} & 0 \\ \alpha^{-1} & \beta^{-1} \end{bmatrix}, \begin{bmatrix} \alpha\beta\alpha^{-1} & 0 \\ 1 & \beta^{-1} \end{bmatrix}, \begin{bmatrix} 0 & -\mu \\ 1 & \beta^{-1} \end{bmatrix},$$

$$\begin{bmatrix} 0 & -\mu \\ 1 & \mu \end{bmatrix}, \begin{bmatrix} 1 & 0 \\ 1 & \mu \end{bmatrix}, \begin{bmatrix} 1 & 0 \\ 0 & \mu \end{bmatrix}$$

each matrix is obtained from its predecessor by one elementary operation. Thus for $n > 1$ the matrix E_n can be brought to the form $d(\mu)$ by elementary operations. This proves the lemma. ∎

2. *The determinant of a matrix over a division ring.* Let $\overline{T^*}$ be the factor group of T^* by its derived group. To $\overline{T^*}$ we adjoin a zero $\bar{0}$ by defining $\bar{0}\overline{T^*} = \overline{T^*}\,\bar{0} = \bar{0}$, and we denote the resulting commutative semigroup by $\overline{T}$. If $T^* \longrightarrow \overline{T^*}$ is the canonical homomorphism, we denote the image of an element $\alpha \in T^*$ by $\bar{\alpha}$. Extending this mapping, we set $\bar{\alpha} = \bar{0}$ for $\alpha = 0$. Then, for $\alpha, \beta \in T$,

$$\overline{\alpha\beta} = \bar{\alpha}\bar{\beta} = \bar{\beta}\bar{\alpha} = \overline{\beta\alpha}.$$

It is clear that the unit element of the semigroup $\overline{T}$ is $\bar{1}$.

Our aim is to prove the existence of a mapping

$$\delta\colon T_n \to \overline{T}, \tag{2}$$

satisfying three conditions:

I. *If b is a matrix obtained from a by multiplication of one row on the left by α, then $\delta(b) = \bar{\alpha}\delta(a)$.*

II. *If b is a matrix obtained from a by adding one row to another, then* $\delta(b) = \delta(a)$.

III. $\delta(E) = \bar{1}$.

If there exists a mapping δ satisfying conditions I, II *and* III, *it also has the following properties*:

1) *If $s \in SL(n, T)$, then $\delta(sa) = \delta(a)$.*

2) *If $a = sd(\mu)$, $s \in SL(n, T)$, then $\delta(a) = \bar{\mu}$.*

3) *If some row of a consists entirely of zeros, then $\delta(a) = \bar{0}$.*

4) $a \in T_n \backslash GL(n, T) \Leftrightarrow \delta(a) = \bar{0}$ (*here* $\Leftrightarrow$ *has the usual meaning of equivalence*).

5) $\delta(ab) = \delta(a)\delta(b)$.

6) *If b is a matrix obtained from a by transposition of two rows, then* $\delta(b) = -\bar{1}\delta(a)$.

PROOF. 1) Since s is a product of elementary matrices $t_{ij}(\lambda)$, it suffices to show that

$$\delta(t_{ij}(\lambda)a) = \delta(a). \tag{3}$$

Let $\lambda \neq 0$, and let b be obtained from a by replacing its jth row a_j by λa_j. Then by I we have $\delta(b) = \bar{\lambda}\delta(a)$. On the other hand, if c is obtained from b by adding the jth row to the ith, then

$$\delta(c) = \delta(b) = \bar{\lambda}\delta(a) = \bar{\lambda}(\delta(t_{ij}(\lambda)a)). \tag{4}$$

Since $\bar{\lambda} \in \overline{T^*}$, it follows that (4) $\Rightarrow$ (3).

2) follows from 1), I and III. 3) follows from I.

4) Let

$$a \in T_n \setminus GL(n, T). \tag{5}$$

Then, by Lemma 8.1b, any row of this matrix is a left linear combination of certain other rows. Hence in $SL(n, T)$ there is a matrix s such that sa contains a row of zeros. Therefore $\delta(a) = \delta(sa) = \bar{0}$. Thus (5) implies that $\delta(a) = \bar{0}$. The reverse implication follows from 2) and Lemma 1.

5) If the matrix a is singular, it follows from Lemma 8.1a that its columns are right linearly dependent. The columns of the matrix ab are right linear combinations of those of a. By Steinitz' theorem they are also linearly dependent, and so ab is singular. Then

$$\delta(ab) = \bar{0} = \delta(a)\delta(b).$$

Now let $a \in GL(n, T)$. Then, by Lemma 1, $a = sd(\mu)$, $s \in SL(n, T)$ and

$$\delta(ab) = \delta(sd(\mu)b) = \delta(d(\mu)b).$$

The matrix $d(\mu)b$ is obtained from b by multiplying the last row by μ on the left. Therefore $\delta(d(\mu)b) = \bar{\mu}\delta(b) = \delta(a)\delta(b)$.

6) follows from 1) and I. ∎

THEOREM 1. *There exists a unique mapping* (2) *satisfying* I, II *and* III. *This mapping is an epimorphism.*

PROOF. It follows from 2) and 4) that there is at most one mapping (2) satisfying I, II and III. It therefore suffices to construct the mapping and show that it is an epimorphism. We proceed by induction on n. For $n = 1$ we set $\delta(\alpha) = \bar{\alpha}$. Then I, II and III are obviously satisfied. Now let $n > 1$ and suppose that we have already constructed a mapping $\delta: T_{n-1} \to \bar{T}$ satisfying the conditions. We now construct (2). Let $a \in T_n$. If a is singular, we set $\delta(a) = \bar{0}$. If a is invertible, its rows $a_1, \ldots, a_n$ form a basis of T_L^n. Consequently there is a unique vector $(\lambda_1, \ldots, \lambda_n) \in T^n$ such that

$$\sum_{\nu=1}^{n} \lambda_\nu a_\nu = (1, 0, \ldots, 0). \tag{6}$$

We write the row a_i in the form $a_i = (\alpha_{i1}, b_i)$, $b_i \in T_L^{n-1}$. Then from (6) we obtain

$$\sum_{\nu=1}^{n} \lambda_\nu b_\nu = (0, \ldots, 0). \tag{7}$$

Let

$$f = \begin{bmatrix} b_1 \\ b_2 \\ \cdot \\ \cdot \\ \cdot \\ b_n \end{bmatrix},$$

and let $c_i \in T_{n-1}$ be obtained from f by deleting the ith row. If $\lambda_i \neq 0$, $\lambda_j \neq 0$ and $i \neq j$, we define two more matrices g and h in T_{n-1} as follows: g is obtained from c_i by replacing b_j by $\lambda_j b_j$, and h by replacing b_j by b_i. Then, by I

$$\delta(c_i) = \overline{\lambda_j^{-1}} \delta(g).$$

We now add to the row $\lambda_j b_j$ of g each of the other rows b_ν, multiplied on the left by λ_ν. By (7), this is equivalent to replacing the row $\lambda_j b_j$ by $-\lambda_i b_i$. Consequently $\delta(g) = -\overline{\lambda}_i \delta(h)$. Hence

$$\delta(c_i) = \overline{\lambda_j^{-1}} (-\overline{\lambda}_i) \delta(h). \tag{8}$$

We now observe that h can be obtained from c_j by a permutation of rows. Counting the necessary number of transpositions, we find

$$\delta(h) = \overline{(-1)^{j-i-1}} \delta(c_j).$$

Hence and from (8) we get

$$\overline{(-1)^{i+1} \lambda_i^{-1}} \delta(c_i) = \overline{(-1)^{j+1} \lambda_j^{-1}} \delta(c_j).$$

This shows that the quantity

$$\overline{(-1)^{i+1} \lambda_i^{-1}} \delta(c_i)$$

assumes the same value for all i for which $\lambda_i \neq 0$.

We set

$$\delta(a) = \overline{(-1)^{i+1} \lambda_i^{-1}} \delta(c_i), \qquad \lambda_i \neq 0. \tag{9}$$

This completes construction of the mapping (2): $\delta(a) = \overline{0}$ *if a is singular*; $\delta(a)$ *is defined by* (9) *if a is invertible.* We now show that this mapping satisfies I, II and III.

Let $a \in T_n$, $\mu \in T^*$, and suppose that b is obtained from a by multiplying its ith row on the left by μ. If a is singular, so is b, and $\delta(b) = \overline{0} = \overline{\mu}\delta(a)$. If a is invertible, so is b, and hence $\delta(a)$ and $\delta(b)$ must be computed from (9). It is clear that in going from $\delta(a)$ to $\delta(b)$ the factor λ_i must be replaced by $\lambda_i \mu^{-1}$, the other λ_ν's remaining unchanged. If $\lambda_i \neq 0$, then

$$\delta(b) = \overline{(-1)^{i+1} \mu \lambda_i^{-1}} \delta(c_i) = \overline{\mu} \delta(a),$$

because c_i does not change on going from a to b. But if $\lambda_i = 0$ and $\lambda_\nu \neq 0$, then $\delta(c_\nu)$ is multiplied by $\overline{\mu}$ when we go from a to b, and so $\delta(b) = \overline{\mu}\delta(a)$.

Now let b be obtained from a by replacing the row a_i by $a_i + a_j$. If a is singular, so is b, and $\delta(b) = \overline{0} = \delta(a)$. If a is invertible, so is b. Since $\lambda_i(a_i + a_j) + (\lambda_j - \lambda_i)a_j = \lambda_i a_i + \lambda_j a_j$, the transition from a to b replaces λ_j by $\lambda_j - \lambda_i$ and leaves the other λ_ν's unchanged. Obviously the $\delta(c_\nu)$, $\nu \neq j$, are also left unchanged. Consequently, if $\lambda_\nu \neq 0$, then

$$\delta(a) = (-1)^{\nu+1}\lambda_\nu^{-1}\delta(c_\nu) = \delta(b).$$

But if $\lambda_\nu = 0$ for all $\nu \neq j$, then, by (7), $\lambda_j b_j = (0, \ldots, 0)$, and so $b_j = (0, \ldots, 0)$. Therefore the transition from a to b changes b_i into $b_i + b_j = b_i$, and c_j is unchanged. Since $\lambda_i = 0$, it follows that λ_j is also unchanged, and again $\delta(b) = \delta(a)$.

It remains to check III. Let $a = E_n$. Then $\lambda_1 = 1$, $\lambda_\nu = 0$ for $\nu > 1$, and $c_1 = E_{n-1}$. By (9), we find $\delta(E_n) = \delta(a) = \delta(E_{n-1}) = \overline{1}$.

Thus the mapping δ satisfies I, II and III. By 5), δ is a homomorphism. Surjectivity of δ follows from 2) and 4). ∎

If T is commutative (in this case $\overline{T^*} = T^*$ and we may assume that $\overline{T} = T$), *then $\delta(a)$ coincides with the classical determinant* det a *of a matrix a over a field.* For let $a = sd(\mu)$, $s \in SL(n, T)$. Then obviously $\det a = \mu$. On the other hand, $\delta(a) = \overline{\mu} = \mu$.

For an arbitrary division ring T we call $\delta(a)$ the *determinant of the matrix* $a \in T_n$, and write det a.

We consider a few examples.

1. Let

$$a = \begin{bmatrix} \alpha_{11} & 0 \ldots 0 \\ \alpha_{21} & \\ \vdots & A_{n-1} \\ \alpha_{n1} & \end{bmatrix} \in GL(n, T).$$

Then, by (9), $\det a = \overline{\alpha}_{11} \det A_{n-1}$. Hence, if

$$a = \begin{bmatrix} \alpha_{11} & 0 & \ldots & 0 \\ \alpha_{21} & \alpha_{22} & \ldots & 0 \\ \cdot & \cdot & \cdot & \cdot \\ \alpha_{n1} & \alpha_{n2} & \ldots & \alpha_{nn} \end{bmatrix}$$

$$\det a = \overline{\alpha_{11} \ldots \alpha_{nn}}.$$

then $\det a = \overline{\alpha_{11} \cdots \alpha_{nn}}$.

2. Let $c \in GL(n, T)$ be a monomial matrix (that is, a matrix with exactly one nonzero element in each row and each column), and let $\alpha_{i_1 1}, \ldots, \alpha_{i_n n}$ be all its nonzero elements. Then

$$\det c = \overline{(-1)^t \alpha_{i_1 1} \ldots \alpha_{i_n n}},$$

where t is the number of inversions in the permutation $i_1, \ldots, i_n$.

THEOREM 2. *If $n \geqslant 2$, then $SL(n, T)$ is the set of all matrices in T_n with determinant $\bar{1}$.*

PROOF. If $a \in SL(n, T)$, then $\det a = \bar{1}$ by virtue of 1) and III. Now let $\det a = \bar{1}$. Then, by 4), a is an invertible matrix, and therefore $a = sd(\mu)$, where $s \in SL(n, T)$ and $\bar{\mu} = \bar{1}$. It follows that $\mu = \mu_1 \cdots \mu_k$, where each μ_j is a commutator in T^*. By Lemma 2, $d(\mu_j) \in SL(n, T)$, and so

$$d(\mu) = d(\mu_1) \ldots d(\mu_n) \in SL(n, T), \qquad a \in SL(n, T). \blacksquare$$

It is obvious that $\delta(GL(n, T)) = \bar{T}^*$. Hence and from Theorem 2 we derive

THEOREM 3. *If $n \geqslant 2$, then $SL(n, T)$ is a normal subgroup of $GL(n, T)$, and the factor group is isomorphic to $\overline{T^*}$.*

THEOREM 4. *Let $n > 2$, or $n = 2$ but $T \neq GF(2)$. Then $SL(n, T)$ is the derived group of $GL(n, T)$.*

PROOF. By Theorem 3, $SL(n, T)$ contains the derived group of $GL(n, T)$. But by Theorem 8.5 and Corollary 1 to Lemma 8.4 $SL(n, T)$ is contained in the derived group. $\blacksquare$

COROLLARY 1. *If $n \geqslant 2$, every subgroup of $GL(n, T)$ containing $SL(n, T)$ is normal.*

If M is a right space over a division ring T and h an endomorphism of M, we set $\det h = \det a$, where a is the matrix of h relative to some basis. The matrix of h relative to any other basis has the form $c^{-1}ac$, and $\det(c^{-1}ac) = \det a$; hence $\det h$ is independent of the choice of basis. $SL(M)$ can now be defined as the group of all automorphisms of M with determinant $\bar{1}$.

§10. Normal subgroups of $GL(M)$

Let M be an n-dimensional right space over a division ring T. By the results of §8, the group $GL(M)$ of all automorphisms of M is isomorphic to the group $GL(n, T)$ of all invertible $n \times n$ matrices over T. It follows from the properties of the determinant that $GL(n, T)$ is the set of all matrices $a \in T_n$ for which $\det a \neq \bar{0}$, and $SL(n, T)$ is the set of all matrices in T_n with determinant $\bar{1}$. But $\det h$ is also defined for any endomorphism h of M. Let $SL(M)$ be the group of all h in End M with $\det h = \bar{1}$. If $\psi\colon GL(M) \longrightarrow GL(n, T)$ is an isomorphism such that $\psi(h)$ is the matrix of the automorphism h, then

$$\psi(SL(M)) = SL(n, T), \qquad \psi^{-1}(SL(n, T)) = SL(M).$$

In what follows we frequently identify $GL(n, T)$ with $GL(M)$ and $SL(n, T)$ with $SL(M)$.

LEMMA 1. *The set of all transvections of M generates the group $SL(M)$.*

PROOF. By Lemma 7.7, any nonidentity transvection τ in $GL(M)$ is conjugate to a transvection $\tau_{ij}(\lambda)$, $\lambda \in T^*$. Now $\det(\tau_{ij}(\lambda)) = \det(t_{ij}(\lambda)) = \bar{1}$; therefore $\det \tau = \bar{1}$ and $\tau \in SL(M)$.

On the other hand, the set of all elementary matrices $t_{ij}(\lambda)$ generates $SL(n, T)$. Consequently the set of all transvections $\tau_{ij}(\lambda)$ generates $SL(M)$. ■

LEMMA 2. *If $n > 2$, the centralizer of any transvection τ in $GL(M)$ contains a transversal to $SL(M)$ in $GL(M)$.*

PROOF. By Lemma 7.7, any two nonidentity transvections are conjugate in $GL(M)$, so that we need only prove the lemma for one transvection, $\tau_{12}(1)$ say. It is clear that for any $\mu \in T^*$ the matrix $d(\mu)$ commutes with $t_{12}(1)$, and therefore the centralizer of $t_{12}(1)$ in $GL(n, T)$ contains a transversal to $SL(n, T)$ in $GL(n, T)$. ■

LEMMA 3. *If $n > 2$, any two nonidentity transvections are conjugate in $SL(M)$.*

PROOF. Let τ_1 and τ_2 be nonidentity transvections. By Lemma 7.7, $\tau_1 = g\tau_2 g^{-1}$, $g \in GL(M)$. Let c be an element of the centralizer of τ_2 in $GL(M)$ in the same coset of $SL(M)$ in $GL(M)$ as g. Then

$$\tau_1 = (gc^{-1})\,\tau_2\,(gc^{-1})^{-1}, \qquad gc^{-1} \in SL(M). \ ■$$

THEOREM 1. *If $n > 2$, then every subgroup Γ of $GL(M)$ normalized by $SL(M)$ either contains $SL(M)$ or is contained in the center of $GL(M)$.*

PROOF. Suppose that Γ is not in the center of $GL(M)$. By Lemma 3 it suffices to prove that Γ contains a nonidentity transvection. By Corollary 7.2.3 and Lemma 7.6, there exist a $\sigma \in \Gamma$ and an $a \in M$ such that a and $b = \sigma(a)$ are linearly independent. We can write M in the form $M = P \dotplus bT$, where P is a hyperplane containing a. We consider the transvection τ defined by $\tau(p) = p$, $p \in P$; $\tau(b) = b + a$. Let $\tau_1 = \sigma\tau\sigma^{-1}$. Then $\tau_1(b) = b$, $\tau_1 \neq \tau$, and so $\rho = \tau_1\tau^{-1} \neq e$. Since $\rho = \sigma\tau\sigma^{-1}\tau^{-1}$ and Γ is invariant under $SL(M)$, it follows that $\rho \in \Gamma$. If $x \in M$, then $x = p + b\lambda$, $\tau(x) = p + (b + a)\lambda = x + a\lambda$ and

$$\tau(x) - x \in aT. \tag{1}$$

Similarly,

$$\tau_1(x) - x \in bT. \tag{2}$$

Now if $Q = aT + bT$, then by (1) and (2) for $x \in M$ we have

$$\rho(x) - x \in Q. \tag{3}$$

Since $n > 2$, Q is contained in some hyperplane H, and it follows from (3) that

$$\rho(x) - x \in H, \quad \rho(H) = H. \tag{4}$$

There are now two possible cases:

1) ρ commutes with every transvection in $K(H)$.

2) $K(H)$ contains a transvection φ that does not commute with ρ.

In the first case, by Theorem 7.2 we have $\rho = \rho_\alpha \tau_\alpha$, where $\rho_\alpha \in H(M)$, $\tau_\alpha \in K(H)$, $\tau_\alpha(c) = c + h$ and $M = H \dot{+} cT$. Next, $\rho(c) - c = (c + h)\alpha - c = c(\alpha - 1) + h\alpha \in H$. Hence $c(\alpha - 1) = 0$, $\alpha - 1 = 0$, $\rho_\alpha = e$ and $\rho = \tau_\alpha \in K(H)$, and so τ_α is a nonidentity transvection in Γ.

In the second case, $\varphi_1 = \rho\varphi\rho^{-1}\varphi^{-1} \in \Gamma$. Then $\rho\varphi\rho^{-1} \in K(\rho(H)) = K(H)$. Therefore also $\varphi_1 \in K(H)$. ■

It is easy to verify that Theorem 1 fails for the groups $GL(2, 2)$ and $GL(2, 3)$. However, these are the only exceptions; to prove this we need three lemmas.

LEMMA 4. *Let M be a two-dimensional space over a division ring T, and let P and Q be one-dimensional subspaces of M. Then the groups $K(P)$ and $K(Q)$ are conjugate in $SL(M)$.*

PROOF. We need only consider the case $P \neq Q$. Let $P = pT$ and $Q = qT$. Then

$$p, \; q \tag{5}$$

is a basis of M. Now let σ be the automorphism defined by $\sigma(p) = q$, $\sigma(q) = p(-1)$. The matrix of σ in the basis (5) is

$$\begin{bmatrix} 0 & -1 \\ 1 & 0 \end{bmatrix},$$

$\det \sigma = \bar{1}$, $\sigma \in SL(M)$. If τ is an arbitrary transvection in $K(P)$, we have

$$\tau(p) = p, \quad \tau(q) = q + p\lambda, \quad \lambda \in T, \quad \tau = \tau_\lambda.$$

Then $\sigma\tau\sigma^{-1}(q) = q$ and

$$\sigma\tau\sigma^{-1}(p) = \sigma\tau(q(-1)) = \sigma(q(-1) + p(-\lambda)) = p + q(-\lambda).$$

Consequently $\sigma\tau_\lambda\sigma^{-1} \in K(Q)$. When λ ranges over T, $\sigma\tau_\lambda\sigma^{-1}$ ranges over $K(Q)$; that is, $\sigma K(P)\sigma^{-1} = K(Q)$. ■

COROLLARY 1. *Let M be a two-dimensional space over a division ring T, and P a one-dimensional subspace of M. If a subgroup Γ of $GL(M)$ containing $K(P)$ is normalized by $SL(M)$, then $\Gamma \supset SL(M)$.*

LEMMA 5. *Let M be a two-dimensional space over a division ring T. Suppose either that T is not a field of characteristic 2 or, if it is, that none of its elements are transcendental over its prime field. If a subgroup Γ of $GL(M)$ is normalized by $SL(M)$ and contains a nonidentity transvection, then $\Gamma \supset SL(M)$.*

PROOF. Let $\tau \in \Gamma$ be a nonidentity transvection. For some basis a, b of M we can write

$$\tau(a) = a, \quad \tau(b) = a\lambda + b, \quad \lambda \in T. \tag{6}$$

Let $\sigma \in GL(M)$ be defined by $\sigma(a) = a\alpha$ and $\sigma(b) = b\gamma$. Then $\sigma^{-1}(a) = a\alpha^{-1}$, $\sigma^{-1}(b) = b\gamma^{-1}$ and

$$\sigma\tau\sigma^{-1}(a) = a, \quad \sigma\tau\sigma^{-1}(b) = a\alpha\lambda\gamma^{-1} + b. \tag{7}$$

If

$$\gamma = \lambda^{-1}\alpha^{-1}\lambda, \tag{8}$$

then

$$\det\sigma = \overline{\alpha\lambda^{-1}\alpha^{-1}\lambda} = \bar{1}, \quad \sigma \in SL(2, T).$$

Substituting (8) in (7), we obtain

$$\sigma\tau\sigma^{-1}(a) = a, \quad \sigma\tau\sigma^{-1}(b) = a\alpha^2\lambda + b. \tag{9}$$

We set $\tau = \tau_\lambda$. Then, by (9),

$$\sigma\tau_\lambda\sigma^{-1} = \tau_{\alpha^2\lambda} \in \Gamma. \tag{10}$$

Let $L = \{\lambda \in T | \tau_\lambda \in \Gamma\}$. Obviously L is a subgroup of the additive group of T. It follows from (10) that $\alpha^2 L \subset L$ for any $\alpha \in T$. Consequently

$$\alpha_1, \ldots, \alpha_k \in T \Rightarrow \alpha_1^2 \ldots \alpha_k^2 L \subset L.$$

If S is the subgroup of the additive group of T generated by the products of squares in T, then $SL \subset L$. By Proposition 6.2, $S = T$. Thus $TL \subset L$, and so $L = T$. Clearly the set of all τ_λ such that $\lambda \in T$ is the group $K(aT)$. Thus $K(aT) \subset \Gamma$. By Lemma 4, Γ also contains all transvections, and so $\Gamma \supset SL(M)$. ■

LEMMA 6. *Let M be a two-dimensional space over a division ring T, a, b a basis of M, σ an element of the form $\sigma(a) = a\alpha$, $\sigma(b) = a\beta + b\gamma$, and Γ a subgroup of $GL(M)$ normalized by $SL(M)$ and containing σ. If γ is not in the center $Z(T)$ of T or if $\gamma \in Z(T)$ but $\gamma \neq \alpha$, then $SL(M) \subset \Gamma$.*

PROOF. Let τ be a transvection of the form (6). Then

$$\rho = \sigma\tau\sigma^{-1}\tau^{-1} \in \Gamma, \quad \tau^{-1}(a) = a, \quad \tau^{-1}(b) = a(-\lambda) + b.$$

Hence $\rho(a) = a$ and $\rho(b) = a(\alpha\lambda\gamma^{-1} - \lambda) + b$; that is, $\rho \in K(aT)$.

We now consider two cases: 1) T is not commutative, 2) T is a field.

In the first case, according to Lemma 5 we need only choose λ so that

$$\alpha\lambda\gamma^{-1} - \lambda \neq 0. \tag{11}$$

If $\alpha \neq \gamma$, (11) is satisfied for $\lambda = 1$. If $\alpha = \gamma$, by the assumptions of the lemma $\gamma \notin Z(T)$. Then there exists a λ in T such that $\gamma\lambda \neq \lambda\gamma$, and so $\gamma\lambda\gamma^{-1} - \lambda \neq 0$.

Now for the second case. By assumption, $\alpha \neq \gamma$ and

$$\alpha\lambda\gamma^{-1} - \lambda = (\alpha\gamma^{-1} - 1)\lambda \neq 0.$$

When λ ranges over T, so does $(\alpha\gamma^{-1} - 1)\lambda$. Therefore $K(aT) \subset \Gamma$. By the corollary to Lemma 4, $SL(M) \subset \Gamma$. ■

THEOREM 2. *Let M be a two-dimensional space over a division ring T containing at least four elements. Then any subgroup of $GL(M)$ normalized by $SL(M)$ either contains $SL(M)$ or is contained in the center of $GL(M)$.*

PROOF. Let Γ be a subgroup of $GL(M)$, normalized by $SL(M)$ and not lying in the center of $GL(M)$. Then there exist $a \in M$ and $\sigma \in \Gamma$ such that

$$a, \; b = \sigma(a) \tag{12}$$

is a basis for M. The element σ is of the form

$$\sigma(a) = b, \quad \sigma(b) = a\beta + b\gamma, \quad \beta \neq 0.$$

Let τ be the automorphism defined by

$$\tau(a) = a\gamma\mu + b(-\mu), \quad \tau(b) = a\mu^{-1}, \quad \mu \in T^*.$$

Its matrix in the basis (12) is

$$\begin{bmatrix} \gamma\mu & \mu^{-1} \\ -\mu & 0 \end{bmatrix}.$$

Hence $\tau \in SL(M)$ and $\rho = (\tau^{-1}\sigma^{-1}\tau)\sigma \in \Gamma$. Next we find that

$$\rho(a) = a(-\mu^{-1}\beta^{-1}\mu^{-1}),$$
$$\rho(b) = a\mu^{-1}\beta^{-1}(-\gamma\mu\beta - \mu^{-1}\gamma) + b(-\mu^2\beta).$$

We now try to choose the parameter μ in such a way that Lemma 6 is applicable.

If $\beta \notin Z(T)$ and $\mu = 1$, then $-\mu^2\beta \notin Z(T)$, and we can apply Lemma 6.

If $\beta \in Z(T)$ but T is not commutative, then by Proposition 6.2 there exists a $\mu \in T$ such that $\mu^2 \notin Z(T)$. Then $-\mu^2\beta \notin Z(T)$, and we can again use Lemma 6.

It remains to consider the case where T is a field. To apply Lemma 6, we must choose μ so that $-\mu^{-1}\beta^{-1}\mu^{-1} \neq -\mu^2\beta$. This is equivalent to

$$\mu^4 \neq \beta^{-2}. \tag{13}$$

Now the equation

$$x^4 = \beta^{-2} \tag{14}$$

has at most four solutions in T. Consequently, if there are more than four elements in T^*, then, given any β in T^*, we can find a μ satisfying (13). Thus there remain only two cases: 1) $T = GF(4)$; 2) $T = GF(5)$. In $GF(4)$ equation (14) has only one solution $x = \beta^{-2}$, and any element $\mu \neq \beta^{-2}$ will do. Now let $T = GF(5)$. If $\beta \in T^*$, then $\mu^4 = 1$. Thus for $\beta \neq \pm 1$ condition (13) holds for any $\mu \in T^*$. But if $\beta = \pm 1$, condition (13) cannot be satisfied. We then choose μ so that $\mu^2\beta = 1$. Then ρ takes the form $\rho(a) = a(-1)$, $\rho(b) = a(-2\gamma) + b(-1)$. Thus $\rho^2(a) = a$ and $\rho^2(b) = a(4\gamma) + b$. If $\gamma \neq 0$, then ρ^2 is a nonidentity transvection in Γ. By Lemma 5, $\Gamma \supset SL(M)$. The only remaining case is $\gamma = 0$, $\beta = \pm 1$. If $\beta = 1$, then $\sigma(a) = b$ and $\sigma(b) = a$. We define a new basis c, d by $c = a + b$, $d = a - b$. Then

$$\sigma(c) = c, \quad \sigma(d) = d(-1). \tag{15}$$

But if $\beta = -1$, we set $c = a + 2b$ and $d = a - 2b$. Then

$$\sigma(c) = c(-2), \quad \sigma(d) = d2. \tag{16}$$

Lemma 6 is applicable in both cases (15) and (16) ($\gamma \neq \alpha$). ∎

Combining Theorems 1 and 2 and Corollary 9.4.1, we obtain

THEOREM 3. *Let M be an n-dimensional space over a division ring T, $n > 1$. Then every subgroup of $GL(M)$ containing $SL(M)$ or contained in $H(M)$ is normal. If $n > 2$, or $n = 2$ but $T \neq GF(2)$ and $T \neq GF(3)$, then any normal subgroup of $GL(M)$ contains $SL(M)$ or is contained in the center of $GL(M)$.*

This theorem enables us to define a certain series of simple groups. Let $Z(SL(M))$ denote the center of $SL(M)$, and let $PSL(M)$ be the factor group $SL(M)/Z(SL(M))$.

THEOREM 4. *Suppose that M is an n-dimensional space over a division ring T, and that either $n > 2$, or $n = 2$ but T contains more than three elements. Then $PSL(M)$ is a simple group.*

PROOF. Let Γ be a proper normal subgroup of $PSL(M)$. Then $\Gamma = U/Z(SL(M))$, where U is a normal subgroup of $SL(M)$ containing $Z(SL(M))$. By Theorem 3, U is in the center of $GL(M)$. But $U \subset SL(M)$, and therefore $U = Z(SL(M))$. Thus $PSL(M)$ is simple. ∎

The results of §§9 and 10 are due to Dieudonné [1]. Our exposition follows that of Artin [1].

To end the chapter we quote without proof a few theorems of Rosenberg [1] on the structure of $GL(M)$ for an infinite-dimensional space over a division ring T.

Let $M:T = \aleph_\delta$, $\delta \geqslant 0$, $L = \operatorname{End} M$ and $G = GL(M)$. Let e be the identity of the ring L. We set $\Sigma = \{s \in L \mid (s-e)^2 = 0\}$. Then Σ is a system of generators for G, and G is equal to its derived group.

For any cardinal $\aleph_\nu$, $0 \leqslant \nu \leqslant \delta$, we set

$$F_\nu = \{f \in L \mid \operatorname{Im} f : T < \aleph_\nu\};$$

F_ν is a two-sided ideal in L. We define groups G_0 and $G_\nu(1)$ by $G_0 = Z^*e + F^0$, where Z is the center of T, and $G_\nu(1) = e + F_\nu$.

Let N be a proper normal subgroup of G. Then either $N \subset G_0$ or $N = HG_\nu(1)$, where H is a subgroup of Z^*e.

THIRD CHAPTER

THE NORMAL STRUCTURE OF THE GROUPS $GL(\Delta)$ AND $GL(n, Z)$, $n > 2$

§11. Normal subgroups of the infinite general linear group

1. *The infinite general linear group.* Let Δ be an associative ring with unit element 1. Identifying any matrix $a \in GL(n, \Delta)$ with $\mathrm{diag}[a, 1] \in GL(n+1, \Delta)$, we obtain an infinite ascending chain of subgroups $GL(1, \Delta) \subset GL(2, \Delta) \subset \cdots$. We call the union of the groups in this chain the limiting or, following Bass [1], the *stable general linear group over* Δ, and denote it by $GL(\Delta)$.

Thus, *every element* $a \in GL(\Delta)$ *belongs to* $GL(n, \Delta)$ *for some* $n = n(a)$.

The subgroup $SL(\Delta)$ of $GL(\Delta)$ generated by all elementary matrices $t_{ij}(\lambda) = E + e_{ij}(\lambda)$, $\lambda \in \Delta$, is called the *stable special linear group over* Δ.

THEOREM 1 (WHITEHEAD'S LEMMA). *The derived group of* $GL(\Delta)$ *is* $SL(\Delta)$.

PROOF. Let K be the derived group of $GL(\Delta)$. Since every elementary matrix is a commutator (Lemma 8.3), $SL(\Delta) \subset K$. Now let $x, y \in GL(\Delta)$. Then $x, y \in GL(n, \Delta)$ for some n. Obviously

$$\begin{aligned} xyx^{-1}y^{-1} &= \mathrm{diag}[xyx^{-1}y^{-1}, E_n] \\ &= \mathrm{diag}[x, x^{-1}]\,\mathrm{diag}[y, y^{-1}]\,\mathrm{diag}[(yx)^{-1}, yx]. \end{aligned} \tag{1}$$

Next

$$\mathrm{diag}[x, x^{-1}] = \begin{bmatrix} E_n & -x \\ 0 & E_n \end{bmatrix} \begin{bmatrix} E_n & 0 \\ x^{-1} - E_n & E_n \end{bmatrix} \begin{bmatrix} E_n & E_n \\ 0 & E_n \end{bmatrix} \begin{bmatrix} E_n & 0 \\ x - E_n & E_n \end{bmatrix}.$$

On the other hand, all the matrices of the form

$$\begin{bmatrix} E_n & a \\ 0 & E_n \end{bmatrix}, \quad \begin{bmatrix} E_n & 0 \\ a & E_n \end{bmatrix}$$

are products of elementary matrices, because

$$\begin{bmatrix} E_n & a_1 \\ 0 & E_n \end{bmatrix} \begin{bmatrix} E_n & a_2 \\ 0 & E_n \end{bmatrix} = \begin{bmatrix} E_n & a_1 + a_2 \\ 0 & E_n \end{bmatrix}.$$

Thus

$$\begin{bmatrix} E_n & a \\ 0 & E_n \end{bmatrix}, \quad \begin{bmatrix} E_n & 0 \\ a & E_n \end{bmatrix} \in SL(\Delta). \tag{2}$$

By (1) and (2), $xyx^{-1}y^{-1} \in SL(\Delta)$, and so $K \subset SL(\Delta)$. Consequently $K = SL(\Delta)$. ∎

2. *Congruence subgroups.* Let N be a two-sided ideal of Δ. It is clear that the canonical ring homomorphism $\gamma_N\colon \Delta \to \Delta/N = \bar{\Delta}$ can be extended to an epimorphism of matrix rings

$$h_N\colon \Delta_n \to \bar{\Delta}_n, \quad \|\alpha_{ij}\| \mapsto \|\bar{\alpha}_{ij}\|, \quad \bar{\alpha}_{ij} = \gamma_N(\alpha_{ij}).$$

This epimorphism, in turn, determines a group homomorphism

$$g_N\colon GL(n, \Delta) \to GL(n, \bar{\Delta}), \quad x \mapsto h_N(x).$$

We call g_N a *Minkowski homomorphism.* Its kernel $CL(n, \Delta, N)$ is called the *N-congruence subgroup of* $GL(n, \Delta)$.

If $N = (0)$, then $\bar{\Delta} = \Delta$ and g_N is the identity automorphism. Hence $CL(n, \Delta, (0)) = (1)$.

When $N = \Delta$, we set $GL(n, \bar{\Delta}) = 1$. Then $CL(n, \Delta, \Delta) = GL(n, \Delta)$.

Let $E(n, \Delta, N)$ denote the group generated by all elementary matrices in $CL(n, \Delta, N)$. The next three lemmas are due to Bass [1].

LEMMA 1. *If $n > 2$, then*

(i) $E(n, \Delta, N) \lhd SL(n, \Delta)$;

(ii) $E(n, \Delta, N) = (SL(n, \Delta), E(n, \Delta, N))$.

PROOF. By definition, $E(n, \Delta, N)$ is generated by all elementary matrices $t_{ij}(\mu)$, where $\mu \in N$. By Lemma 8.3,

$$t_{ik}(\lambda)\, t_{kj}(\mu)\, t_{ik}^{-1}(\lambda)\, t_{kj}^{-1}(\mu) = t_{ij}(\lambda\mu) \in E(n, \Delta, N), \tag{3}$$

where i, j, k are pairwise distinct. Consequently, for $\lambda \in \Delta$ and $\mu \in N$,

$$t_{ik}(\lambda)\, t_{kj}(\mu)\, t_{ik}^{-1}(\lambda) \in E(n, \Delta, N). \tag{4}$$

Similarly,

$$t_{ji}(\lambda)\, t_{kj}(\mu)\, t_{ij}^{-1}(\lambda) \in E(n, \Delta, N). \tag{5}$$

Now, if $\alpha \neq \beta$ and $(\alpha - j)(\beta - k) \neq 0$, then

$$t_{\alpha\beta}(\lambda)\, t_{kj}(\mu)\, t_{\alpha\beta}^{-1}(\lambda) = t_{kj}(\mu) \in E(n, \Delta, N). \tag{6}$$

It follows from (4)–(6) that for any $a \in SL(n, \Delta)$

$$a t_{kj}(\mu) a^{-1} \in E(n, \Delta, N), \quad \mu \in N,$$

and so $E(n, \Delta, N) \lhd SL(n, \Delta)$.

We now prove (ii). By (i),

$$K = (SL(n, \Delta), E(n, \Delta, N)) \subset E(n, \Delta, N).$$

Setting $\lambda = 1$ in (3), we find that $t_{ij}(\mu) \in K$. Thus $K = E(n, \Delta, N)$. ■

LEMMA 2. $(GL(n, \Delta), CL(n, \Delta, N)) \subset E(2n, \Delta, N)$.

PROOF. Let $x \in GL(n, \Delta)$ and $y \in CL(n, \Delta, N)$. Then $y = E_n + y_1$, where y_1 is a matrix over N. We define matrices t_1, t_2, t_3 by

$$t_1 = \begin{bmatrix} E_n & (yx)^{-1} y_1 \\ 0 & E_n \end{bmatrix}, \qquad t_2 = \begin{bmatrix} E_n & -x^{-1}y_1 \\ 0 & E_n \end{bmatrix},$$

$$t_3 = \begin{bmatrix} E_n & 0 \\ -y^{-1}y_1 x & E_n \end{bmatrix}.$$

Clearly, $t_1, t_2, t_3 \in E(2n, \Delta, N)$. If

$$s = \begin{bmatrix} E_n & 0 \\ x & E_n \end{bmatrix},$$

then $s \in SL(2n, \Delta)$. Hence, by Lemma 1,

$$t = t_1 s^{-1} t_2 s t_3 \in E(2n, \Delta, N). \tag{7}$$

Now we set

$$a = \operatorname{diag}[yx, E_n], \qquad b = \operatorname{diag}[x, y].$$

Then

$$at_1 = \begin{bmatrix} yx & y_1 \\ 0 & E_n \end{bmatrix}, \qquad at_1 s^{-1} = \begin{bmatrix} x & y_1 \\ -x & E_n \end{bmatrix},$$

$$at_1 s^{-1} t_2 = \begin{bmatrix} x & 0 \\ -x & y \end{bmatrix}, \qquad at_1 s^{-1} t_2 s = \begin{bmatrix} x & 0 \\ y_1 x & y \end{bmatrix},$$

$$at = \operatorname{diag}[x, y] = b. \tag{8}$$

In particular, it follows from (7) and (8) that

$$u = \operatorname{diag}[y^{-1}, y] \in E(2n, \Delta, N). \tag{9}$$

This is proved by putting $x = y^{-1}$. Now let $c = \operatorname{diag}[xy, E_n]$. Then

$$cu = b. \tag{10}$$

By (9) and (10),

$$y^{-1}x^{-1}yx = \operatorname{diag}[y^{-1}x^{-1}yx, E_n] = c^{-1}a = ut^{-1} \in E(2n, \Delta, N),$$

and this proves the lemma. ■

We now set

$$CL(\Delta, N) = \bigcup_n CL(n, \Delta, N), \qquad E(\Delta, N) = \bigcup_n E(n, \Delta, N)$$

LEMMA 3. *$E(\Delta, N) = (SL(\Delta), E(\Delta, N)) = (GL(\Delta), CL(\Delta, N)) = (GL(\Delta), E(\Delta, N))$.*

PROOF. By Lemmas 1 and 2,

$$\begin{aligned} E(n, \Delta, N) &= (SL(n, \Delta), E(n, \Delta, N)) \\ &\subset (GL(n, \Delta), CL(n, \Delta, N)) \subset E(2n, \Delta, N). \end{aligned}$$

Passing now to the stable groups $E(\Delta, N)$, $SL(\Delta)$, $GL(\Delta)$ and $CL(\Delta)$, we get the required equalities. ∎

3. *Normal subgroups of $GL(\Delta)$.*

LEMMA 4. *Let $n > 2$; let H be a subgroup of $GL(n, \Delta)$ normalized by $SL(n, \Delta)$, and N the set of all $\nu \in \Delta$ such that at least one elementary matrix $t_{ij}(\nu)$ is in H. Then N is a two-sided ideal of Δ and $H \supset E(n, \Delta, N)$.*

PROOF. $t_{ij}(0) = E_n \in H$, so that $0 \in N$, $N \neq \varnothing$. Now let $t_{kl}(\alpha) \in H$. If $\delta \in \Delta$ and $j \neq l, k$, then by Lemma 8.3

$$t_{kj}(\alpha\delta) = t_{kl}(\alpha)\, t_{lj}(\delta)\, t_{kl}^{-1}(\alpha)\, t_{lj}^{-1}(\delta) \in H. \tag{11}$$

For $\gamma \in \Delta$ and $i \neq k, j$

$$t_{ij}(\gamma\alpha\delta) = t_{ik}(\gamma)\, t_{kj}(\alpha\delta)\, t_{ik}^{-1}(\gamma)\, t_{kj}^{-1}(\alpha\delta) \in H. \tag{12}$$

Hence we have the implication

$$t_{kl}(\alpha) \in H \Rightarrow t_{ij}(\gamma\alpha\delta) \in H; \quad i \neq k; \quad j \neq l,\ k; \quad \gamma,\ \delta \in \Delta. \tag{13}$$

This, in turn, gives

$$\alpha \in N \Rightarrow \gamma\alpha\delta \in N, \qquad \gamma,\ \delta \in \Delta. \tag{14}$$

Moreover, setting $\gamma = \delta = 1$ in (13), we obtain

$$t_{kl}(\alpha) \in H \Rightarrow t_{ij}(\alpha) \in H, \tag{15}$$

where $i \neq j$. Now let $\alpha, \beta \in N$. Then $t_{kl}(\alpha)$, $t_{ij}(\beta) \in H$. By (15),

$$t_{kl}(\beta) \in H, \quad t_{kl}(\alpha - \beta) = t_{kl}(\alpha)\, t_{kl}^{-1}(\beta) \in H, \quad \alpha - \beta \in N. \tag{16}$$

It follows from (14) and (16) that N is a two-sided ideal of Δ. By (15),

$$E(n, \Delta, N) \subset H. \ ∎$$

THEOREM 2. (i) *Let H be a subgroup of $GL(\Delta)$ normalized by $SL(\Delta)$. Then there exists a unique two-sided ideal $N = N(H)$ of Δ such that*

$$E(\Delta, N) \subset H \subset CL(\Delta, N). \tag{17}$$

(ii) *Every subgroup H of $GL(\Delta)$ satisfying* (17) *is normal in $GL(\Delta)$.*

PROOF. Let H be a subgroup of $GL(\Delta)$ normalized by $SL(\Delta)$, and N the two-sided ideal of Δ generated by the elements of all the matrices $h - 1$, $h \in H$.

Then obviously $H \subset CL(\Delta, N)$. We claim that

$$E(\Delta, N) \subset H. \tag{18}$$

If $H = (1)$, then $N = (0)$, $CL(\Delta, (0)) = (1) = E(\Delta, (0))$ and (18) is trivially true. We assume that $H \neq (1)$ and consider the group

$$H_n = H \cap GL(n, \Delta)$$

and the two-sided ideal N_n generated by the elements of the matrices $h - E_n$, $h \in H_n$. Now $H_n \neq (E_n)$ and $N_n \neq (0)$ for sufficiently large n. Obviously $H_n \subset \mathrm{Aff}(n, \Delta)$, and since H is normalized by $SL(\Delta)$, it follows that $H_n = \langle H_n, (0)\rangle$ is normalized by the group $SL(n, \Delta) = \langle SL(n, \Delta), (0)\rangle$ (see §8). By Lemma 8.9,

$$K = (\langle H_n, (0)\rangle, \langle E_n, \Delta^n\rangle) = \langle E_n, \Phi^n\rangle,$$

where Φ is a left ideal of Δ. On the other hand, by the same lemma,

$$K = \langle E_n, A\rangle, \quad A = \Sigma(h - E_n)\Delta^n, \quad h \in H_n.$$

Therefore Φ contains any element of the matrix $h - E_n$, $h \in H_n$. Moreover, since H is normalized by the group $SL(\Delta) \supset \langle E_n, \Delta^n\rangle$, it follows that

$$K \subset H \cap GL(n+1, \Delta) = H_{n+1}.$$

Thus H_{n+1} contains the elementary matrices $t_{ij}(\nu)$, where ν are elements of the matrices $h - E_n$, $h \in H_n$. By Lemma 4,

$$H_{n+1} \supset E(n+1, \Delta, N_n).$$

Passing to the limit as $n \to \infty$, we obtain (17).

It remains to prove (ii) and the uniqueness of $N = N(H)$. Let H be a subgroup of $GL(\Delta)$ satisfying (17). Then

$$(GL(\Delta), E(\Delta, N)) \subset (GL(\Delta), H) \subset (GL(\Delta), CL(\Delta, N)).$$

Hence and by Lemma 3,

$$(GL(\Delta), H) = E(\Delta, N) \subset H. \tag{19}$$

This implies that N is unique and that H is normal in $GL(\Delta)$. ■

Concerning the structure of Aut M, where M is a free module with a basis of arbitrary infinite cardinality over a commutative ring, see Maxwell [1].

4. *Minkowski's Theorem.* We now return to the congruence subgroup $CL(n, \Delta, N)$.

THEOREM 3 (MINKOWSKI'S THEOREM). *Let Δ be a Gaussian domain of characteristic zero, q an indecomposable element of Δ such that $q \nmid 2$ and $q^2 \nmid p$ for any rational prime p; let $N = q\Delta$. Then $CL(n, \Delta, N)$ is torsion free.*

PROOF. Suppose that $CL(n, \Delta, N)$ contains elements of finite order other than E_n. Then it contains an element a of prime order p. Clearly $a = E_n + qb$, where $b \in \Delta_n$, $b \neq 0$. Now

$$a^p = (E_n + qb)^p$$
$$= E_n + pqb + \frac{p(p-1)}{2} q^2b^2 + \ldots + q^p b^p = E_n.$$

Hence

$$pb + \frac{p(p-1)}{2} qb^2 + \ldots + q^{p-1}b^p = 0. \tag{20}$$

There are two possibilities: $q \nmid p$ or $q | p$. In the first case it follows from (20) that $q | b$; that is, q divides every element of the matrix b. Let

$$q^\alpha | b, \quad q^{\alpha+1} \nmid b, \quad \alpha > 0. \tag{21}$$

By (21), the terms in (20), from the second on, are divisible by $q^{2\alpha+1}$. Consequently pb and b are divisible by $q^{2\alpha+1}$. This contradicts (21). In the second case it follows from the assumptions of the theorem that $p > 2$ and $q^2 \nmid p$. Thus $p = qr$, $r \in \Delta$, $(r, q) = 1$. Now (20) can be rewritten as

$$qrb = q^2b^2c, \tag{22}$$

where $c \in \Delta_n$. Since $(r, q) = 1$, it follows from (22) that $q | b$. Let $q^\alpha | b$ and $q^{\alpha+1} \nmid b$. Then the right-hand side of (22) is divisible by $q^{2\alpha+2}$, but the left-hand side is not. This contradiction completes the proof. ■

COROLLARY 1. *Let Δ and $N = q\Delta$ be as in Theorem* 3. *Then $CL(\Delta, N)$ is torsion free.*

COROLLARY 2. *Periodic subgroups of $GL(n, Z)$, where Z is the ring of rational integers, are finite. For any prime $p > 2$ and any finite subgroup H of $GL(n, Z)$, there is a subgroup of $GL(n, Z/p)$ isomorphic to H.*

COROLLARY 3. *Let Σ be a field of characteristic zero and x an indeterminate. Then every periodic subgroup of $GL(n, \Sigma[x])$ is isomorphic to some subgroup of $GL(n, \Sigma)$.*

To prove this it is enough to put $\Delta = \Sigma[x]$ and $q = q(x) = x$ in Theorem 3.

COROLLARY 4. *Let Σ be a field of characteristic zero and $x_1, \ldots, x_m$ independent indeterminates. Then every finite subgroup of $GL(n, \Sigma(x_1, \ldots, x_m))$ is isomorphic to some subgroup of $GL(n, \Sigma)$.*

PROOF. We first consider a finite subgroup G of $GL(n, \Sigma(x))$. By Burnside's Theorem (Theorem 8.3) G is isomorphic to some subgroup of $GL(n, \Sigma[x])$, and, by Corollary 3, to some subgroup of $GL(n, \Sigma)$. The rest follows by induction. ■

THEOREM 4. *Let Δ be a Dedekind domain of characteristic zero, and $\mathfrak{P}$ a prime ideal of Δ such that $\mathfrak{P} \not\supset (2)$ and $\mathfrak{P}^2 \not\supset (p)$ for any prime $p \in Z$. Then $CL(n, \Delta, \mathfrak{P})$ is torsion free.*

PROOF. Let $a \in CL(n, \Delta, \mathfrak{P})$ be an element of prime order p. Then

$$a = E_n + b, \quad b \in \mathfrak{P}_n, \quad b = \|\beta_{ij}\| \neq 0, \quad \beta_{ij} \in \mathfrak{P}$$

and

$$pb + \frac{p(p-1)}{2} b^2 + \dots + b^p = 0, \tag{23}$$

$b^k = \|\beta_{ij}^k\|$, $\beta_{ij}^k \in \mathfrak{P}$. Let l be the number such that $\beta_{ij} \in \mathfrak{P}^l$ for all i, j, but $\beta_{\mu\nu} \notin \mathfrak{P}^{l+1}$ for some pair μ, ν. By (23),

$$p\beta_{\mu\nu} + C_p^2 \beta_{\mu\nu}^2 + \dots + \beta_{\mu\nu}^p = 0. \tag{24}$$

There are two possibilities: $\mathfrak{P} \not\supset (p)$ or $\mathfrak{P} \supset (p)$. In the first case,

$$p\beta_{\mu\nu} \in \mathfrak{P}^l \setminus \mathfrak{P}^{l+1},$$

which contradicts (24). In the second case,

$$p \neq 2, \quad p\beta_{\mu\nu} \in \mathfrak{P}^{l+1} \setminus \mathfrak{P}^{l+2},$$

and the remaining terms of (24) belong to the ideal $\mathfrak{P}^{2l+1}$. This contradiction completes the proof. ∎

In connection with Theorems 3 and 4 it would be interesting to clarify for what rings Δ and ideals N the group $CL(n, \Delta, N)$ is torsion free.

§12. Normal subgroups of $GL(n, Z)$, $n > 2$. Subgroups of finite index

1. *Preliminary lemmas.* First let Δ be an arbitrary associative ring with unit element. We prove two lemmas (see Bass [1]).

LEMMA 1. *Let $n > 2$, and let H be a subgroup of $GL(n, \Delta)$ that is normalized by $SL(n, \Delta)$ and satisfies one of the following conditions (see the end of* §8):

(i) *$H \cap \langle E_{n-1}, \Delta^{n-1}\rangle \neq (E_n)$;*

(ii) *$H \cap Z(\Delta)^*\langle E_{n-1}, \Delta^{n-1}\rangle \not\subset Z(\Delta)^*E_n$, where $Z(\Delta)^*$ is the multiplicative group of the center of Δ.*

(iii) *There exist an $h \in H$ and a $\mu \in Z(\Delta)^*$ such that the matrix $h - \mu E_n$ has a row or a column of zeros, but $h - \mu E_n \neq 0$.*

(iv) *There exists a matrix $h \in H$, $h \neq \gamma E_n$, $\gamma \in Z(\Delta)^*$, such that the matrices $t_{12}(\delta)$ and h commute for any $\delta \in \Delta$.*

Then $H \supset E(n, \Delta, N)$, $N \neq (0)$.

PROOF. (i) It is obvious that

$$H \cap \langle E_{n-1}, \Delta^{n-1}\rangle = A = \langle E_{n-1}, B\rangle,$$

where $B \neq (0)$ is a subgroup of the additive group of Δ^{n-1}. Since H is normalized by $SL(n, \Delta) \supset \langle SL(n-1, \Delta), (0)\rangle$ and $\langle E_{n-1}, \Delta^{n-1}\rangle$ is normalized by $\langle SL(n-1, \Delta), (0)\rangle$, it follows that $A = \langle E_{n-1}, B\rangle$ is also normalized by

$$\langle SL(n-1, \Delta), (0)\rangle.$$

By Lemmas 8.7 and 8.8 we see that $B = \Phi^{n-1}$, where $\Phi \neq 0$ is a left ideal of Δ. Hence $t_{1n}(\varphi) \in A \subset H$ for $\varphi \in \Phi$. Then, by Lemma 11.4, $H \supset E(n, \Delta, N)$, $N \neq (0)$.

(ii) Let $\mu \in Z(\Delta)^*$ and $d = \mu\langle E_{n-1}, a\rangle \in H$, $a \neq (0)$. Then for $s \in SL(n-1, \Delta)$

$$\langle s, 0\rangle d \langle s, 0\rangle^{-1} = \mu \langle E_{n-1}, sa\rangle = d_1 \in H.$$

The matrix s can clearly be chosen so that $sa \neq a$. Then

$$dd_1^{-1} \neq E_n, \quad dd_1^{-1} \in H \cap \langle E_{n-1}, \Delta^{n-1}\rangle;$$

that is, H satisfies (i).

(iii) We may assume that the matrix $h - \mu E_n$ has a row of zeros, and that $h \in H$, $\mu \in Z(\Delta)^*$ and $h \neq \mu E_n$. For if $h - \mu E_n$ has a column of zeros, we consider instead the group of transposed matrices over the anti-isomorphic ring. We may thus assume that $h - \mu E_n$ has a row of zeros. Transforming $h - \mu E_n$ by an element of $SL(n, \Delta)$, we can achieve that this is the last row. We may therefore assume that $h \in Z(\Delta)^*\mathrm{Aff}(n-1, \Delta)$. We consider the intersection

$$H \cap Z(\Delta)^* \mathrm{Aff}(n-1, \Delta) = A.$$

Since $h \neq \mu E_n$, it follows that $A \not\subset Z(\Delta)^* E_n$. If $A \subset Z(\Delta)^*\langle E_{n-1}, \Delta^{n-1}\rangle$, then H satisfies (ii). But if $A \not\subset Z(\Delta)^*\langle E_{n-1}, \Delta^{n-1}\rangle$, we consider the commutator subgroup

$$K = (A, \langle E_{n-1}, \Delta^{n-1}\rangle) = (A, Z(\Delta)^* \langle E_{n-1}, \Delta^{n-1}\rangle).$$

It is clear that $K \subset H$, because $\langle E_{n-1}, \Delta^{n-1}\rangle \subset SL(n, \Delta)$. By Theorem 7.2, $Z(\Delta)^*\langle E_{n-1}, \Delta^{n-1}\rangle$ is a maximal abelian subgroup of $GL(n, \Delta)$, and so $K \neq (E_n)$. Obviously

$$K = (A_1, \langle E_{n-1}, \Delta^{n-1}\rangle),$$

where A_1 is a subgroup of $\mathrm{Aff}(n-1, \Delta)$. Consequently $K \subset \langle E_{n-1}, \Delta^{n-1}\rangle$ and $K \neq (E_n)$; that is, H satisfies (i).

(iv) Let $h \in H$ and suppose $t_{12}(\delta)h = ht_{12}(\delta)$ for any $\delta \in \Delta$. Then the first column of h is of the form

$$\begin{bmatrix} \gamma \\ 0 \\ \cdot \\ \cdot \\ \cdot \\ 0 \end{bmatrix}, \qquad \gamma \in Z(\Delta)^*,$$

that is, H satisfies (iii). ∎

LEMMA 2. *Let $n > 2$, and let H be a subgroup of $GL(n, \Delta)$ that is normalized by $SL(n, \Delta)$ and contains a matrix h not in the center of $GL(n, \Delta)$, such that at least one of its elements is zero. Then $H \supset E(n, \Delta, N)$, $N \neq (0)$.*

PROOF. Let $h = \|\gamma_{ij}\|$, $\gamma_{kl} = 0$. There are two possibilities: $k \neq l$ or $k = l$. In the first case it follows from Lemma 8.5 that $SL(n, \Delta)$ contains a matrix c such that

$$chc^{-1} = \|\alpha_{ij}\|, \qquad \alpha_{n1} = 0.$$

But if $k = l$, then $SL(n, \Delta)$ contains a matrix c such that

$$chc^{-1} = \|\alpha_{ij}\|, \qquad \alpha_{11} = 0.$$

Consequently we may assume that $\gamma_{n1} = 0$ or $\gamma_{11} = 0$. Bearing Lemma 1 (iv) in mind, it suffices to consider the case when there is an $\alpha \in \Delta$ such that

$$r = ht_{12}(\alpha)h^{-1}t_{12}^{-1}(\alpha) \neq E_n.$$

Obviously $r \in H$. Now

$$ht_{12}(\alpha)h^{-1} = E_n + he_{12}\alpha h^{-1} = E_n + u\alpha v,$$

where u is the first column of h and v the second row of h^{-1}. Consequently

$$r = t_{12}^{-1}(\alpha) + u\alpha vt_{12}^{-1}(\alpha) = t_{12}^{-1}(\alpha) + uc, \quad c = \alpha vt_{12}^{-1}(\alpha).$$

We now distinguish between the two cases $\gamma_{n1} = 0$ and $\gamma_{11} = 0$. In the first case it is clear that the last row of uc is zero. Thus the last row of r has the form $[0 \cdots 01]$, and H satisfies (iii) of Lemma 1. But if $\gamma_{11} = 0$, the first row of r is the same as the first row of $t_{12}^{-1}(\alpha)$; that is, it has the form $[1 - \alpha, 0, \ldots, 0]$. Clearly $r = \|\rho_{ij}\|$ is a noncentral element and $\rho_{13} = 0$. This brings us back to the case $k \neq l$. ∎

From now on Δ is a *commutative Euclidean domain.*

LEMMA 3. *For any column vector*

$$a = \begin{bmatrix} \alpha_1 \\ \cdot \\ \cdot \\ \cdot \\ \alpha_n \end{bmatrix} \in \Delta^n, \qquad n > 1,$$

$SL(n, \Delta)$ contains a matrix s such that

$$sa = \begin{bmatrix} \mu \\ 0 \\ \cdot \\ \cdot \\ \cdot \\ 0 \end{bmatrix}, \qquad \mu\Delta = \alpha_1\Delta + \ldots + \alpha_n\Delta.$$

PROOF. If $a = (0)$, then $\mu = 0$ and $s = E_n$. Now let $a \neq (0)$. Obviously

$$t_{ij}(\delta)\, a = \begin{bmatrix} \alpha_1 \\ \cdot \\ \cdot \\ \cdot \\ \alpha_{i-1} \\ \alpha_i + \delta\alpha_j \\ \alpha_{i+1} \\ \cdot \\ \cdot \\ \cdot \\ \alpha_n \end{bmatrix}. \tag{1}$$

Let O be the orbit of $SL(n, \Delta)$ containing a. From all the nonzero coordinates of vectors in O, let μ be one of minimum norm. If b is a vector in O one of whose coordinates is μ, then by (1) all its coordinates are divisible by μ. Thus O contains

$$\begin{bmatrix} 0 \\ \cdot \\ \cdot \\ \cdot \\ 0 \\ \mu \\ 0 \\ \cdot \\ \cdot \\ \cdot \\ 0 \end{bmatrix},$$

and with it also the vector

$$\begin{bmatrix} \mu \\ 0 \\ \cdot \\ \cdot \\ \cdot \\ 0 \end{bmatrix}.$$

In addition, it follows from (1) that for $s \in SL(n, \Delta)$ the coordinates of the vectors a and sa generate the same ideal of Δ. Thus $\alpha_1\Delta + \cdots + \alpha_n\Delta = \mu\Delta$. ∎

COROLLARY 1. *Vectors a and b of Δ^n are in the same orbit of $SL(n, \Delta)$ if and only if the coordinates of a and those of b generate the same ideal of Δ.*

PROOF. Necessity follows in an obvious manner from Lemma 3. To prove sufficiency, let

$$a = \begin{bmatrix} \alpha_1 \\ \cdot \\ \cdot \\ \cdot \\ \alpha_n \end{bmatrix}, \quad b = \begin{bmatrix} \beta_1 \\ \cdot \\ \cdot \\ \cdot \\ \beta_n \end{bmatrix},$$

$$\alpha_1\Delta + \dots + \alpha_n\Delta = \beta_1\Delta + \dots + \beta_n\Delta.$$

Then, by Lemma 3, $SL(n, \Delta)$ contains matrices s_1 and s_2 such that

$$s_1 a = \begin{bmatrix} \mu_1 \\ 0 \\ \cdot \\ \cdot \\ \cdot \\ 0 \end{bmatrix}, \quad s_2 b = \begin{bmatrix} \mu_2 \\ 0 \\ \cdot \\ \cdot \\ \cdot \\ 0 \end{bmatrix}, \quad (\mu_1) = (\mu_2).$$

If $\mu_1 = 0$, then $\mu_2 = 0$ and $a = b = 0$. Let $\mu_1 \neq 0$. Then $\mu_2 = \epsilon\mu_1$, $\epsilon \in \Delta^*$. Let $O(a)$ and $O(b)$ be the orbits of $SL(n, \Delta)$ containing a and b, respectively. Now,

$$t_{21}(\varepsilon) \begin{bmatrix} \mu_1 \\ 0 \\ \cdot \\ \cdot \\ \cdot \\ 0 \end{bmatrix} = \begin{bmatrix} \mu_1 \\ \mu_2 \\ 0 \\ \cdot \\ \cdot \\ \cdot \\ 0 \end{bmatrix} \in O(a);$$

$\mu_1 = \epsilon^{-1}\mu_2$, $\epsilon^{-1} \in \Delta^*$. Consequently,

$$\begin{bmatrix} \mu_2 \\ 0 \\ \cdot \\ \cdot \\ \cdot \\ 0 \end{bmatrix} \in O(a);$$

that is, $O(a) = O(b)$. ∎

LEMMA 4. *Let $n > 2$ and $h = \|\gamma_{ij}\| \in \Delta_n$. Then there is a matrix s in $SL(n, \Delta)$ such that $shs^{-1} = g = \|\beta_{ij}\|$, where*

$$\begin{bmatrix} \beta_{21} \\ \beta_{31} \\ \cdot \\ \cdot \\ \cdot \\ \beta_{n1} \end{bmatrix} = \begin{bmatrix} \mu \\ 0 \\ \cdot \\ \cdot \\ \cdot \\ 0 \end{bmatrix}, \quad \mu\Delta = \gamma_{21}\Delta + \dots + \gamma_{n-1}\Delta, \quad \beta_{11} = \gamma_{11}. \tag{2}$$

PROOF. By Lemma 3 there exists a matrix $s_1 \in SL(n-1, \Delta)$ such that

$$s_1 \begin{bmatrix} \gamma_{21} \\ \cdot \\ \vdots \\ \gamma_{n1} \end{bmatrix} = \begin{bmatrix} \mu \\ 0 \\ \vdots \\ 0 \end{bmatrix}.$$

Setting $s = \operatorname{diag}[1, s_1]$, we obtain (2), and $s \in SL(n, \Delta)$. ∎

THEOREM 1. *Let $n > 2$, and let H be a subgroup of $GL(n, \Delta)$ that is normalized by $SL(n, \Delta)$ and is not contained in $Z(\Delta)^*E_n$. Then $H \supset E(n, \Delta, N)$ for some nontrivial ideal N of Δ.*

PROOF. Let $g \in H$ and $g \notin Z(\Delta)^*E_n$. Then, by Lemma 4, H contains a matrix

$$h = \| \gamma_{ij} \| = sgs^{-1}, \quad s \in SL(n, \Delta),$$

such that $\gamma_{n1} = 0$. Clearly $h \notin Z(\Delta)^*E_n$. It then follows from Lemma 2 that $H \supset E(n, \Delta, N), N \neq (0)$. ∎

2. *The group $CSL(n, Z, N)$*. Let Δ be an associative ring with unit element, and N a two-sided ideal of Δ. We define a new group $CSL(n, \Delta, N)$ by

$$CSL(n, \Delta, N) = CL(n, \Delta, N) \cap SL(n, \Delta).$$

$CSL(n, \Delta, N)$ is obviously a normal subgroup of both $SL(n, \Delta)$ and $GL(n, \Delta)$. It follows from the definition that $E(n, \Delta, N) \subset CSL(n, \Delta, N)$.

We now consider the case when Δ is the ring Z of rational integers. By definition, $SL(n, Z)$ is the group generated by all the elementary matrices $t_{ij}(\lambda)$, $\lambda \in Z$. Since $t_{ij}(\lambda) = (t_{ij}(1))^\lambda$, it follows that *$SL(n, Z)$ is generated by the $n(n-1)$ elementary matrices $t_{ij}(1), i, j = 1, \ldots, n$.*

THEOREM 2. *$SL(n, Z)$ is the group of all $n \times n$ matrices over Z with determinant 1.*

PROOF. Since $\det t_{ij}(\lambda) = 1$, we have $\det g = 1$ for any $g \in SL(n, Z)$. Conversely, let $h \in Z_n$ and $\det h = 1$. Using elementary transformations (that is, multiplying on the right and left by elementary matrices), we can bring h to the form

$$t_1 h t_2 = \operatorname{diag}[\pm 1, \ldots, \pm 1] = b, \qquad t_1, t_2 \in SL(n, Z),$$

where -1 appears an even number of times on the diagonal of b. We claim that $b \in SL(n, Z)$. To prove this, we consider the matrix $c = \operatorname{diag}[-1, 1, \ldots, 1, -1]$. Successively applying elementary transformations, we obtain

$$\begin{bmatrix} -1 & 0 & \dots & 0 & 1 \\ 0 & 1 & \dots & 0 & 0 \\ \cdot & \cdot & \cdot & \cdot & \cdot \\ 0 & 0 & \dots & 1 & 0 \\ 0 & 0 & \dots & 0 & -1 \end{bmatrix}, \quad \begin{bmatrix} 0 & 0 & \dots & 0 & 1 \\ 0 & 1 & \dots & 0 & 0 \\ \cdot & \cdot & \cdot & \cdot & \cdot \\ 0 & 0 & \dots & 1 & 0 \\ -1 & 0 & \dots & 0 & -1 \end{bmatrix},$$

$$\begin{bmatrix} 0 & 0 & \dots & 0 & 1 \\ 0 & 1 & \dots & 0 & 0 \\ \cdot & \cdot & \cdot & \cdot & \cdot \\ 0 & 0 & \dots & 1 & 0 \\ -1 & 0 & \dots & 0 & 0 \end{bmatrix}, \quad \begin{bmatrix} 1 & 0 & \dots & 0 & 1 \\ 0 & 1 & \dots & 0 & 0 \\ \cdot & \cdot & \cdot & \cdot & \cdot \\ 0 & 0 & \dots & 1 & 0 \\ -1 & 0 & \dots & 0 & 0 \end{bmatrix},$$

$$\begin{bmatrix} 1 & 0 & \dots & 0 & 0 \\ 0 & 1 & \dots & 0 & 0 \\ \cdot & \cdot & \cdot & \cdot & \cdot \\ 0 & 0 & \dots & 1 & 0 \\ -1 & 0 & \dots & 0 & 1 \end{bmatrix}, \quad E_n.$$

It follows that $c \in SL(n, Z)$. Hence also $h \in SL(n, Z)$. ∎

We now consider $CSL(n, Z, N) = CL(n, Z, N) \cap SL(n, Z)$. We define a binary relation R on $CSL(n, Z, N)$ by setting gRh for two elements $g, h \in CSL(n, Z, N)$ if h is obtained from g by finitely many operations of the following three types: multiplication on the right or left by an element of $E(n, Z, N)$, and transformation by an element of $SL(n, Z)$. *Clearly, R is an equivalence relation.*

LEMMA 5 (BRENNER [1]). *Let $n > 2$. Then for any matrix $a = \|\alpha_{ij}\| \in CSL(n, Z, N)$ there is a matrix*

$$b = \operatorname{diag}[c, 1], \qquad c \in CSL(n-1, Z, N), \tag{3}$$

such that aRb.

PROOF. If $N = Z$, then $CL(n, Z, N) = GL(n, Z)$, $CSL(n, Z, N) = SL(n, Z) = E(n, Z, N)$, and the lemma is trivial. It is also trivial for $N = (0)$, for $CL(n, Z, (0)) = CSL(n, Z, (0)) = E(n, Z, (0)) = (E_n)$. Now let $N = (\nu)$, $\nu \in Z$, $\nu > 1$. The lemma is almost obvious if $\alpha_{ii} = 1$ for some i, $1 \leqslant i \leqslant n$. For then $\alpha_{ki} \in N$ for $k \neq i$, and so

$$t_{ki}(\alpha_{ki}) \in E(n, Z, N).$$

On the other hand, let $d = t_{ki}(-\alpha_{ki})a$. It is clear that the (k, i)-element of d is zero. Thus, multiplying a on the left by matrices in $E(n, Z, N)$, we can make all the nondiagonal elements of the ith column vanish. The same can be done for the nondiagonal elements of the ith row. Finally, by Lemma 8.5, $SL(n, Z)$ contains a matrix transforming the resulting matrix to the form (3).

Now let $r = \|\rho_{ij}\| \in CSL(n, Z, N)$ be a matrix such that

$$(\rho_{21}\nu^{-1}, \rho_{32}\nu^{-1}) = 1, \qquad \rho_{31} = 0 \tag{4}$$

(on the left of the first equality we have used the standard notation for the greatest common divisor of two numbers). Then there is a matrix r_1 with 1 at the (2, 2)-position such that rRr_1. For $\rho_{22} = 1 + \nu\lambda$, and it follows from (4) that there exist integers δ_1 and δ_2 such that

$$\nu\lambda = \rho_{21}\delta_1 + \rho_{32}\delta_2. \tag{5}$$

Now, if $t = t_{12}(\delta_1)t_{23}(-\delta_2)$ and $r_1 = trt^{-1}$, then rRr_1, and by (5) the (2, 2)-element of r_1 is 1.

To complete the proof, it remains to prove the existence of a matrix $r = \|\rho_{ij}\|$ satisfying (4) and such that aRr. We may assume that $\alpha_{11} \neq \pm 1$. Then the g. c. d. of the nondiagonal elements of the first column in a is some number $\mu = \lambda\nu > 0$. Since $\det a = 1$, we have $(\alpha_{11}, \lambda) = 1$. By Lemma 4, a is conjugate in $SL(n, Z)$ to the matrix $f = \|\gamma_{ij}\|$, $\gamma_{11} = \alpha_{11}$, $\gamma_{21} = \lambda\nu$, $\gamma_{i1} = 0$ for $i > 2$. There obviously exists an $i > 1$ such that $\gamma_{3i} \neq 0$. We may therefore assume that $\gamma_{32} \neq 0$. Since $(\gamma_{11}, \lambda) = (\alpha_{11}, \lambda) = 1$, it follows from Dirichlet's theorem on primes in an arithmetical progression that there exists a $\gamma \in Z$ such that

$$(\gamma_{11}\gamma + \lambda, \ \gamma_{32}\nu^{-1}) = 1.$$

Now we set $r = \|\rho_{ij}\| = t_{21}(\gamma_\nu)f$. Then fRr and

$$\rho_{21} = \gamma_{21} + \gamma_{11}\gamma\delta = \lambda\nu + \gamma_{11}\gamma\nu = (\gamma_{11}\gamma + \lambda)\,\nu,$$
$$\rho_{32} = \gamma_{32}, \qquad \rho_{31} = 0.$$

By the choice of γ we have $(\rho_{32}\nu^{-1}, \rho_{21}\nu^{-1}) = 1$. Consequently aRr, and r satisfies (4). ∎

LEMMA 6. *Let $n > 2$. Then*

$$E(n, Z, N) \neq CSL(n, Z, N) \Rightarrow E(n-1, Z, N) \neq CSL(n-1, Z, N).$$

PROOF. Let

$$CSL(n, Z, N) \neq E(n, Z, N). \tag{6}$$

Then $D = CSL(n, Z, N) \setminus E(n, Z, N) \neq \varnothing$. Let $a \in D$, aRb. Since both sides of (6) are normal in $SL(n, Z)$, it follows that $b \in D$. By Lemma 5, there exists a matrix $b = \operatorname{diag}[c, 1]$ such that aRb. Since $b \notin E(n, Z, N)$, we have $c \notin E(n-1, Z, N)$. Consequently $c \in CSL(n-1, Z, N) \setminus E(n-1, Z, N)$. ∎

3. *Equality of the groups $E(3, Z, N)$ and $CSL(3, Z, N)$.* We introduce the canonical homomorphism

$$f\colon SL(3, Z) \to SL(3, Z)/E(3, Z, N)$$

and set $\Gamma = CSL(3, Z, N)$.

LEMMA 7. *$f(\Gamma)$ is contained in the center of* Im f.

PROOF. If gRh and $f(h)$ is in the center C of Im f, then $f(g) = f(h) \in C$. Let $g \in \Gamma$. By Lemma 5 there is a matrix

$$h = \begin{bmatrix} \alpha & \beta & 0 \\ \gamma & \delta & 0 \\ 0 & 0 & 1 \end{bmatrix} \in \Gamma,$$

such that gRh. We consider

$$h_1 = ht_{13}(1)h^{-1}t_{13}^{-1}(1) = \begin{bmatrix} 1 & 0 & \alpha \\ 0 & 1 & \gamma \\ 0 & 0 & 1 \end{bmatrix} t_{13}(-1) = t_{13}(\alpha - 1)\, t_{23}(\gamma).$$

Since $\alpha - 1, \gamma \in N$, it follows that $h_1 \in E(3, Z, N)$. In similar fashion one proves that the other commutators $\bar{h}$ of the type $ht_{ij}(1)h^{-1}t_{ij}^{-1}(1)$, where $(i, j) \neq (1, 2)$ or $(2, 1)$, are also in $E(3, Z, N)$. Therefore $f(\bar{h}) = E(3, Z, N)$, and $f(h)$ commutes with $f(t_{ij}(1))$. Furthermore,

$$t_{12}(1) = t_{13}(1)\, t_{32}(1)\, t_{13}(-1)\, t_{32}(-1),$$
$$t_{21}(1) = t_{23}(1)\, t_{31}(1)\, t_{23}(-1)\, t_{31}(-1),$$

so that $SL(3, Z)$ is generated by the matrices $t_{ij}(1)$, $(i, j) \neq (1, 2), (2, 1)$. Therefore $f(h) \in C$, and so also $f(g) \in C$. ∎

LEMMA 8. *Let*

$$x = \begin{bmatrix} \alpha & \beta & 0 \\ \gamma & \delta & 0 \\ 0 & 0 & 1 \end{bmatrix} \in \Gamma, \qquad y = f(x),$$

and let $k \geqslant 1$ be an integer. Then Γ contains a matrix

$$x_k = \begin{bmatrix} \alpha & \beta^k & 0 \\ (-1)^{k+1}\gamma^k & \delta_k & 0 \\ 0 & 0 & 1 \end{bmatrix},$$

such that $f(x_k) = y^k$.

PROOF. We use induction on k. For $k = 1$ the lemma is true with $x_1 = x$ and $\delta_1 = \delta$. Supposing that the lemma is true for k, we prove it for $k + 1$. We set

$$\bar{x} = \begin{bmatrix} 0 & 1 & 0 \\ 0 & 0 & -1 \\ -1 & 0 & 0 \end{bmatrix} x \begin{bmatrix} 0 & 1 & 0 \\ 0 & 0 & -1 \\ -1 & 0 & 0 \end{bmatrix}^{-1} = \begin{bmatrix} \delta & 0 & -\gamma \\ 0 & 1 & 0 \\ -\beta & 0 & \alpha \end{bmatrix}.$$

By Lemma 7, $f(x) \in C$. Thus $f(\bar{x}) = y$. By the inductive hypothesis, $f(x_k\bar{x}) = f(x_k)f(\bar{x}) = y^{k+1}$. Multiplying the matrix

$$x_k\bar{x} = \begin{bmatrix} \alpha\delta & \beta^k & -\alpha\gamma \\ (-1)^{k+1}\gamma^k\delta & \delta_k & (-1)^{k+2}\gamma^{k+1} \\ -\beta & 0 & \alpha \end{bmatrix}$$

successively by elementary matrices in $E(3, Z, N)$, we obtain

$$x_k\bar{x}, \quad \begin{bmatrix} 1 & \beta^k & 0 \\ (-1)^{k+1}\gamma^k\delta & \delta_k & (-1)^{k+2}\gamma^{k+1} \\ -\beta & 0 & \alpha \end{bmatrix},$$

$$\begin{bmatrix} 1 & 0 & 0 \\ (-1)^{k+1}\gamma^k & \delta_{k+1} & (-1)^{k+2}\gamma^{k+1} \\ -\beta & \beta^{k+1} & \alpha \end{bmatrix}, \quad \begin{bmatrix} 1 & 0 & 0 \\ 0 & \delta_{k+1} & (-1)^{k+2}\gamma^{k+1} \\ 0 & \beta^{k+1} & \alpha \end{bmatrix}.$$

Transforming the last matrix by an element of $SL(3, Z)$ (Lemma 8.5), we obtain x_{k+1}. By construction, $f(x_{k+1}) = y^{k+1}$. ∎

LEMMA 9. *Let*

$$x = \begin{bmatrix} \alpha & \beta & 0 \\ \gamma & \delta & 0 \\ 0 & 0 & 1 \end{bmatrix} \in \Gamma,$$

and let k be an integer such that $\beta^k \equiv \epsilon (\mathrm{mod}\ \alpha)$, where $\epsilon = \pm 1$. Then $x^k \in E(3, Z, N)$.

PROOF. We consider $u_k = x_k t_{12}(\lambda\nu)$, where $(\nu) = N$:

$$u_k = \begin{bmatrix} \alpha & \alpha\lambda\nu + \beta^k & 0 \\ \gamma' & \delta' & 0 \\ 0 & 0 & 1 \end{bmatrix}.$$

Since $x \in CSL(3, Z)$, it follows that $\alpha = 1 + \mu\nu$, $\nu | \beta$. Consequently $\nu | \beta^k + \epsilon\mu\nu$. On the other hand,

$$\beta^k = \varepsilon + \rho\alpha, \quad \beta^k + \varepsilon\mu\nu = \varepsilon + \rho\alpha + \varepsilon(\alpha - 1) = (\rho + \varepsilon)\alpha.$$

Consequently $\nu | (\rho + \epsilon)\alpha$, and since $(\nu, \alpha) = 1$ we have that $\nu | \rho + \epsilon$. Then $\nu\alpha | (\rho + \epsilon)\alpha$; that is, $\nu\alpha | \beta^k + \epsilon\mu\nu$. Thus the equation $\alpha\lambda\nu + \beta^k = -\epsilon\mu\nu$ is solvable for λ in Z. Substituting its solution in u_k, we obtain

$$u_k = \begin{bmatrix} \alpha & -\varepsilon\mu\nu & 0 \\ \gamma' & \delta' & 0 \\ 0 & 0 & 1 \end{bmatrix}.$$

Now,

$$t_{21}(-\varepsilon)\, u_k t_{21}(\varepsilon) = \begin{bmatrix} 1 & -\varepsilon\mu\nu & 0 \\ \gamma'' & \delta'' & 0 \\ 0 & 0 & 1 \end{bmatrix}.$$

This matrix is evidently in $E(3, Z, N)$. Consequently $x_k \in E(3, Z, N)$. By Lemma 8,

$$f(x^k) = f(x_k) = E(3, Z, N), \qquad x^k \in E(3, Z, N). \blacksquare$$

LEMMA 10 (MENNICKE [1]). $CSL(3, Z, N) = E(3, Z, N)$.

PROOF. Let $x \in \Gamma$. We claim that $x \in E(3, Z, N)$. If xRy, then $x \in E(3, Z, N) \Leftrightarrow y \in E(3, Z, N)$; hence by Lemma 5 we may assume that

$$x = \begin{bmatrix} \alpha & \beta & 0 \\ \gamma & \delta & 0 \\ 0 & 0 & 1 \end{bmatrix}.$$

For $\tau_i \in Z$,

$$t_{21}^{-1}(\tau_i)\, x t_{21}(\tau_i) = \begin{bmatrix} \alpha + \tau_i\beta & \beta & 0 \\ \gamma' & \delta' & 0 \\ 0 & 0 & 1 \end{bmatrix} = x_i.$$

If $\alpha_i = \alpha + \tau_i\beta$ and k_i is an integer such that

$$\beta^{k_i} \equiv \pm 1 \pmod{\alpha_i}, \tag{7}$$

then $x_i^{k_i} \in E(3, Z, N)$ by Lemma 9. But

$$x_i^{k_i} = t_{21}^{-1}(\tau_i)\, x^{k_i} t_{21}(\tau_i),$$

so that $x_i^{k_i} R x^{k_i}$, whence

$$x^{k_i} \in E(3, Z, N). \tag{8}$$

Now let $i = 1, 2$, and let (k_1, k_2) be the g.c.d. of k_1 and k_2. By (8),

$$x^{(k_1, k_2)} \in E(3, Z, N).$$

It now remains to prove the existence of numbers α_1 and α_2 in the progression

$$\alpha + \tau\beta \tag{9}$$

and relatively prime k_1 and k_2 in Z satisfying (7).

We first prove the existence of a prime p such that

$$p \equiv -1 \pmod 4 \tag{10}$$

and either p or $-p$ is in the progression (9). Since $\det x = 1$, we have $(\alpha, \beta) = 1$.

Let $\alpha \equiv -1 \pmod 4$. Then $(\alpha, 4\beta) = 1$, and therefore, again by Dirichlet's theorem, the progression $\alpha + 4\beta\tau$ contains a prime p.

Similarly, if $\alpha \equiv 1 \pmod 4$ there is a prime p in the progression $-\alpha + \tau(-\beta)$ such that $p \equiv -\alpha \equiv -1 \pmod 4$. Then $-p$ is a member of the progression (9).

Now suppose that α is even. Then $\beta \equiv 1 \pmod 2$ and the congruence $\alpha + k\beta \equiv -1 \pmod 4$ has a unique solution k. Consider the progression $\alpha + \tau\beta$, where $\tau \equiv k \pmod 4$:

$$\alpha + \tau\beta = \alpha + (4l + k)\beta = (\alpha + k\beta) + 4\beta l. \tag{11}$$

Since $(\alpha + k\beta, 4\beta) = 1$, the progression (11) contains a prime p.

Suppose that p satisfies (10) and either p or $-p$ belongs to (9). Let

$$p - 1 = 2^l q_1^{l_1} \ldots q_s^{l_s}$$

be the canonical representation, and set $m = \beta q_1 \cdots q_s$, or $m = \beta$ for $p - 1 = 2^l$. By Dirichlet's theorem there is a prime n such that $n \equiv -1 \pmod m$. Further, since $(p, m) = 1$, there exists a prime r such that $r \equiv -p \pmod m$. Let a be the exponent of β modulo p and k_2 the exponent of β modulo nr. Then $a | p - 1$ and $k_2 | \varphi(nr)$, where $\varphi(x)$ is Euler's function. By the choice of n and r, this shows that a and k_2 have no odd prime factors in common. Since p satisfies (10), either a is odd, or a is even and $a/2$ is odd. Now we take $\alpha_1 = \pm p$ and $\alpha_2 = \pm nr$ in such a way that α_1 and α_2 belong to (9). Let $k_1 = a$ if a is odd and $k_1 = a/2$ if a is even. Then α_1 and α_2 are in (9), and k_1 and k_2 satisfy (7). ∎

Lemmas 6 and 10 imply

THEOREM 3. *$CSL(n, Z, N) = E(n, Z, N)$ for $n > 2$.*

4. *Normal subgroups of $GL(n, Z)$ and $SL(n, Z)$, $n > 2$. Subgroups of finite index.* Applying Theorem 1 to the case $\Delta = Z$ and using Theorem 3, we obtain the following proposition:

THEOREM 4 (MENNICKE [1]). *Let $n > 2$; let H be a subgroup of $GL(n, Z)$ normalized by $SL(n, Z)$ and not contained in Z^*E_n. Then $H \supset CSL(n, Z, N)$ for some nontrivial ideal N of Z.*

COROLLARY 1. *Let $n > 2$, $\Gamma = GL(n, Z)$ or $\Gamma = SL(n, Z)$, $H \lhd \Gamma$ and H not a subgroup of the center of Γ. Then H is a subgroup of finite index.*

Since every subgroup of finite index contains a normal subgroup of finite index, we infer from Theorem 4:

THEOREM 5 (BASS, LAZARD AND SERRE [1]; MENNICKE [1]). *Let $n > 2$, and let G be a subgroup of finite index in $SL(n, Z)$. Then $G \supset CSL(n, Z, N)$ for some $N \neq (0)$.*

The condition $n > 2$ in Theorems 4 and 5 is essential. Examples were known as far back as 1887 of normal subgroups of finite index in $SL(2, Z)$ that do not contain $CSL(2, Z, N)$ for any $N \neq (0)$. A number of such examples were constructed by Reiner in [1].

Subgroups of finite index in $SL(n, \Delta)$, $n > 2$, where Δ is the ring of algebraic integers of a finite extension of the rationals, are studied in Bass, Milnor and Serre [1]. In this case Theorem 5 remains valid if and only if, for any ideal N in

Δ, the factor group $C_N = CSL(n, \Delta, N)/E(n, \Delta, N)$ is trivial. If K has a real embedding then $C_N = 1$ for any ideal N. But if K is completely imaginary, then C_N is isomorphic to some subgroup of order $r = r(N)$ of the group of roots of unity in K. In particular, if m is the number of roots of unity in K and m^2 divides the ideal N, then $r(N) = m$.

FOURTH CHAPTER

REDUCIBILITY AND IMPRIMITIVITY

§13. Abelian groups with operators. The structure of semisimple algebras

In this section we present, mainly without proof, the properites of abelian operator groups and the theorems on the structure of associative semisimple algebras that we need later.

1. *Abelian operator groups.* Let A be an additive abelian group with operator domain Ω. Then, by definition, for any $\omega \in \Omega$ and $x \in A$, an element $\omega x \in A$ is defined, and $\omega(a + b) = \omega a + \omega b$ for $a, b \in A$. A subgroup B of A is said to be *admissible* if $\omega b \in B$ for any $\omega \in \Omega$ and $b \in B$. It is evident that A and (0) are admissible subgroups of A. If a nontrivial group A has only these two trivial admissible subgroups, then A is said to be *operator-simple.*

If B is an admissible subgroup of A, the factor group A/B can be made into an operator group with the same operator domain Ω by setting $\omega(a + B) = \omega a + B$ for $a \in A$.

A *composition series* of an operator group A is a finite series

$$A_0 = A \supset A_1 \supset \ldots \supset A_k = (0)$$

of admissible subgroups whose factors $F_i = A_i/A_{i+1}$, $i = 0, 1, \ldots, k - 1$, are operator-simple.

Let A and $\bar{A}$ be groups with the same operator domain Ω. We say that A and $\bar{A}$ are *operator-isomorphic* if there exists a bijection $f: A \longrightarrow \bar{A}$ such that $f(x + y) = f(x) + f(y)$ and $f(\omega x) = \omega f(x)$ for any $x, y \in A$ and $\omega \in \Omega$.

If an operator group A has a composition series, then its factors F_i are uniquely determined to within operator isomorphism and order (Jordan-Hölder theorem).

PROPOSITION 1. *Suppose that a group A is a direct sum*

$$A = B_1 \dotplus \ldots \dotplus B_k \tag{1}$$

of operator-simple admissible subgroups B_i. Then every nontrivial admissible subgroup B of A is operator-isomorphic to a direct sum $B_{i_1} \dotplus \ldots \dotplus B_{i_r}$, and A can *be represented as a direct sum*

$$A = B \dotplus B_{i_{r+1}} \dotplus \dots \dotplus B_{i_k},$$

where $i_1, \dots, i_r, i_{r+1}, \dots, i_k$ *is some permutation of* $1, \dots, k$.

PROPOSITION 2. *Let*

$$A = B_1 \dotplus \dots \dotplus B_k = C_1 \dotplus \dots \dotplus C_l, \tag{2}$$

where $B_1, \dots, B_k$ *and* $C_1, \dots, C_l$ *are operator-simple admissible subgroups of* A. *Then* $k = l$ *and, with a suitable numbering of the summands,* B_j *and* C_j *are operator-isomorphic* ($j = 1, \dots, k$).

These two propositions imply

PROPOSITION 3. *Let* (2) *be two decompositions of* A *into a direct sum of operator-simple admissible subgroups,* H *an operator-simple admissible subgroup of* A, $B(H)$ *the direct sum of all* B_j *in* (2) *that are operator-isomorphic to* H, *and* $C(H)$ *the direct sum of all* C_j *in* (2) *that are operator-isomorphic to* H. *Then* $H \subset B(H)$ *and* $B(H) = C(H)$.

PROPOSITION 4. *A group* A *having a composition series can be represented as a direct sum of operator-simple admissible subgroups if and only if every operator-simple admissible subgroup of* A *has an admissible direct complement.*

A group A is said to be *decomposable* if it can be expressed as the direct sum of two nontrivial admissible subgroups. Otherwise A is said to be *indecomposable.*

PROPOSITION 5. *An operator group* A *having a composition series is either indecomposable or the direct sum of finitely many indecomposable admissible subgroups. If* (2) *are two representations of* A *as a direct sum of indecomposable admissible subgroups, then* $k = l$ *and, with a suitable numbering of the summands,* B_j *and* C_j *are operator-isomorphic* ($j = 1, \dots, k$).

2. *Homomorphisms of abelian operator groups.* Let A and B be additive abelian groups with the same operator domain Ω. The symbol $\mathrm{Hom}(A, B)$ denotes the set of all operator homomorphisms of A into B; that is, the set of all mappings $h: A \longrightarrow B$ such that

$$h(x + y) = h(x) + h(y), \qquad h(\omega x) = \omega h(x)$$

for arbitrary $x, y \in A$ and $\omega \in \Omega$. We define an addition of elements g and h in $\mathrm{Hom}(A, B)$ by setting $(g + h)(x) = g(x) + h(x)$. Then $\mathrm{Hom}(A, B)$ evidently becomes an additive abelian group. It is clear that if $h \in \mathrm{Hom}(A, B)$, then $\mathrm{Ker}\, h$ is an admissible subgroup of A, and $\mathrm{Im}\, h$ an admissible subgroup of B.

When $B = A$, the elements of $\mathrm{Hom}(A, B) = \mathrm{Hom}(A, A)$ may also be multiplied together as mappings. A direct check shows that $\mathrm{Hom}(A, A)$ *is an associative ring with unit element.*

LEMMA 1. (SCHUR'S LEMMA). *Let A and B be operator-simple abelian groups with the same domain of operators, and $h \in \mathrm{Hom}(A, B)$. Then h is either the zero of* $\mathrm{Hom}(A, B)$ *or an operator isomorphism.*

PROOF. Let $h \neq 0$. Then $\mathrm{Ker}\, h \neq A$. Since A is operator-simple, $\mathrm{Ker}\, h = (0)$. Since $h \neq 0$, we have $\mathrm{Im}\, h \neq (0)$. But then, since B is operator-simple, $\mathrm{Im}\, h = B$. Consequently h is an operator isomorphism. ∎

COROLLARY 1. *If A is an operator-simple group, then* $\mathrm{Hom}(A, A)$ *is a division ring. If A and B are nonisomorphic operator-simple groups, then* $\mathrm{Hom}(A, B) = (0)$.

LEMMA 2. *Let A be a direct sum* (1), *where the B_j are operator-simple pairwise operator-isomorphic groups. Then the ring* $\mathrm{Hom}(A, A)$ *is isomorphic to the full matrix ring T_k over the division ring* $T = \mathrm{Hom}(B_1, B_1)$.

PROOF. We may assume that A is the direct sum of k copies of a group $B = B_1$, and write $a \in A$ as $a = (a_1, \ldots, a_k)$, $a_j \in B$, so that B_j contains all the elements of the form $(0, \ldots, a_j, \ldots, 0)$, with a_j in the jth place. We define k^2 operator endomorphisms $\epsilon_{\mu\nu}$, $\mu, \nu = 1, \ldots, k$, by

$$\varepsilon_{\mu\nu}(a_1, \ldots, a_k) = (0, \ldots, a_\nu, \ldots, 0),$$

where a_ν is in the μth place. For example, $\epsilon_{21}(a_1, \ldots, a_k) = (0, a_1, 0, \ldots, 0)$. Let $T = \mathrm{Hom}(B, B)$. For $t \in T$ we set $t(a_1, \ldots, a_k) = (ta_1, \ldots, ta_k)$. Thus t may be treated as an endomorphism of A. Hence it follows that any element h of the form

$$h = \sum_{\alpha, \beta=1}^{k} t_{\alpha\beta}\varepsilon_{\alpha\beta}, \qquad t_{\alpha\beta} \in T, \tag{3}$$

is an operator endomorphism of A.

By the definition of $\epsilon_{\mu\nu}$ we have $\epsilon_{\alpha\beta}\epsilon_{\mu\nu} = \delta^\beta_\mu \epsilon_{\alpha\nu}$, where δ^β_μ is the Kronecker symbol. It is obvious that $t\epsilon_{\mu\nu} = \epsilon_{\mu\nu}t$, $t \in T$.

The endomorphisms $\epsilon_{\alpha\beta}$ are linearly independent over T. For suppose that $h = 0$. Then, multiplying (3) by $\epsilon_{\nu\nu}$ on the right and by $\epsilon_{\mu\mu}$ on the left, we obtain $t_{\mu\nu}\epsilon_{\mu\nu} = 0$. Hence $t_{\mu\nu} = 0$. Thus the endomorphisms (3) form a ring isomorphic to the full matrix ring T_k.

It remains to prove that every element $g \in \mathrm{Hom}(A, A)$ has the form (3). For fixed g, i, j, we define a mapping $t_{ij}: B \longrightarrow B$ as follows. Let $b \in B$,

$$b_j = (0, \ldots, b, \ldots, 0), \qquad gb_j = (c_1, \ldots, c_k).$$

Then we set $t_{ij}b = c_i$. Obviously, $t_{ij} \in T$. We claim that

$$g = \sum_{i,j} t_{ij}\varepsilon_{ij}. \tag{4}$$

For $a = (a_1, \ldots, a_k) \in A$ we can write

$$ga = g(a_1, \ldots, a_k) = \sum_{j=1}^{k} g(0, \ldots, a_j, \ldots, 0),$$

where the jth term contains the element a_j in the jth place. Now,

$$g(0, \ldots, a_j, \ldots, 0) = (t_{1j}a_j, \ldots, t_{kj}a_j).$$

Therefore

$$ga = \sum_{i,j} (0, \ldots, t_{ij}a_j, \ldots, 0),$$

with $t_{ij}a_j$ in the ith place. On the other hand,

$$t_{ij}\varepsilon_{ij}(a) = t_{ij}(0, \ldots, a_j, \ldots, 0) = (0, \ldots, t_{ij}a_j, \ldots, 0),$$

with a_j and $t_{ij}a_j$ in the ith place. Thus,

$$ga = \sum_{i,j} t_{ij}\varepsilon_{ij}(a).$$

This implies (4). ■

THEOREM 1. *Let A be a direct sum*

$$A = B_{11} \dotplus \ldots \dotplus B_{1r_1} \dotplus B_{21} \dotplus \ldots \dotplus B_{2r_2} \dotplus \ldots \dotplus B_{s1} \dotplus \ldots \dotplus B_{sr_s}$$

of operator-simple admissible groups B_{ij} such that $B_{i\alpha}$ and $B_{i\beta}$ are operator-isomorphic for any i, α and β, while $B_{i\alpha}$ and $B_{j\beta}$ are not operator-isomorphic for any $i \neq j$. Then the ring Hom(A, A) *is the direct sum of s rings $R_1, \ldots, R_s$ such that $R_iR_j = (0)$ for $i \neq j$, and R_j ($j = 1, \ldots, s$) is isomorphic to the full matrix algebra $T^j_{r_j}$ of degree r_j over the division ring $T^j =$* Hom(B_{j1}, B_{j1}).

PROOF. Let $A_j = B_{j1} \dotplus \ldots \dotplus B_{jr_j}$, $j = 1, \ldots, s$. Then $A = A_1 \dotplus \ldots \dotplus A_s$. For any $h \in R =$ Hom(A, A),

$$h(A_j) \subset A_j \tag{5}$$

For let us consider the mapping

$$\bar{h}\colon B_{j\alpha} \to h(B_{j\alpha}), \qquad x \mapsto h(x), \qquad x \in B_{j\alpha}.$$

Since h is a homomorphism and $B_{j\alpha}$ is operator-simple, $h(B_{j\alpha})$ is either (0) or a group operator-isomorphic to $B_{j\alpha}$. By Proposition 3, $h(B_{j\alpha}) \subset A_j$, and this proves (5).

We now define endomorphisms π_j, $j = 1, \ldots, s$, by

$$\pi_j\colon x \mapsto x_j, \quad x \in A, \quad x = x_1 + \ldots + x_s, \quad x_j \in A_j.$$

It is obvious that $\pi_1 + \ldots + \pi_s = 1$, where 1 is the identity automorphism of A. For $R = \mathrm{Hom}(A, A)$ we can write

$$R = (\pi_1 + \ldots + \pi_s)R = \pi_1 R + \ldots + \pi_s R = R_1 + \ldots + R_s,$$

where $R_j = \pi_j R$. It follows from (5) that

$$R(A_j) = A_j, \quad \pi_i R(A_i) = A_i, \quad \pi_i R(A_j) = (0) \quad \text{for} \quad i \neq j.$$

Hence $R_i \cap R_j = (0)$, $R_i R_j = (0)$ for $i \neq j$. Hence $R = R_1 \dotplus \ldots \dotplus R_s$. It is obvious that $R_j | A_j = \mathrm{Hom}(A_j, A_j)$. Since $R_j(A_i) = (0)$ for $i \neq j$, we have

$$R_j \cong \mathrm{Hom}(A_j, A_j).$$

Hence and from Lemma 2, the assertion of the theorem follows. ∎

3. *Semisimple algebras over a field.* Let A be a linear space over a field Δ, in which a multiplication of vectors is defined, so that any a and b in A have a product $ab \in A$. If A is a ring under addition and multiplication of vectors and if $(ab)\lambda = (a\lambda)b = a(b\lambda)$ for any $a, b \in A$ and $\lambda \in \Delta$, then A is called an *algebra over* Δ. If the multiplication of vectors is associative, A is called an *associative algebra.* The dimension of A as a linear space is called the *dimension* (or *rank*) of the algebra A. Any ideal of the ring A that admits multiplication by field elements is called an *ideal* of the algebra A.

In this subsection, A always denotes an *associative algebra of finite dimension* over a field Δ, having a unit element e. An ideal N (left, right or two-sided) of A is said to be *nilpotent* if there is a natural number l such that any product of l elements in N is zero. Since A has a unit element, every ideal of A is a subspace of the underlying linear space of A.

Let R be an ideal of maximal dimension over Δ among the nilpotent ideals of an algebra A. Then R is a two-sided ideal of A and contains all nilpotent ideals of A. Hence R is uniquely determined by A. R is known as the *radical* of A. An algebra A is said to be *semisimple* if $R = (0)$ and *simple* if its only two-sided ideals are A and (0).

It is sometimes convenient to regard an algebra A as an abelian operator group $W(A)$ with A as operator domain and with ax as the result of operating with a in A on an element x of A. It is clear that the admissible subgroups of $W(A)$ are simply the left ideals of A, and the operator-simple subgroups of $W(A)$ are the minimal left ideals of A. The structure of semisimple algebras is fully described by the following theorems.

THEOREM 2. *An algebra A is semisimple if and only if*

$$A = L_1 \dotplus \ldots \dotplus L_t, \tag{6}$$

where L_j, $j = 1, \ldots, t$, are minimal left ideals of A. In other words, A is semisimple if and only if the operator group $W(A)$ is the direct sum of operator-simple admissible subgroups.

THEOREM 3. *An algebra A is semisimple if and only if it is a direct sum*

$$A = A_1 \dotplus \ldots \dotplus A_s, \tag{7}$$

where A_j are two-sided ideals of A such that $A_iA_j = (0)$ for $i \neq j$, $A_j \cong T^j_{r_j}$, and T^j is a division ring whose center contains Δ.

By Proposition 2, the summands L_j in (6) are uniquely determined by A (to within operator isomorphism and order). The decomposition (7) is uniquely determined by A to within the order of the summands A_j. The decompositions (6) and (7) are related as follows. The number s of summands in (7) is the number of pairwise operator-isomorphic summands in (6). The dimensions over Δ of the summands in (6) are

$$n_i = (T^i : \Delta)\, r_i, \qquad i = 1, \ldots, s,$$

because A_i is the direct sum of r_i operator-isomorphic minimal left ideals of dimension n_i.

In particular, we have the following criteria for simplicity:

1) *An algebra A is simple if and only if $A = L_1 \dotplus \ldots \dotplus L_r$, where the L_j are pairwise operator-isomorphic minimal left ideals of A.*

2) *An algebra A is simple if and only if $A \cong T_r$, where T is a division ring.*

4. *On nilpotent rings and nilpotent groups.* Let A be an associative ring, B and C two-sided ideals of A. The *product* BC is the ideal in A consisting of all sums $b_1c_1 + \ldots + b_tc_t$, where t is any positive integer, $b_i \in B$ and $c_i \in C$. It is easy to see that multiplication of ideals is associative. This enables us to introduce powers A^ν, $\nu \geqslant 1$, of the ring A: $A^2 = AA$, $A^3 = AAA$, and so on.

Suppose A is *nilpotent*; that is, $A^k = 0$ for some positive integer k. The smallest positive integer k for which $A^k = 0$ is called the *nilpotency class* of A. It is easy to verify that the following strict inclusions hold:

$$A \supset A^2 \supset \ldots \supset A^{k-1} \supset 0.$$

Now let A be an arbitrary associative ring. We define an operation $\circ$ on the set A by setting $a \circ b = a + b + ab$, $a, b \in A$. Obviously, *under this operation A is a semigroup $\langle A, \circ\rangle$, whose unit element is the zero of A.* $\langle A, \circ\rangle$ is known as the *adjoint semigroup* of A.

THEOREM 4. *Let A be a nilpotent ring of class k. Then the adjoint semigroup $G = \langle A, \circ\rangle$ is a nilpotent group of class at most $k - 1$.*

PROOF. For any $a \in A$, we have $a^\nu = 0$, $\nu \leqslant k$. We set

$$a_1 = \sum_{j=1}^{\nu-1} (-1)^j a^j.$$

Then $a \circ a_1 = a_1 \circ a = 0$, so that a is an invertible element of G. Thus $G = \langle A, \circ\rangle$ is a group. We now show that G is nilpotent of class at most $k - 1$. We set $G_0 = G$ and $G_j = (G, G_{j-1})$, $j > 0$, where (G, G_{j-1}) denotes the commutator subgroup. We must show that G_{k-1} is trivial; that is, $G_{k-1} = (0)$. The theorem will be proved if we can show that $G_j \subset A^{j+1}$ for any j. This is obvious for $j = 0$. Suppose that $G_{j-1} \subset A^j$, $j > 0$. Now $G_j = (G, G_{j-1})$ is generated by the commutators $c = a \circ b \circ a_1 \circ b_1$, where $a \in G$, $b \in G_{j-1}$, $a \circ a_1 = 0$ and $b \circ b_1 = 0$. For this c we have

$$\begin{aligned} c = (a + b + ab) \circ (a_1 + b_1 + a_1 b_1) &= a + b + ab \\ + a_1 + b_1 + a_1 b_1 &+ (a + b + ab)(a_1 + b_1 + a_1 b_1) \\ &\equiv a + b + a_1 + b_1 + (a + b)(a_1 + b_1) \\ &\equiv a + a_1 + aa_1 + b + b_1 + bb_1 \pmod{A^{j+1}}. \end{aligned}$$

Consequently $a \circ b \circ a_1 \circ b_1 = c \in A^{j+1}$, and so $G_j \subset A^{j+1}$. ■

Now let Δ be an arbitrary associative ring with unit element. We single out in Δ_n the subring $N(n, \Delta)$ of all matrices of the form

$$a = \begin{bmatrix} 0 & \alpha_{12} & \alpha_{13} & \ldots & \alpha_{1n} \\ 0 & 0 & \alpha_{23} & \ldots & \alpha_{2n} \\ \cdot & \cdot & \cdot & \cdot & \cdot \\ 0 & 0 & 0 & \ldots & \alpha_{n-1, n} \\ 0 & 0 & 0 & \ldots & 0 \end{bmatrix}, \qquad \alpha_{ij} \in \Delta.$$

It is easy to see that the product of any n matrices in $N(n, \Delta)$ is zero. On the other hand, $e_{12}e_{23} \ldots e_{n-1,n} = e_{1n} \neq 0$. Thus $N(n, \Delta)$ is nilpotent of class n.

THEOREM 5. *$G = E_n + N(n, \Delta)$ is a nilpotent group of class $n - 1$.*

PROOF. The mapping $f\colon N(n, \Delta) \to E_n + N(n, \Delta)$ defined by $f(x) = E_n + x$ is an isomorphism of the adjoint group of $N(n, \Delta)$ onto the group G. For f is bijective, and

$$\begin{aligned} f(x \circ y) &= f(x + y + xy) = E_n + x + y + xy \\ &= (E_n + x)(E_n + y) = f(x)f(y) \end{aligned}$$

Hence, by Theorem 4, G is nilpotent of class at most $n - 1$. On the other hand, if we set $t_{ij}(1) = t_{ij}$, then $t_{13} = (t_{12}, t_{23}) \in G_1$, $t_{14} = (t_{13}, t_{34}) \in G_2, \ldots, t_{1n}$

$= (t_{1,n-1}, t_{n-1,n}) \in G_{n-2}$, where $G_0 = G$ and $G_j = (G, G_{j-1})$. Consequently $G_{n-2} \neq E_n$. Thus G is nilpotent of class $n - 1$. ∎

Now we take for Δ a division ring T.

THEOREM 6. *Let N be a nilpotent subring of T_n. Then there is an element d in $GL(n, T)$ such that $d^{-1}Nd \subset N(n, T)$.*

PROOF. We treat T_n as the ring of endomorphisms of the space T^n. Let $N^k = 0$ and $N^{k-1} \neq 0$, $k > 1$. We define a subspace Q_ν, $0 < \nu < k$, of T^n as the linear T-hull of the set of vectors $N^\nu T^n$. Obviously

$$Q_{k-1} \subset Q_{k-2} \subset \ldots \subset Q_1 \subset T^n = Q_0.$$

It follows from the definition of Q_ν that for $g \in N$ and $\nu < k - 1$

$$g(Q_\nu) \subset Q_{\nu+1}, \qquad g(Q_{k-1}) = 0, \qquad N^{k-\nu} Q_\nu = 0.$$

It can be shown that all the inclusions $Q_\nu \subset Q_{\nu-1}$, $\nu = 1, \ldots, k - 1$, are strict. For suppose that $Q_\nu = Q_{\nu-1}$. Then $N(Q_{\nu-1}) = N(Q_\nu) \subset Q_{\nu+1}$, so that

$$N^{k-1}T^n = N^{k-\nu}(N^{\nu-1}T^n) \subset N^{k-\nu}Q_{\nu-1} = N^{k-\nu}Q_\nu = 0.$$

But $N^{k-1}T^n \neq 0$, because k is the nilpotency class of N. It is now easy to prove the theorem. We construct a T-basis of T^n as follows. We complete a basis of Q_{k-1} to a basis of Q_{k-2}, then complete the resulting basis of Q_{k-2} to a basis of Q_{k-3}, etc. In this basis the matrix of an element g of N is of the form

$$\begin{bmatrix} 0 & a_{12} & \ldots & a_{1k} \\ 0 & 0 & \ldots & a_{2k} \\ \cdot & \cdot & \cdot & \cdot \\ 0 & 0 & \ldots & 0 \end{bmatrix}.$$

This completes the proof. ∎

COROLLARY 1. *$N(n, T)$ is a maximal nilpotent subring of T_n. If N_1 and N_2 are both maximal nilpotent subrings of T_n, there is an element d in $GL(n, T)$ such that $d^{-1}N_1d = N_2$.*

§14. Linear representations. Reducibility and irreducibility of linear groups

1. *Linear representations.* Let M be a right free module over a ring Δ, and Y an arbitrary nonempty set. Any mapping

$$d\colon Y \to \operatorname{Hom}(M, M) \tag{1}$$

is called a *linear representation of Y.*

For a given representation (1), the module M may be regarded as an additive abelian group $M(\Omega)$ with operator domain $\Omega = Y \cup \Delta$, where the action of $\omega \in \Omega$ on an element $u \in M$ is defined as follows. If $\omega = y \in Y$, we set $\omega u = \sigma u$, where $\sigma = d(y)$. If $\omega = \lambda \in \Delta$, we set $\omega u = u\lambda$. It is evident that every admissible subgroup of $M(\Omega)$ is a submodule of M. Let M_1 and M_2 be free right modules over Δ. Two representations

$$d_1\colon Y \to \operatorname{Hom}(M_1, M_1), \qquad d_2\colon Y \to \operatorname{Hom}(M_2, M_2) \tag{2}$$

of a set Y are said to be *equivalent* if the groups $M_1(\Omega)$ and $M_2(\Omega)$ are operator-isomorphic.

LEMMA 1. *The representations* (2) *are equivalent if and only if there exists a module isomorphism*

$$\psi\colon M_1 \to M_2 \tag{3}$$

such that for any $y \in Y$

$$d_2(y) = \psi d_1(y)\psi^{-1}. \tag{4}$$

PROOF. Suppose that the representations (2) are equivalent. Then there exists an operator isomorphism $\psi\colon M_1(\Omega) \longrightarrow M_2(\Omega)$. Obviously ψ is an isomorphism of M_1 and M_2. We now prove (4). Let x_1 range over M_1. Then $x_2 = \psi(x_1)$ ranges over M_2. For any $y \in Y$ we can write

$$d_2(y)(x_2) = y\psi(x_1) = \psi(yx_1) = \psi(d_1(y)(x_1)).$$

Hence

$$(d_2(y)\psi)(x_1) = (\psi d_1(y))(x_1), \qquad d_2(y)\psi = \psi d_1(y). \tag{5}$$

This implies (4), and we have proved necessity.

Now for sufficiency. Suppose that there exists a module isomorphism (3) such that (4) is satisfied. We claim that ψ is an operator isomorphism of the groups $M_1(\Omega)$ and $M_2(\Omega)$. It is clear that (4) implies (5). For any $x \in M$, we infer from (5) that

$$\psi d_1(y)(x) = d_2(y)\psi(x), \qquad \psi(yx) = y\psi(x).$$

Thus ψ is an operator isomorphism. ∎

In case the modules M_1 and M_2 have finite bases, Lemma 1 may be restated as follows:

The representations (2) *are equivalent if and only if the modules* M_1 *and* M_2 *have bases* B_1 *and* B_2, *respectively, such that for any* $y \in Y$ *the matrix of the endomorphism* $d_1(y)$ *in* B_1 *is the same as that of* $d_2(y)$ *in* B_2.

When the set Y in (1) is a group, we stipulate two more conditions. First,

$d(y_1 y_2) = d(y_1)d(y_2)$ for any $y_1, y_2 \in Y$; and second, Im d is a subgroup of $GL(M)$. A representation $d\colon G \longrightarrow \mathrm{Hom}(M, M)$ of a group G is said to be *faithful* if the mapping d is injective.

Let M_1 and M_2 be free modules over a ring Δ, H_1 a subgroup of $GL(M_1)$ and H_2 a subgroup of $GL(M_2)$. We say that H_1 and H_2 are *similar as linear groups* if there exists a module isomorphism $\varphi\colon M_1 \longrightarrow M_2$ for which $H_2 = \varphi H_1 \varphi^{-1}$.

If d_1 and d_2 are equivalent representations of a group G, it follows from Lemma 1 that Im d_1 and Im d_2 are similar as linear groups.

We now consider the special case of a representation (1) in which $M = V$ is a linear space over a division ring T,

$$d\colon Y \to \mathrm{Hom}(V, V). \tag{6}$$

We write $V(\Omega)$ instead of $M(\Omega)$. The representation (6) is said to be *irreducible* if $V(\Omega)$ is operator-simple. But if $V(\Omega)$ has a nontrivial admissible subgroup, we call the representation *reducible*. The representation (6) is said to be *completely reducible* if $V(\Omega)$ is a direct sum of operator-simple admissible subgroups. It is said to be *decomposable* (*indecomposable*) if $V(\Omega)$ is decomposable (indecomposable).

Now suppose that V is of finite dimension n over T. Since every admissible subgroup of $V(\Omega)$ is a subspace of V, the operator group $V(\Omega)$ has a composition series

$$V(\Omega) = V_0(\Omega) \supset V_1(\Omega) \supset \ldots \supset V_k(\Omega) = 0, \quad k \leqslant n. \tag{7}$$

We set $U_j(\Omega) = V_{j-1}(\Omega)/V_j(\Omega)$, $j = 1, \ldots, k$. The groups $U_j(\Omega)$ are operator-simple, and the representation (6) together with the composition series (7) determines k irreducible linear representations of Y:

$$d_j\colon Y \to \mathrm{Hom}(U_j, U_j), \quad j = 1, \ldots, k,$$

where

$$d_j(y) = \sigma_j, \quad \sigma_j(u) = yu, \quad u \in U_j, \quad y \in Y. \tag{8}$$

The representations (8) are called the *irreducible components* of (6). By the Jordan-Hölder theorem, the irreducible components are uniquely determined, to within equivalence and order.

Now let H be a subset of $\mathrm{Hom}(V, V)$. We define the trivial linear representation $D\colon H \longrightarrow \mathrm{Hom}(V, V)$, $h \mapsto h$. The irreducible components of D are called the *irreducible components* of H.

If Y is an algebra over a field Δ and V a linear space over Δ, we assume that d in (6) is an algebra homomorphism of Y into $\mathrm{Hom}(V, V)$.

2. *Mal'cev's local theorem.* A set Σ of subgroups of a group G is called a *local system of* G if it satisfies the following two conditions:

1) $G = \bigcup H_\alpha,\ H_\alpha \in \Sigma$.

2) For any H_α and H_β in Σ there exists a subgroup H_γ in Σ such that $H_\alpha, H_\beta \subset H_\gamma$.

For example, the collection of all finitely generated subgroups of G is a local system.

The aim of this subsection is to prove the following theorem:

THEOREM 1 (MAL'CEV [1]). *Let Σ be a local system of a group G. If every H_α in Σ has a faithful linear representation of degree n over a field Δ_α, then G also has a faithful linear representation of degree n over some field Δ.*

We precede the proof with some definitions and lemmas concerning the properties of ultraproducts.

Let I be a nonempty set. A collection $\mathfrak{F}$ of subsets of I is called a *filter* over I if it satisfies the following conditions:

1) $\varnothing \notin \mathfrak{F}$,

2) $X, Y \in \mathfrak{F} \Rightarrow X \cap Y \in \mathfrak{F}$,

3) $(I \supset Y \supset X \in \mathfrak{F}) \Rightarrow Y \in \mathfrak{F}$.

A filter $\mathfrak{F}$ over I is called an *ultrafilter* over I if it satisfies

4) For every $X \subset I$, either $X \in \mathfrak{F}$ or $I \backslash X \in \mathfrak{F}$.

It follows from 1) and 4) that $I \in \mathfrak{F}$ for any ultrafilter $\mathfrak{F}$ over I. A collection $\mathfrak{F}$ of subsets of I satisfying only conditions 1) and 2) is called a *centered system* over I. It is clear that every centered system over I is contained in some filter over I. And it is not hard to show that *every filter over I is contained in some ultrafilter over I* (see Mal'cev [5], §8.1, Theorem 3).

Let $\mathfrak{F}$ be an ultrafilter over I and suppose that to each $\alpha \in I$ there corresponds a set A_α. We construct the cartesian product

$$\prod A_\alpha, \qquad \alpha \in I, \tag{9}$$

whose elements we write in the form $a = \{a_\alpha\}$, $a_\alpha \in A_\alpha$. On (9) we define a binary relation $\equiv_\mathfrak{F}$ by setting

$$a \equiv_\mathfrak{F} b \Leftrightarrow I_1 = \{\alpha \in I \mid a_\alpha = b_\alpha\} \in \mathfrak{F}.$$

Since $I \in \mathfrak{F}$, the relation $\equiv_\mathfrak{F}$ is reflexive. Symmetry follows from the definition of $\equiv_\mathfrak{F}$. To show that $\equiv_\mathfrak{F}$ is also transitive, let $a \equiv_\mathfrak{F} b$ and $b \equiv_\mathfrak{F} c$. Then $I_2 = \{\alpha \in I \mid b_\alpha = c_\alpha\} \in \mathfrak{F}$. By 2), $I_1 \cap I_2 \in \mathfrak{F}$. On the other hand,

$$I_1 \cap I_2 \subset \{\alpha \in I \mid a_\alpha = c_\alpha\} = I_3.$$

Hence $I_3 \in \mathfrak{F}$ and $a \equiv_{\mathfrak{F}} c$. Thus $\equiv_{\mathfrak{F}}$ is an equivalence relation on (9). We denote the set of equivalence classes by

$$\prod A_\alpha/\mathfrak{F} \tag{10}$$

and call it an ultraproduct of the sets A_α. The class containing an element $a = \{a_\alpha\}$ is denoted by $\bar{a} = \overline{\{a_\alpha\}}$. If all the A_α are groups (rings, modules), we can define operations on the ultraproduct (10) in a natural manner and turn it into a group (ring, module). For example, let $A_\alpha = \Delta_\alpha$ be a ring. If $\lambda = \{\lambda_\alpha\}$ and $\mu = \{\mu_\alpha\}$, $\lambda_\alpha, \mu_\alpha \in \Delta_\alpha$, we set

$$\bar{\lambda} + \bar{\mu} = \overline{\{\lambda_\alpha + \mu_\alpha\}}, \qquad \bar{\lambda}\bar{\mu} = \overline{\{\lambda_\alpha\mu_\alpha\}}.$$

The classes $\bar{\lambda} + \bar{\mu}$ and $\bar{\lambda}\bar{\mu}$ are independent of the choice of representatives λ and μ of the classes $\bar{\lambda}$ and $\bar{\mu}$. For let $\bar{\lambda} = \bar{\rho}$ and $\bar{\mu} = \bar{\sigma}$. Then

$$X_1 = \{\alpha \in I \mid \lambda_\alpha = \rho_\alpha\} \in \mathfrak{F}, \qquad X_2 = \{\alpha \in I \mid \mu_\alpha = \sigma_\alpha\} \in \mathfrak{F},$$
$$Y = X_1 \cap X_2 \in \mathfrak{F}.$$

For $\alpha \in Y$ we have $\lambda_\alpha + \mu_\alpha = \rho_\alpha + \sigma_\alpha$ and $\lambda_\alpha\mu_\alpha = \rho_\alpha\sigma_\alpha$. Hence

$$Y \subset \{\alpha \in I \mid \lambda_\alpha + \mu_\alpha = \rho_\alpha + \sigma_\alpha\} \in \mathfrak{F},$$
$$Y \subset \{\alpha \in I \mid \lambda_\alpha\mu_\alpha = \rho_\alpha\sigma_\alpha\} \in \mathfrak{F};$$

that is, $\bar{\lambda} + \bar{\mu} = \bar{\rho} + \bar{\sigma}$ and $\bar{\lambda}\bar{\mu} = \bar{\rho}\bar{\sigma}$. It is obvious that

$$\Delta = \prod \Delta_\alpha/\mathfrak{F} \tag{11}$$

is a ring, the so-called *ultraproduct of the rings* Δ_α. Similarly one defines ultraproducts of groups, modules, and other algebraic systems.

If every ring Δ_α, $\alpha \in I$, *is a field, then* (11) *is also a field.* For Δ is a ring with unit element $\bar{1} = \overline{\{1_\alpha\}}$ and zero $\bar{0} = \overline{\{0_\alpha\}}$, where 1_α and 0_α are the unit element and zero of Δ_α. Let $\bar{\lambda} = \overline{\{\lambda_\alpha\}} \in \Delta$, $\bar{\lambda} \neq \bar{0}$. Then

$$X = \{\alpha \in I \mid \lambda_\alpha = 0_\alpha\} \notin \mathfrak{F}, \qquad Y = I \setminus X \in \mathfrak{F}.$$

We set $\bar{\mu} = \overline{\{\mu_\alpha\}}$, where $\mu_\alpha = \lambda_\alpha^{-1}$ for $\alpha \in Y$, and $\mu_\alpha = 1_\alpha$ for $\alpha \in X$. Then

$$\bar{\lambda}\bar{\mu} = \overline{\{\lambda_\alpha\mu_\alpha\}}, \qquad \{\alpha \in I \mid \lambda_\alpha\mu_\alpha = 1_\alpha\} = Y \in \mathfrak{F}.$$

Therefore $\bar{\lambda}\bar{\mu} = \bar{1}$, so that Δ is indeed a field. ∎

Now suppose that to each $\alpha \in I$ there correspond a field Δ_α, an n-dimensional linear space V_α over Δ_α, and a subgroup G_α of $GL(V_\alpha)$. We construct the ultraproducts

$$\Delta = \prod \Delta_\alpha/\mathfrak{F}, \qquad V = \prod V_\alpha/\mathfrak{F}, \qquad G = \prod G_\alpha/\mathfrak{F}.$$

We have already seen that Δ is a field. We now show that V is an n-dimensional linear space over Δ.

For $\bar{\lambda} = \overline{\{\lambda_\alpha\}} \in \Delta$, $\bar{u} = \overline{\{u_\alpha\}} \in V$ and $\bar{v} = \overline{\{v_\alpha\}} \in V$ we have

$$\bar{u} + \bar{v} = \overline{\{u_\alpha + v_\alpha\}}, \qquad \bar{u}\bar{\lambda} = \overline{\{u_\alpha \lambda_\alpha\}},$$

whence it is clear that V is a linear space over Δ. Let

$$b_{\alpha 1}, \ldots, b_{\alpha n} \tag{12}$$

be a basis of V_α over Δ_α, and set $\bar{b}_1 = \overline{\{b_{\alpha 1}\}}, \ldots, \bar{b}_n = \overline{\{b_{\alpha n}\}}$. Then

$$\bar{b}_1, \ldots, \bar{b}_n \tag{13}$$

is a basis of V over Δ. In fact, for any $\bar{v} \in V$

$$\bar{v} = \overline{\{b_{\alpha 1}\lambda_{\alpha 1} + \ldots + b_{\alpha n}\lambda_{\alpha n}\}} = \bar{b}_1\bar{\lambda}_1 + \ldots + \bar{b}_n\bar{\lambda}_n.$$

On the other hand, the system (13) is linearly independent over Δ, because it follows from $\bar{b}_1\bar{\lambda}_1 + \ldots + \bar{b}_n\bar{\lambda}_n = \bar{0}$ that

$$\{\alpha \in I \mid b_{\alpha 1}\lambda_{\alpha 1} + \ldots + b_{\alpha n}\lambda_{\alpha n} = 0_\alpha\} \in \mathfrak{F},$$

where 0_α is the zero of V_α. Since (12) is a basis for V_α, for $j = 1, \ldots, n$ we have

$$X_j = \{\alpha \in I \mid \lambda_{\alpha j} = 0_\alpha\} \in \mathfrak{F},$$

where 0_α is the zero of Δ_α. Consequently $\bar{\lambda}_j = \overline{\{\lambda_{\alpha j}\}} = \bar{0}$. Thus V is an n-dimensional linear space over Δ.

It is easy to see that $G = \Pi G_\alpha / \mathfrak{F}$ is a subgroup of $GL(V)$. For let $\bar{g} = \overline{\{g_\alpha\}} \in G$. Then for $\bar{\lambda} \in \Delta$ and $\bar{u}, \bar{v} \in V$, we can write

$$\bar{g}(\bar{u}\bar{\lambda}) = \overline{\{g_\alpha(u_\alpha\lambda_\alpha)\}} = \overline{\{g_\alpha(u_\alpha)\lambda_\alpha\}} = \overline{\{g_\alpha\}}\,\overline{\{u_\alpha\}}\,\bar{\lambda} = \bar{g}(\bar{u})\bar{\lambda},$$
$$\bar{g}(\bar{u} + \bar{v}) = \bar{g}(\bar{u}) + \bar{g}(\bar{v}).$$

Hence $G \subset \mathrm{Hom}(V, V)$. Since G is a group and its unit element is the same as that of $\mathrm{Hom}(V, V)$, it follows that G is a subgroup of $GL(V)$. We have thus proved

LEMMA 2. *An ultraproduct of linear groups G_α of degree n over fields Δ_α is a linear group of degree n over some field Δ.*

LEMMA 3. *Let G be an arbitrary group and Σ a local system of G. Then G is isomorphic to a subgroup of the ultraproduct of all subgroups in Σ for a suitable ultrafilter over Σ.*

PROOF. We first construct a certain centered system of subsets of Σ. For each finite subset F of G we set

$$\Sigma_F = \{H \in \Sigma \mid H \supset F\}. \tag{14}$$

If F_1 and F_2 are two finite subsets of G, then by (14)

$$\Sigma_{F_1 \cup F_2} = \Sigma_{F_1} \cap \Sigma_{F_2}. \tag{15}$$

Since $G = \bigcup H_\alpha$, $H_\alpha \in \Sigma$, it follows that for singletons $F = \{g\}$

$$\Sigma_g = \Sigma_F \neq \varnothing.$$

Now let card $F = m > 1$ and suppose we have already proved that $\Sigma_{F_j} \neq \varnothing$ for subsets F_j of G of cardinality less than m. We write F in the form $F = F_1 \cup F_2$, where card $F_j < m$. Then

$$\Sigma_F = \Sigma_{F_1} \cap \Sigma_{F_2}, \quad \Sigma_{F_j} \neq \varnothing.$$

Let $R \in \Sigma_{F_1}$ and $S \in \Sigma_{F_2}$. Since Σ is a local system, it contains a group H such that $H \supset R$ and $H \supset S$. Obviously $H \supset F = F_1 \cup F_2$. Hence

$$\Sigma_F \neq \varnothing. \tag{16}$$

Let Φ be the set of all Σ_F, where F is an arbitrary finite subset of G. It follows from (15) and (16) that Φ is a centered system over Σ. As we mentioned above, Φ is contained in some ultrafilter $\mathfrak{F}$ over Σ. We claim that G can be isomorphically embedded in the ultraproduct $\Pi S_\alpha/\mathfrak{F}$, $S_\alpha \in \Sigma$. Consider the mapping

$$\varphi\colon\ G \to \prod S_\alpha/\mathfrak{F}, \quad g \longmapsto \overline{\{g_\alpha\}},$$

where $g_\alpha = g$ for $g \in S_\alpha$ and $g_\alpha = 1$ for $g \notin S_\alpha$. Let us show that φ is a homomorphism. Indeed,

$$\varphi(gh) = \overline{\{(gh)_\alpha\}}.$$

Clearly, for g and h in G we have

$$\Sigma_{\{g,h\}} \subset \{S_\alpha \in \Sigma \mid (gh)_\alpha = g_\alpha h_\alpha\} = X.$$

Since $\Sigma_{\{g,h\}} \in \mathfrak{F}$, it follows that $X \in \mathfrak{F}$. Therefore

$$\varphi(gh) = \overline{\{(gh)_\alpha\}} = \overline{\{g_\alpha\}}\,\overline{\{h_\alpha\}} = \varphi(g)\,\varphi(h).$$

It remains to prove that φ is injective. Let $g \in G$, $g \neq 1$, but $\varphi(g) = \overline{1}$. Then

$$\{S_\alpha \in \Sigma \mid g_\alpha = 1\} \in \mathfrak{F}, \quad X = \{S_\alpha \in \Sigma \mid g \notin S_\alpha\} \in \mathfrak{F},$$
$$\Sigma_g = \Sigma \setminus X \notin \mathfrak{F}.$$

But by construction $\mathfrak{F}$ contains all the Σ_F, where F is a finite subset of G. ∎

Mal′cev's theorem now follows from Lemmas 2 and 3 (see Kegel [1]).

3. *Semisimplicity and complete reducibility.* Let A be an algebra of finite dimension over a field Δ with unit element e, and V a finite-dimensional linear space over Δ. We consider linear representations $d: A \longrightarrow \mathrm{Hom}(V, V)$ such that $d(e)$ is the unit element of the ring $\mathrm{Hom}(V, V)$. An example is the *left regular representation* ρ of A:

$$\rho\colon A \to \mathrm{Hom}(W, W),$$

where W is the underlying linear space of A, $\rho(a) = \sigma_a$, $\sigma_a: A \longrightarrow A$, $\sigma_a(x) = ax$. It is readily seen that

$$\rho(a+b) = \sigma_{a+b} = \sigma_a + \sigma_b, \quad \rho(ab) = \sigma_{ab} = \sigma_a\sigma_b,$$
$$\mathrm{Ker}\,\rho = (0).$$

Therefore ρ is a faithful linear representation of A. Clearly $\rho(e)$ is the unit element of $\mathrm{Hom}(W, W)$. The additive operator group defined by ρ is denoted by $W(A)$. It is clear that the admissible subgroups of $W(A)$ are the left ideals of A. It follows from Theorem 13.2 that *ρ is completely reducible if and only if A is semisimple.*

The following theorem is well known from representation theory.

THEOREM 2. *Let A be a semisimple algebra. Then every representation of the form d is completely reducible, and any irreducible component of d is equivalent to some irreducible component of the left regular representation* (*see, for example, van der Waerden* [2], §123).

In particular, if A is a simple algebra over Δ, then the minimal left ideals of A are pairwise operator-isomorphic. Together with Theorem 2, this implies

COROLLARY 1. *Let A be a simple algebra over a field Δ. Then, up to equivalence, A has a unique irreducible representation over Δ.*

THEOREM 3. *Let V be a finite-dimensional space over a field Δ and A a subalgebra of* $\mathrm{Hom}(V, V)$ *whose unit element e is the same as that of* $\mathrm{Hom}(V, V)$. *Then, if A is irreducible, it is simple, and if A is completely reducible, it is semisimple.*

PROOF. Suppose that a subalgebra A of $\mathrm{Hom}(V, V)$ is completely reducible as a set of endomorphisms of V. Let W be the underlying linear space of A. We consider two operator groups: the space $W(A)$ of the left regular representation ρ of A, and the space $V(A)$ of the trivial representation D of A. Since A is of finite dimension over Δ, $W(A)$ contains operator-simple admissible subgroups; let W_1 be any one of these. Since $V(A)$ is a direct sum of admissible operator-simple subgroups, it contains an admissible operator-simple subgroup

V_1 such that $W_1 V_1 \neq (0)$. Consequently there is a vector $v_1 \in V_1$ such that

$$W_1 v_1 \neq (0). \tag{17}$$

Consider the mapping $\gamma_1 \colon W_1 \longrightarrow V_1$, $x \mapsto xv_1$. For $x, y \in W$ and $a \in A$ we have

$$\gamma_1 (x + y) = \gamma_1 (x) + \gamma_1 (y), \qquad \gamma_1 (ax) = a\gamma_1 (x).$$

Consequently γ_1 is an operator homomorphism. Since W_1 and V_1 are operator-simple, it follows from (17) that γ_1 is an operator isomorphism, and

$$V_1 = W_1 v_1. \tag{18}$$

If $W_1 = W(A)$, we are done, because in this case A is simple (see the end of §13).

Now suppose that

$$W_1 \neq W(A). \tag{19}$$

By (18), there is an element e_1 in W_1 such that $e_1 v_1 = v_1$. Since W_1 is operator-simple, $W_1 = Ae_1$. Hence it follows from (19) that $e_1 \neq e$. For any x in W_1 we can write

$$\gamma_1 (xe_1) = xe_1 v_1 = xv_1 = \gamma_1 (x).$$

Since γ_1 is injective, it follows that $xe_1 = x$, for any $x \in W_1$. In particular, $e_1^2 = e_1$. Therefore $W(A) = W$ can be written in the form

$$W(A) = A = Ae = Ae_1 \dotplus A(e - e_1) = W_1 \dotplus A(e - e_1).$$

Clearly $A(e - e_1)$ is an admissible subgroup of $W(A)$. Thus any operator-simple admissible subgroup W_1 of $W(A)$ has an admissible complement. By Proposition 13.4, $W(A)$ is a direct sum of operator-simple admissible subgroups; that is, A is semisimple. In particular, if A is irreducible, then $V_1 = V$ in (17) and any operator-simple admissible subgroup W_1 is operator-isomorphic to $V(A)$, so that A is simple. ■

4. *The linear hull of a completely reducible group.* Let V be an n-dimensional linear space over a field Δ, and G a set of endomorphisms in $\mathrm{Hom}(V, V)$. The *linear Δ-hull* $[G]_\Delta$ of G is defined as the set of all linear combinations $g_1\lambda_1 + \ldots + g_k\lambda_k$, $g_j \in G$, $\lambda_j \in \Delta$. In particular, *if G is a semigroup under multiplication, then obviously $[G]_\Delta$ is an algebra over Δ.* The space V can be regarded as an additive group $V(\Omega)$ with operator domain $\Omega = \Delta \cup G$. The admissible subgroups of $V(\Omega)$ are then the G-invariant subspaces of V. It is clear that the system of admissible subgroups of $V(\Omega)$ remains unchanged if we replace the operator domain $\Omega = \Delta \cup G$ by $\Omega_1 = \Delta \cup [G]_\Delta$; that is, every admissible subgroup of $V(\Omega)$ is an admissible subgroup of $V(\Omega_1)$ and vice versa. The pas-

sage from $V(\Omega)$-admissible to $V(\Omega_1)$-admissible subgroups also preserves the relation of operator isomorphism, so that two admissible subgroups of $V(\Omega_1)$ are operator-isomorphic if and only if they are operator-isomorphic as subgroups of $V(\Omega)$.

In particular, if G is a completely reducible subgroup of $GL(V)$, then $[G]_\Delta$ is a completely reducible subalgebra of $\mathrm{Hom}(V, V)$.

THEOREM 4. *Let G be a completely reducible subgroup of $GL(V)$, where V is a linear space over a field Δ; let $n_1, \ldots, n_s$ be the degrees of the pairwise inequivalent irreducible components of G and $A = [G]_\Delta$. Then $A = R_1 \dotplus \ldots \dotplus R_s$ where $R_j \cong T^j_{r_j}$, T^j is a division ring whose center contains Δ, $(T^j : \Delta)r_j = n_j$, and $R_iR_j = (0)$ for $i \neq j$. In particular, if G is irreducible, then $A = T_r$ and $(T: \Delta)r = n$, where T is a division ring and n the dimension of V over Δ.*

PROOF. Since $A = [G]_\Delta$ is completely reducible, A is semisimple by Theorem 3. By the preceding remarks, the numbers

$$n_1, \ldots, n_s \tag{20}$$

are the degrees of all the pairwise inequivalent components of A. By Theorem 2 every irreducible component of A is equivalent to some irreducible component of the left regular representation of A. Hence A contains at least s operator-simple left ideals that are not operator-isomorphic. On the other hand, if L is any operator-simple left ideal of $A = [G]_\Delta = \mathrm{Hom}(V, V)$, then at least one of the admissible subgroups of $V(\Omega_1)$ $(\Omega_1 = \Delta \cup A)$ is operator-isomorphic to L. For since $L \neq (0)$, V contains a vector v such that $V_1 = L(v) \neq (0)$; V_1 is the required admissible subgroup. Thus A contains exactly s operator-simple left ideals that are not operator-isomorphic; their degrees are given by (20). This and Theorem 13.3 imply our theorem. ■

Clearly, in a suitable basis of V the matrices g of the group G in Theorem 4 all take the form

$$g = \mathrm{diag}\,[g_1, \ldots, g_1, \ldots, g_j, \ldots, g_j, \ldots, g_s, \ldots, g_s],$$

where $\gamma_j: g \mapsto g_j$ is an irreducible representation of G of degree n_j, $n = n_1k_1 + \ldots + n_sk_s$, k_j being the multiplicity of g_j. For $i \neq j$ the representations γ_i and γ_j are inequivalent. Let $G_j = \mathrm{Im}\,\gamma_j$ and $A_j = [G_j]_\Delta$, $j = 1, \ldots, s$. Then Theorem 4 can be stated as follows:

THEOREM 4a. *The algebra $A = [G]_\Delta$ consists of all matrices of the form*

$$\mathrm{diag}\,[a_1, \ldots, a_1, \ldots, a_s, \ldots, a_s],$$

where $a_1, \ldots, a_s$ range independently over $A_1, \ldots, A_s$, respectively. Thus $A: \Delta = A_1: \Delta + \ldots + A_s: \Delta$.

It is well known that the only division ring of finite dimension over an algebraically closed field P is P itself. Consequently, if the field Δ in Theorem 4 is algebraically closed, then $T^i\colon \Delta = 1$ and $n_i = r_i$. If G is irreducible and Δ algebraically closed, then $A = [G]_\Delta = \Delta_n$ and $A\colon \Delta = n^2$.

Let Δ be an arbitrary field and Σ an extension of Δ. Then obviously every subgroup of $GL(n, \Delta)$ is a subgroup of $GL(n, \Sigma)$. A subgroup G of $GL(n, \Delta)$ is said to be *absolutely irreducible* if it is an irreducible subgroup of $GL(n, \Sigma)$ for any extension Σ of Δ.

COROLLARY 1. *A subgroup G of $GL(n, \Delta)$ is absolutely irreducible if and only if its linear Δ-hull is Δ_n.*

PROOF. Let Σ be an extension of Δ. Then $[G]_\Delta\colon \Delta = [G]_\Sigma\colon \Sigma$, because a system of matrices in Δ_n that is linearly independent over Δ remains linearly independent over Σ. By a standard theorem in field theory, Δ can be embedded in an algebraically closed field P. If G is absolutely irreducible, then $n^2 = [G]_P\colon P = [G]_\Delta\colon \Delta$; hence

$$[G]_\Delta = \Delta_n; \tag{21}$$

that is, (21) follows from the absolute irreducibility of G. The converse assertion is obvious. ∎

This corollary can be stated in another form.

COROLLARY 2. *A subgroup G of $GL(n, \Delta)$ is absolutely irreducible if and only if G contains n^2 linearly independent matrices over Δ.*

We state a few more properties of irreducible groups and algebras.

LEMMA 4. *Let G be an irreducible subgroup of $GL(l, \Sigma)$, where Σ is an extension of Δ such that $\Sigma\colon \Delta = m$, $\Sigma \subset \Delta_m$. Setting $n = ml$ and replacing in the matrices of G the elements σ of Σ by matrices of Δ_m, we can regard G as a subgroup of $GL(n, \Delta)$. If now*

$$[G]_\Delta \supset \Sigma E_l, \tag{22}$$

then G is an irreducible subgroup of $GL(n, \Delta)$.

PROOF. The n-dimensional space Δ^n over Δ on which G acts can be regarded as an l-dimensional space Σ^l over Σ. Let $L \neq (0)$ be a $[G]_\Delta$-invariant subspace of Δ^n. By (22), L is also a subspace of Σ^l. By assumption, G is an irreducible subgroup of $GL(l, \Sigma)$, and so $L = \Sigma^l = \Delta^n$; that is, G is an irreducible subgroup of $GL(n, \Delta)$. ∎

LEMMA 4a. *Let T be a division ring contained in Δ_m, where Δ is a field, the center of T contains ΔE_m, $T\colon \Delta = m$, and*

$$B = T_r. \tag{23}$$

Set $n = mr$, and regard B as a subalgebra of Δ_n. Then B is an irreducible subalgebra of Δ_n.

PROOF. It follows from (23), by Theorem 13.3, that B is a simple algebra over Δ and the dimension of an operator-simple left ideal over Δ is $rm = n$. Consequently, by Theorem 2, the degree of an irreducible representation of B by matrices over Δ is n. But $B \subset \Delta_n$. Hence the trivial representation $D: B \longrightarrow B$, $x \mapsto x$, of degree n, is irreducible. Thus B is an irreducible subalgebra of Δ_n. ∎

Later we also need the right regular representation of an algebra A with unit element e over a field Δ.

Let W be the underlying space of A. The *right regular representation* of A is the mapping

$$\rho_1\colon A \to \operatorname{Hom}(W, W),$$
$$\rho_1(a) = \tau_a, \qquad \tau_a\colon A \to A, \qquad \tau_a(x) = xa.$$

As is easy to verify,

$$\begin{aligned} &\rho_1(a+b) = \tau_{a+b} = \tau_a + \tau_b,\\ &\rho_1(ab) = \tau_{ab} = \tau_b\tau_a,\\ &\rho_1(a\lambda) = \tau_{a\lambda} = \tau_a\lambda, \ \lambda \in \Delta,\\ &\operatorname{Ker} \rho_1 = (0). \end{aligned}$$

Thus ρ_1 is an injective antihomomorphism of A into $\operatorname{Hom}(W, W)$. By Theorem 1.3, $\operatorname{Im} \rho_1$ is the centralizer of $\operatorname{Im} \rho$, and $\operatorname{Im} \rho$ the centralizer of $\operatorname{Im} \rho_1$, in $\operatorname{Hom}(W, W)$. Hence

$$\operatorname{Im} \rho \cap \operatorname{Im} \rho_1 = Z(\operatorname{Im} \rho) = Z(\operatorname{Im} \rho_1), \tag{24}$$

where $Z(\operatorname{Im} \rho)$ and $Z(\operatorname{Im} \rho_1)$ are the centers of $\operatorname{Im} \rho$ and $\operatorname{Im} \rho_1$.

THEOREM 5. *Let A be an irreducible subalgebra with unit element E_n in Δ_n, where Δ is a field. Then there is a matrix g in $GL(n, \Delta)$ such that $A = gT_rg^{-1}$, where T is a division ring such that $\Delta_m \supset T \supset \Delta E_m$, $T\colon \Delta E_m = m$ and $n = mr$. Moreover, the centralizer of A in Δ_n is $gT'E_rg^{-1}$, where T' is a division ring anti-isomorphic to T, whose center is the same as that of T, namely $T' \cap T$.*

PROOF. By Theorem 3, A is a simple subalgebra of Δ_n, and so, by Theorem 13.3, A is isomorphic to Φ_r, where Φ is a division ring of finite dimension m over Δ. We consider the left and the right regular representations ρ and ρ_1 of Φ:

$$\rho\colon \Phi \to \Delta_m, \qquad \rho_1\colon \Phi \to \Delta_m.$$

We set $T = \operatorname{Im}\rho$ and $T' = \operatorname{Im}\rho_1$. Then the division rings T and T' are anti-isomorphic, and the center of T' is also that of T, namely $T \cap T'$. (See (24)). If we replace in the matrices of Φ_r the elements φ of Φ by the elements $\rho(\varphi) \in T$, we obtain an algebra T_r isomorphic to A. Obviously T_r is a subalgebra of Δ_{mr}, and it is irreducible by Lemma 4a. Since the simple algebra A has (up to equivalence) a unique irreducible representation, we see that $mr = n$ and $A = gT_rg^{-1}$, where $g \in GL(n, \Delta)$. This proves the first part. Since T' is the centralizer of T in Δ_m, it follows that $T'E_r$ is the centralizer of T_r in Δ_n, and $gT'E_rg^{-1}$ the centralizer of A in Δ_n. ∎

THEOREM 6. *Let G be an irreducible subgroup of $GL(n, \Delta)$, Z' the center of G. Then $[Z']_\Delta = \Sigma$ is an extension of Δ, and the degree $\Sigma\colon \Delta$ divides n.*

PROOF. Let V be the linear space over Δ on which G acts. Since G is irreducible, the additive group $V(\Omega)$ with operator domain $\Omega = G \cup \Delta$ is operator-simple. The elements $\sigma \in \Sigma$ are endomorphisms of $V(\Omega)$. By Corollary 1 to Lemma 13.1, Σ is contained in some division ring $T \supset \Delta$. Therefore Σ is a commutative algebra without divisors of zero, of finite dimension over Δ. Hence Σ is a field. It remains to show that the degree $m = \Sigma\colon \Delta$ divides n. But we can regard V as a linear space over Σ. If $k = V\colon \Sigma$, then $mk = n$. ∎

In connection with the proof of Theorem 6, we observe that for any $\sigma \in \Sigma$, $x \in V$ and $g \in G$, we have $g(\sigma(x)) = \sigma g(x)$. Thus G can be regarded as a subgroup of $GL(k, \Sigma)$.

5. *θ-subgroups.* Let θ be an abstract group-theoretic property satisfying the following condition: if a subgroup Γ of a group G has the property θ and C is an abelian subgroup of G lying in the centralizer of Γ in G, then ΓC also has the property θ. Examples of such properties are solvability, commutativity, local nilpotency, and nilpotency of a fixed class. A subgroup Γ of a group G having the property θ is called a θ-subgroup.

THEOREM 7. *Let Δ be an arbitrary field and G a maximal irreducible θ-subgroup of $GL(n, \Delta)$. Then there exists an extension Σ of Δ, of degree m, such that m divides n, G is an absolutely irreducible maximal θ-subgroup of $GL(r, \Sigma)$, $r = nm^{-1}$, and the center of G is Σ^*E_r.*

PROOF. By Theorem 5, we may assume that $[G]_\Delta = T_r$, where T is a division ring, $(T\colon \Delta)r = n$, and the centralizer of T_r in Δ_n is $T'E_r$, where T' is a division ring anti-isomorphic to T, with $T' = T$ if and only if T is commutative. We claim that T is indeed a field. Suppose T and hence T' is noncommutative.

If a is a noncentral element of T', then $a \notin T$ and $aE_r \notin G \subset T_r$. On the other hand, by the definition of a θ-group, $(a)E_rG$ is a θ-group, and $(a)E_rG \neq G$. This contradicts the assumption that G is a maximal θ-subgroup of $GL(n, \Delta)$. Hence $T = \Sigma$, where Σ is a field. By the maximality of G, $\Sigma^*E_r \subset G$. Thus $[G]_\Delta = \Sigma_r$, and Σ^*E_r is the center of G. By Corollary 4.1, G is an absolutely irreducible subgroup of $GL(r, \Sigma)$. It is obvious that $m = \Sigma: \Delta = nr^{-1}$. ∎

§15. Primitivity and imprimitivity

1. *Definition of imprimitivity and simplest properties of imprimitive groups.* Let V be a linear space over a division ring T and G a subgroup of $GL(V)$. If V can be represented as a direct sum

$$V = \dot{\sum} Q_\nu, \qquad \nu \in I, \tag{1}$$

of subspaces $Q_\nu \neq (0)$, where card $I > 1$ and for any $\alpha \in I$ and $g \in G$ there exists $\beta \in I$ such that

$$g(Q_\alpha) \subset Q_\beta, \tag{2}$$

then G is said to be *imprimitive*. (1) is called a *decomposition of V into systems of imprimitivity*, namely the subspaces Q_ν, *for* G. Otherwise the group is said to be *primitive*.

LEMMA 1. *Let* (1) *be a decomposition of V into systems of imprimitivity for a group G. Then it follows from* (2) *that* $g(Q_\alpha) = Q_\beta$.

PROOF. By (2) and the definition of imprimitivity,

$$Q_\alpha \subset g^{-1}(Q_\beta) \subset Q_\gamma, \qquad \gamma \in I. \tag{3}$$

Since the Q_ν are direct summands, we have from (3)

$$Q_\alpha = Q_\gamma, \qquad g^{-1}(Q_\beta) = Q_\alpha, \qquad g(Q_\alpha) = Q_\beta. \ ∎$$

Let (1) be a decomposition of V into systems of imprimitivity for G. It follows from Lemma 1 that each element of G induces a certain permutation $\bar{g}$ on the systems Q_ν, $\nu \in I$; that is, (1) defines an epimorphism

$$\gamma\colon G \to \Gamma, \qquad g \mapsto \bar{g}, \tag{4}$$

where $\bar{g}\colon Q_\nu \mapsto g(Q_\nu)$, $\nu \in I$, and Γ is the set of all $\bar{g}$. Clearly Γ is a subgroup of the symmetric group $S(X)$, where X is the set of all Q_ν in (1). Now let ν be a fixed index in I and Γ_{Q_ν} the stabilizer of $Q_\nu \in X$ in Γ. Using the epimorphism (4), we define a subgroup H_ν of G by setting

$$H_\nu = \gamma^{-1}(\Gamma_{Q_\nu}). \tag{5}$$

Obviously, $H_\nu = \{h | h \in G, h(Q_\nu) = Q_\nu\}$. With this notation we have the following theorem.

THEOREM 1. *Let G be an irreducible imprimitive subgroup of $GL(V)$ and* (1) *a decomposition of V into systems of imprimitivity for G. Then the following assertions are true*:

(i) *The group Γ is transitive.*

(ii) *The set of all distinct subgroups H_ν, $\nu \in I$, is a class of conjugate subgroups of G.*

(iii) *For any $\nu \in I$, the restriction $G_\nu = H_\nu | Q_\nu$ is an irreducible subgroup of $GL(Q_\nu)$.*

(iv) *For any $\alpha, \beta \in I$, G_α and G_β are similar as linear groups.*

PROOF. Let Y be an orbit of Γ. Then the sum W of all the subspaces Q_ν contained in Y is a G-invariant subspace of V. Hence and from the irreducibility of G it follows that $W = V$ and $Q_\nu \in Y$ for every $\nu \in I$; that is, Γ is transitive and (i) is proved. Since Γ is transitive, the set of all distinct stablizers Γ_x is a conjugacy class of subgroups of Γ (Corollary 2.3.1). This and (5) imply (ii). To prove (iii), we can construct for fixed $\alpha \in I$ a set

$$\{g_\nu\}, \quad \nu \in I, \tag{6}$$

of elements of G such that $g_\alpha = e$ and $g_\nu(Q_\alpha) = Q_\nu$. Since $\{\gamma(g_\nu)\}$, $\nu \in I$, is a left transversal to Γ_{Q_α} in Γ, (6) is a left transversal to H_α in G.

Let R_α be a nontrivial G_α-invariant subspace of Q_α. Then $R_\nu = g_\nu(R_\alpha) \subset Q_\nu$. We set $R = \dot{\Sigma} R_\nu$, $\nu \in I$. For $g \in G$ we have $gg_\nu = g_\mu h$, $h \in H_\alpha$. Therefore

$$g(R_\nu) = gg_\nu(R_\alpha) = g_\mu h(R_\alpha) \subset g_\mu(R_\alpha) = R_\mu.$$

Hence $g(R) \subset R$. Since $R \neq V$, this contradicts the irreducibility of G, and (iii) is proved. It remains to prove (iv). By means of the elements (6) we define a mapping φ: $Q_\alpha \longrightarrow Q_\beta$, $x \mapsto g_\beta(x)$, $x \in Q_\alpha$. Since $g_\beta(Q_\alpha) = Q_\beta$, φ is an isomorphism of Q_α and Q_β. As is easy to verify, $H_\beta = g_\beta H_\alpha g_\beta^{-1}$. For $b \in G_\beta = H_\beta | Q_\beta$ we have

$$b = g_\beta h_\alpha g_\beta^{-1} | Q_\beta = \varphi a \varphi^{-1}, \qquad a \in G_\alpha.$$

Consequently $G_\beta \subset \varphi G_\alpha \varphi^{-1}$. On the other hand, $G_\alpha \subset \varphi^{-1} G \varphi$. Hence

$$G_\beta = \varphi G_\alpha \varphi^{-1}.$$

This completes the proof. ∎

Let G be an irreducible imprimitive subgroup of $GL(V)$ and (1) a decomposition of V into systems of imprimitivity for G. We say that (1) is *nonrefinable* if the groups G_ν in Theorem 1 are primitive.

LEMMA 2. *Let G be an irreducible imprimitive subgroup of $GL(V)$, where V is a linear space over a division ring T. Suppose that there is a decomposition* (1) *of V into systems of imprimitivity for G such that*

$$Q_\nu : T < \infty, \quad \nu \in I. \tag{7}$$

Then there exists a nonrefinable decomposition of V into systems of imprimitivity for G.

PROOF. If the groups G_ν are primitive, (1) is a nonrefinable decomposition. Suppose that some G_α is imprimitive and let $Q_\alpha = Q_{\alpha 1} \dotplus \ldots \dotplus Q_{\alpha m}$ be a decomposition of Q_α into systems of imprimitivity for G_α. By means of the elements (6) we construct a system of subspaces $Q_{\nu j} = g_\nu(Q_{\alpha j})$. Then V can be expressed as

$$V = \dot{\sum} Q_{\nu j}, \quad \nu \in I, \quad j = 1, \ldots, m. \tag{8}$$

For $g \in G$ we have

$$g(Q_{\nu j}) = g g_\nu (Q_{\alpha j}) = g_\mu h (Q_{\alpha j}) = g_\mu (Q_{\alpha k}) = Q_{\mu k}, \qquad h \in H_\alpha.$$

Therefore (8) is also a decomposition of V into systems of imprimitivity for G, and

$$Q_{\alpha j} : T < Q_\alpha : T < \infty.$$

Hence the lemma follows. ∎

COROLLARY 1. *If V is a finite-dimensional space and G an irreducible imprimitive subgroup of $GL(V)$, then V always has nonrefinable decompositions into systems of imprimitivity for G.*

If there exists a decomposition (1) into systems of imprimitivity for G such that all the subspaces Q_ν are one-dimensional, G is said to be *monomial.*

THEOREM 2. *Let G be an irreducible imprimitive subgroup of $GL(V)$, and suppose that there exists a decomposition* (1) *of V into systems of imprimitivity for G such that* card $I = k$, *where k is a natural number. Then G has a normal subgroup N whose index in G divides $k!$ and is divisible by k.*

PROOF. Let $N = \operatorname{Ker} \gamma$ (see (4)). Then, by Theorem 1, $\Gamma \cong G/N$ is a transitive subgroup of the symmetric group S_k. This proves the theorem. ∎

When V is of finite dimension n over a division ring T, we have

THEOREM 3 (IMPRIMITIVITY CRITERION). *An irreducible subgroup G of $GL(V)$ is imprimitive if and only if there exists a subspace Q of V such that $Q\colon T = m < n$, $m | n$, and the index of the subgroup $H = \{h | h \in G,\ h(Q) = Q\}$ in G is n/m.*

PROOF. Let G be an imprimitive irreducible subgroup of $GL(V)$ and (1) a decomposition of V into systems of imprimitivity for G. Setting $Q = Q_\alpha$ and $H = H_\alpha$, we see that Q: $T = m$, $m|n$, and G: $H = n/m$. This proves the necessity. We now prove sufficiency. Let G be irreducible, and suppose that a subspace Q of V satisfies the conditions of the theorem; $k = n/m$ and $b_1 = e, b_2, \ldots, b_k$ is a left transversal to H in G. We consider the subspace

$$R = Q + b_2(Q) + \ldots + b_k(Q). \tag{9}$$

For $g \in G$ we have

$$g(b_\mu(Q)) = b_\nu h(Q) = b_\nu(Q), \qquad h \in H.$$

Consequently $g(R) = R$. Since G is irreducible, $R = V$. But the dimension of R is n only if (9) is the direct sum of the $b_\mu(Q)$. On the other hand, $g(b_\mu(Q)) = b_\nu(Q)$. Thus (9) is a decomposition of V into systems of imprimitivity for G. ∎

2. *The wreath product of a linear group and a permutation group.* Let U be a linear space over a division ring T, G_1 a subgroup of $GL(U)$, and Γ a subgroup of the symmetric group S_k on $\{1, \ldots, k\}$, $k > 1$. The cartesian product $U^k = V_1$ can be regarded as a linear space over T, and we write any vector $v \in V_1$ in the form $v = (u_1, \ldots, u_k)$, $u_j \in U$. For any $f_1, \ldots, f_k \in G_1$ and $s \in \Gamma$, we define a mapping f: $V_1 \longrightarrow V_1$, $f = \langle f_1, \ldots, f_k, s\rangle$, by setting

$$f(v) = f(u_1, \ldots, u_k) = \bar{v} \in V_1, \tag{10}$$

where the $s(\nu)$th component of $\bar{v}$ is $f_\nu(u_\nu)$, $\nu = 1, \ldots, k$. Obviously f is an automorphism of V_1. We denote the set of all such automorphisms by F. Let $g = \langle g_1, \ldots, g_k, r\rangle \in F$. Then $gf(v) = g(\bar{v}) = (\bar{\bar{v}}) \in V_1$, where the $rs(\bar{\bar{v}})$th component of v is $g_{s(\nu)}f_\nu(u_\nu)$, $\nu = 1, \ldots, k$. Hence

$$gf = \langle g_{s(1)}f_1, \ldots, g_{s(k)}f_k, rs\rangle \in F. \tag{11}$$

It follows from (11) that F is a subgroup of $GL(V_1)$. The group F is called the *wreath product of the linear group* G_1 *and the permutation group* Γ, and is denoted by $G_1 \sim \Gamma$.

The group $F = G_1 \sim \Gamma$ *is imprimitive.* For let Q_j be the subspace of $V_1 = U^k$ consisting of all vectors $(0, \ldots, u_j, \ldots, 0)$, where $u_j \in U$ is the jth component ($j = 1, \ldots, k$). Then obviously

$$V_1 = Q_1 \dotplus \ldots \dotplus Q_k. \tag{12}$$

By (10), for $f \in F$ we have $f(Q_j) = Q_{s(j)}$. Thus F is imprimitive and (12) is a decomposition of V_1 into systems of imprimitivity for F.

It is clear that the mapping $\bar{\gamma}$: $F \longrightarrow \gamma$, $\langle f_1, \ldots, f_k, s\rangle \mapsto s$, is an epimorphism, and Ker $\bar{\gamma}$ is the direct product of k copies of G_1. Consequently F

is an extension of the direct product of k copies of G_1 by Γ.

LEMMA 3. *Let G_1 and H be conjugate subgroups of $GL(U)$, Γ and Φ conjugate subgroups of S_k. Then $G_1 \sim \Gamma$ and $H \sim \Phi$ are conjugate subgroups of $GL(U) \sim S_k$.*

PROOF. Let $H = aG_1a^{-1}$ and $\Phi = b\Gamma b^{-1}$, where $a \in GL(U)$ and $b \in S_k$. In $GL(U) \sim S_k$ we choose the element $c = \langle a, \ldots, a, b\rangle$. Then

$$H \sim \Phi = c\,(G_1 \sim \Gamma)c^{-1}.$$

For it follows from (11) that

$$c^{-1} = \langle a^{-1}, \ldots, a^{-1}, b^{-1}\rangle.$$

For $f \in F$ we find by means of (11) that

$$fc^{-1} = \langle f_{b^{-1}(1)}a^{-1}, \ldots, f_{b^{-1}(k)}a^{-1}, sb^{-1}\rangle,$$
$$cfc^{-1} = \langle af_{b^{-1}(1)}a^{-1}, \ldots, af_{b^{-1}(k)}a^{-1}, bsb^{-1}\rangle$$
$$= \langle h_1, \ldots, h_k, \varphi\rangle, \quad h_j \in H, \quad \varphi \in \Phi.$$

Therefore $cfc^{-1} \in H \sim \Phi$ and $c(G_1 \sim \Gamma)c^{-1} \subset H \sim \Phi$. By the same device we prove that $c^{-1}(H \sim \Phi)c \subset G_1 \sim \Gamma$. Hence

$$c(G_1 \sim \Gamma)c^{-1} = H \sim \Phi. \quad \blacksquare$$

LEMMA 4. *Let G_1 be an irreducible subgroup of $GL(U)$, and Γ a transitive subgroup of the symmetric group S_k, $k > 1$. Then the group $F = G_1 \sim \Gamma$ is irreducible, except when U is a one-dimensional space over the field $GF(2)$.*

PROOF. Suppose that the assumptions of the lemma hold, that U is not a one-dimensional space over $GF(2)$, and $R \neq (0)$ an F-invariant subspace of V_1, $v \in R$, $v \neq 0$. Then $v = q_1 + \ldots + q_k$, $q_j \in Q_j$, $q_\alpha \neq 0$. Since F contains the direct product of k copies of G_1 and U contains at least two nonzero vectors, there is an element f in F such that $f(q_\nu) = q_\nu$ for $\nu \neq \alpha$, and $f(q_\alpha) = p_\alpha \in Q_\alpha$, where $p_\alpha \neq q_\alpha$. Now, $f(v) - v \in R$, and so

$$v_\alpha = p_\alpha - q_\alpha \in R, \quad v_\alpha \neq 0, \quad v_\alpha \in Q_\alpha.$$

Since G is irreducible, we have $Q_\alpha \subset R$. By the transitivity of Γ this implies that $Q_\nu \subset R$, $\nu = 1, \ldots, k$. Hence $R = V_1$; that is, F is irreducible.

It remains to consider the case of a one-dimensional space U over $GF(2)$. Here Q_j contains a single nonzero vector q_j, and for $f \in F$ we have $f(q_j) = q_{s(j)}$, $s \in \Gamma$. Consequently $f(v) = v$ for $v = q_1 + \ldots + q_k$. Hence F is reducible. $\blacksquare$

Now let G be an irreducible imprimitive subgroup of $GL(V)$, where V is a finite-dimensional linear space over a division ring T, let

$$V = Q_1 \dotplus \ldots \dotplus Q_k \tag{13}$$

be a decomposition of V into systems of imprimitivity for G, and let

$$\gamma\colon G \to \Gamma \tag{14}$$

be the epimorphism defined by (13) (see (4)). Since Γ is transitive, in G we can choose elements

$$g_1 = e,\ g_2,\ \ldots,\ g_k, \tag{15}$$

such that $g_\nu(Q_1) = Q_\nu$, $\nu = 1, \ldots, k$. As before, we set

$$H_\nu = \{h \mid h \in G,\ h(Q_\nu) = Q_\nu\}, \qquad G_\nu = H_\nu \mid Q_\nu, \qquad \nu = 1, \ldots, k. \tag{16}$$

If $\varphi_\nu = g_\nu | Q_1$, then, as follows from the proof of Theorem 1, (iv),

$$G_\nu = \varphi_\nu G_1 \varphi_\nu^{-1}. \tag{17}$$

Consequently, if

$$v_{11}, \ldots, v_{1m} \tag{18}$$

is any basis of Q_1 and $v_{\nu j} = g_\nu(v_{1j})$, $j = 1, \ldots, m$, then G_ν is represented in the basis

$$v_{\nu 1}, \ldots, v_{\nu m} \tag{19}$$

by the same matrix group A_1 over T as G_1 in the basis (18).

LEMMA 5. *As a linear group, G is similar to a subgroup of the wreath product $G_1 \sim \Gamma$.*

PROOF. We first construct a subgroup B of $GL(V)$ isomorphic to the direct product of $G_1, \ldots, G_k$. We set $B = B_1 \times \ldots \times B_k$, where $B_j | Q_j = G_j$, and let $B_j | Q_i$ be trivial for $i \neq j$. Then $gBg^{-1} = B$ for any $g \in G$. For, by (13), $g(Q_j) = Q_\nu$. If $x \in Q_\nu$ and $b_j \in B_j$, then

$$gb_jg^{-1}(x) = ga_jg^{-1}(x) = a_\nu(x) = b_\nu(x),$$

where $a_j \in G_j$, $a_\nu \in G_\nu$ and $b_\nu \in B_\nu$. But if $x \in Q_\mu$, $\mu \neq \nu$, then $g^{-1}(x) \in Q_i$, $i \neq j$, and so

$$gb_jg^{-1}(x) = gb_j(g^{-1}(x)) = gg^{-1}(x) = x.$$

Hence $gB_jg^{-1} = B_\nu$ and $gBg^{-1} = B$.

We now consider the group $\overline{F} = BG$. Since $b(Q_j) = Q_j$ for $b \in B$, it follows that $\bar{F}$ is also imprimitive and (13) is a decomposition of V into systems of imprimitivity for $\bar{F}$. We claim that $\bar{F}$ and $G_1 \wr \Gamma$ are similar as linear groups. As a T-basis for V we take the system of vectors

$$v_{11}, \ldots, v_{1m}, \ldots, v_{k1}, \ldots, v_{km} \tag{20}$$

(see (18) and (19)). For $g \in \bar{F}$ we have

$$g(v_{\nu\mu}) = \sum_{j=1}^{m} v_{i_\nu j} \alpha^{\nu}_{j\mu}, \quad \alpha^{\nu}_{j\mu} \in T, \tag{21}$$

where $i_1, \ldots, i_k$ is a permutation of $1, \ldots, k$. Owing to the choice of the basis (20) we find from (21) that

$$g_\nu g_{i_\nu}^{-1} g(v_{\nu\mu}) = \sum_{j=1}^{m} v_{\nu j} \alpha^{\nu}_{j\mu} = h^{\nu}(v_{\nu\mu}), \quad h^{\nu} \in H_\nu. \tag{22}$$

(21) and (22) give us

$$a_\nu = \| \alpha^{\nu}_{j\mu} \| \in A_1. \tag{23}$$

Now (21) and (23) show that $\bar{F}$ is similar as a linear group to $G_1 \wr \Gamma$. Since $\bar{F} \supset G$, this proves the lemma. ∎

Now let $\bar{\theta}$ be an abstract group-theoretic property that is S-, Q- and P-closed.

A group having the property $\bar{\theta}$ is called a *$\bar{\theta}$-group* for short. Examples are solvable groups, Π-groups and, in particular, p-groups. It follows from the definition of a $\bar{\theta}$-group that the wreath product $G_1 \wr \Gamma$ is a $\bar{\theta}$-group if and only if both G_1 and Γ are $\bar{\theta}$-groups.

THEOREM 4. *Let G be an irreducible maximal $\bar{\theta}$-subgroup of $GL(V)$, where V is a linear space of finite dimension n over a division ring T. Then G is either primitive or similar as a linear group to a wreath product $G_1 \wr \Gamma$, where G_1 is a maximal primitive $\bar{\theta}$-subgroup of $GL(Q)$, Γ a maximal transitive $\bar{\theta}$-subgroup of the symmetric group S_k, and Q a subspace of V such that $(Q: T)k = n$.*

PROOF. If G is primitive, the theorem is obvious. Let G be imprimitive. Then by Lemma 2 there is a nonrefinable decomposition (13) of V into systems of imprimitivity for G. Since G is a $\bar{\theta}$-group, so are the groups G_1 and Γ induced by (13). By Lemma 5, G is similar as a linear group to $G_1 \wr \Gamma$, and G_1 is primitive, Γ transitive. Since G is a maximal $\bar{\theta}$-subgroup of $GL(V)$ and $G_1 \wr \Gamma$ is a $\bar{\theta}$-subgroup, G is similar to $G_1 \wr \Gamma$. The maximality of G implies that of G_1 in $GL(Q)$ and that of Γ in S_k (with respect to $\bar{\theta}$-subgroups). ∎

3. *The Kronecker product of linear transformations.* Let X be the space of linear forms over a field Δ in the indeterminates

$$x_1, \ldots, x_n, \tag{24}$$

Y the space of linear forms over Δ in

$$y_1, \ldots, y_m, \tag{25}$$

and V the nm-dimensional space of bilinear forms over Δ, with the basis

$$x_1y_1, \ldots, x_ny_1, \ldots, x_1y_m, \ldots, x_ny_m. \tag{26}$$

Let $f \in \mathrm{Hom}(X, X)$, $g \in \mathrm{Hom}(Y, Y)$, $a = \|\alpha_{ij}\|$ the matrix of f in the basis (24) and $b = \|\beta_{kl}\|$ that of g in the basis (25). The *Kronecker product* [or tensor product] of f and g is the endomorphism $f \dot{\times} g$ of V defined by

$$(f \dot{\times} g)(x_iy_j) = f(x_i)\,g(y_j), \quad i = 1, \ldots, n, \quad j = 1, \ldots, m. \tag{27}$$

It is obvious that

$$(f \dot{\times} g)(x_iy_j) = \sum_{k=1}^{n} \sum_{l=1}^{m} x_ky_l\alpha_{ki}\beta_{lj}, \quad i = 1, \ldots, n, \quad j = 1, \ldots, m. \tag{28}$$

As is readily checked, the $nm \times nm$ matrix

$$\begin{bmatrix} a\beta_{11} & \ldots & a\beta_{1m} \\ \cdot & \cdot\;\cdot\;\cdot & \cdot \\ a\beta_{m1} & \ldots & a\beta_{mm} \end{bmatrix} = a \dot{\times} b \tag{29}$$

over Δ is the matrix of $f \dot{\times} g$ in the basis (26). The matrix $a \dot{\times} b$ defined by (29) is called the *Kronecker product of a and b*.

Now let $\alpha \in \Delta$, a and c matrices in Δ_n, b and d in Δ_m. Then (29) implies the following equalities:

$$\left.\begin{aligned} (a \dot{\times} b)\alpha &= a\alpha \dot{\times} b = a \dot{\times} b\alpha, \\ (a + c) \dot{\times} b &= a \dot{\times} b + c \dot{\times} b, \\ a \dot{\times} (b + d) &= a \dot{\times} b + a \dot{\times} d, \\ (a \dot{\times} b)(c \dot{\times} d) &= ac \dot{\times} bd, \\ a \dot{\times} E_m &= \mathrm{diag}\,[a, \ldots, a] \in \Delta_{nm}, \\ E_n \dot{\times} b &= \begin{bmatrix} \beta_{11}E_n & \ldots & \beta_{1m}E_n \\ \cdot & \cdot\;\cdot\;\cdot & \cdot \\ \beta_{m1}E_n & \ldots & \beta_{mm}E_n \end{bmatrix}. \end{aligned}\right\} \tag{30}$$

In particular, if a and b are invertible, then

$$a^{-1} \dot{\times} b^{-1} = (a \dot{\times} b)^{-1}.$$

It follows from (29) that

$$a \dot{\times} b = E_{nm} \Leftrightarrow a = \alpha E_n, \; b = \beta E_m, \; \alpha\beta = 1, \; \alpha, \beta \in \Delta. \tag{31}$$

If A is a subset of Δ_n and B a subset of Δ_m, the *Kronecker product* $A \dot{\times} B$ is defined as the set of all $a \dot{\times} b$ for $a \in A$, $b \in B$.

The Kronecker product determines an epimorphism

$$h\colon\ GL(n, \Delta) \times GL(m, \Delta) \to GL(n, \Delta) \dot{\times} GL(m, \Delta),$$

where $GL(n, \Delta) \times GL(m, \Delta)$ is the direct product of the groups in question and $h\colon \langle a, b\rangle \mapsto a \dot{\times} b$. By (31), Ker h consists of all pairs of the form

$$\langle \alpha E_n, \alpha^{-1} E_m\rangle, \quad \alpha \in \Delta^*;$$

that is, Ker $h \cong \Delta^*$. This leads to the next results.

LEMMA 6. *Let A be a subgroup of $GL(n, \Delta)$ and B a subgroup of $GL(m, \Delta)$. Then $A \dot{\times} B \cong (A \times B)/M$, where $A \times B$ is the direct product and M a subgroup of $A \times B$ isomorphic to a subgroup of Δ^*.*

LEMMA 7. *Let G be an absolutely irreducible subgroup of $GL(n, \Delta)$. Then the centralizer C of $G \dot{\times} E_m$ in $GL(nm, \Delta)$ is $E_n \dot{\times} GL(m, \Delta)$.*

PROOF. We write the matrices of $GL(nm, \Delta)$ as $m \times m$ matrices $c = \|c_{ij}\|$ over Δ_n. If $c \in C$, then by (30) all the c_{ij} are in the centralizer of G in Δ_n. Since G is absolutely irreducible, $c_{ij} = \gamma_{ij} E_n$, $\gamma_{ij} \in \Delta$. Hence and from the last equality in (30) the required result follows. ■

LEMMA 8. *Let A be a subgroup of $GL(n, \Delta)$, B a subgroup of $GL(m, \Delta)$. Then $A \dot{\times} B$ is absolutely irreducible if and only if both A and B are absolutely irreducible.*

PROOF. Consider the linear Δ-hull $[A \dot{\times} B]_\Delta$ of $A \dot{\times} B$. By (30),

$$[A \dot{\times} B]_\Delta \supset [A]_\Delta \dot{\times} [B]_\Delta. \tag{32}$$

Let A and B be absolutely irreducible. Then $[A]_\Delta = \Delta_n$ and $[B]_\Delta = \Delta_m$. If now e_{ij} $(i, j = 1, \ldots, n)$ is the canonical basis of Δ_n, and ϵ_{kl} $(k, l = 1, \ldots, m)$ that of Δ_m, then, as follows from (29), the products $e_{ij} \dot{\times} e_{kl}$ form a basis of the algebra Δ_{nm}. Hence and from (32) it follows that

$$[A \dot{\times} B]_\Delta = \Delta_{nm}.$$

This proves the sufficiency. To prove the necessity, let $a_1, \ldots, a_k$ be a maximal system of linearly independent elements of A, and $b_1, \ldots, b_l$ one for B. Then, by (30), every element of $A \dot{\times} B$ is a linear combination of the kl elements $a_i \dot{\times} b_j$. Consequently the absolute irreducibility of $A \dot{\times} B$ implies that of A and B. ■

For the validity of Lemma 8 it is essential that the groups are absolutely irreducible, and not just irreducible. Irreducibility of $A \dot{\times} B$ always implies that of A and B, but, as the following example shows, $A \dot{\times} B$ may turn out to be reducible even when both A and B are irreducible. Let R be the field of reals, and $A = B$ the group of all real matrices of the form

$$\begin{bmatrix} \alpha & \beta \\ -\beta & \alpha \end{bmatrix}, \qquad \alpha^2 + \beta^2 \neq 0.$$

It is obvious that $A \dot{\times} A$ is an abelian subgroup of $GL(4, R)$. But $GL(4, R)$ contains no irreducible abelian subgroups.

We return now to an arbitrary field Δ.

LEMMA 9. *If $A \dot{\times} B$ is primitive, so are A and B.*

PROOF. Let A be imprimitive, and let $u_{11}, \ldots, u_{1r}; \ldots; u_{s1}, \ldots, u_{sr}$ be bases of the s systems of imprimitivity for A. We construct an rm-dimensional subspace Q_i of V with the basis $u_{ik}y_j$ ($k = 1, \ldots, r; j = 1, \ldots, m$). Obviously $V = Q_1 + \ldots + Q_s$ is a decomposition of V into systems of imprimitivity for $A \dot{\times} B$. By the same device it can be proved that if B is imprimitive, so is $A \dot{\times} B$. ■

§16. On normal subgroups of completely reducible groups

1. *Lemmas on the normalizer.* Let V be a finite-dimensional linear space over a division ring T, H a subgroup of $GL(V)$, and G the normalizer of H in $GL(V)$. We regard V as an additive group $V(\Omega)$ with the operator domain $\Omega = T \cup H$. Thus the admissible subgroups of $V(\Omega)$ are the subspaces W of V such that $h(W) = W$ for h in H. Under these assumptions we prove the following two lemmas.

LEMMA 1. *Let $g \in G$, and let W be an admissible subgroup of $V(\Omega)$. Then the following assertions are true*:

(i) *$g(W)$ is an admissible subgroup of $V(\Omega)$.*

(ii) *$H|g(W)$ is similar to $H|W$.*

(iii) *$g(W)$ is operator-simple if and only if W is.*

PROOF. By assumption, $h(W) = W$ for any $h \in H$. On the other hand, every element $\bar{h}$ of H can be written in the form $\bar{h} = ghg^{-1}$, $h \in H$. Hence

$$\bar{h}(g(w)) = gh(w) = g(w).$$

This proves (i). To prove (ii), we consider the mapping γ: $W \longrightarrow g(W)$, $x \mapsto g(x)$. Since $g \in GL(V)$, γ is an isomorphism of W and $g(W)$ (not necessarily an operator isomorphism!). Now $H|g(W)$ is the set of all mappings $\bar{h}|g(W)$, where $\bar{h} = ghg^{-1}$, $h \in H$. For $x \in W$,

$$\bar{h}(g(x)) = ghg^{-1}g(x).$$

Setting $h|W = h_W$, we conclude that

$$\bar{h}(g(x)) = \gamma h_W \gamma^{-1}(g(x)), \qquad \bar{h} \mid g(W) = \gamma h_W \gamma^{-1}.$$

Consequently $H|g(W)$ is similar to $H|W$. (iii) follows easily from (i). ■

LEMMA 2. *Let $g \in G$, and let W_1 and W_2 be operator-isomorphic admissible subgroups of $V(\Omega)$. Then $g(W_2)$ is operator-isomorphic to $g(W_1)$.*

PROOF. Let $f: W_1 \longrightarrow W_2$ be an operator isomorphism. Then for $\lambda \in T$, $x, y \in W_1$ and $h \in H$ we have

$$f(x+y) = f(x) + f(y), \qquad f(x\lambda) = f(x)\lambda, \qquad f(hx) = hf(x).$$

We consider the mapping $f_1: g(W_1) \longrightarrow g(W_2)$, $g(x) \mapsto g(f(x))$. Clearly f_1 is bijective, and we find that

$$\begin{aligned} f_1(g(x)+g(y)) &= f_1(g(x+y)) = g(f(x+y)) \\ &= g(f(x)+f(y)) = gf(x) + gf(y) \\ &= f_1(g(x)) + f_1(g(y)), \\ f_1(g(x)\lambda) &= f_1(g(x\lambda)) = g(f(x\lambda)) = g(f(x)\lambda) = \\ &= g(f(x))\lambda = f_1(g(x))\lambda. \end{aligned}$$

For any $h \in H$ we set $h_1 = g^{-1}hg$. Then

$$\begin{aligned} f_1[h(g(x))] &= f_1[g(h_1(x))] = g[f(h_1(x))] = g(h_1 f(x)) \\ &= gh_1(f(x)) = hg(f(x)) = hf_1(g(x)). \end{aligned}$$

Hence f_1 is an operator isomorphism. ■

We need a further simple lemma on the centralizer of H in $GL(V)$.

LEMMA 3. *If g is in the centralizer of H in $GL(V)$ and W is an admissible subgroup of $V(\Omega)$, then W and $g(W)$ are operator-isomorphic.*

PROOF. The mapping $\gamma: W \longrightarrow g(W)$ defined by $\gamma(x) = g(x)$ for $x \in W$ is an operator isomorphism, because for any h in H we can write

$$\gamma(h(x)) = gh(x) = hg(x) = h\gamma(x). \quad ■$$

2. *Clifford's theorems.* As before, let V be a finite-dimensional linear space over a division ring T.

THEOREM 1 (CLIFFORD'S FIRST THEOREM). *Let G be an irreducible subgroup of $GL(V)$ and H a normal subgroup of G. Then the following assertions are true:*

(i) *H is completely reducible.*

(ii) *If W_1 and W_2 are two H-invariant irreducible subspaces of V, then $H|W_1$ and $H|W_2$ are similar.*

PROOF. Let L_1 be an H-invariant H-irreducible subspace of V. If $L_1 = V$, there is nothing to prove. If $L_1 \neq V$, there is an element g_2 in G such that

$$g_2(L_1) \neq L_1. \tag{1}$$

By Lemma 1, $g_2(L_1)$ is H-invariant and H-irreducible. By (1), $L_1 + g_2(L_1)$ is a direct sum. Now let

$$g_2, \dots, g_r \tag{2}$$

be elements of G such that $L = L_1 + g_2(L_1) + \dots + g_r(L_1)$ is a direct sum, but for every $g \in G$ the sum $L + g(L_1)$ is no longer direct. Since V is of finite dimension, such elements (2) exist. It follows from the construction of L and the H-irreducibility of $g(L_1)$ that

$$g(L_1) \cap L \neq (0), \quad g(L_1) \subset L, \quad gg_i(L_1) \subset L, \quad g(L) = L.$$

Hence, since G is irreducible, $L = V$. Consequently

$$V = L_1 \dotplus g_2(L_1) \dotplus \dots \dotplus g_r(L_1), \tag{3}$$

so that (i) is proved. By Proposition 13.1, any operator-simple subgroup W of a completely reducible group $V(\Omega)$, $\Omega = T \cup H$, is operator-isomorphic to one of the summands $g_i(L_1)$ in (3). Hence and from Lemma 1 we obtain (ii). ∎

COROLLARY 1. *A normal subgroup of a completely reducible subgroup of $GL(V)$ is completely reducible.*

Retaining the notation of Theorem 1 we turn to the decomposition (3) and set $L_\nu = g_\nu(L_1)$, $\nu = 2, \dots, r$. Obviously

$$L_1, \dots, L_r \tag{4}$$

are operator-simple admissible subgroups of $V(\Omega)$, $\Omega = T \cup H$. We divide the system of subspaces (4) into classes, putting L_μ and L_ν into the same class if and only if they are operator-isomorphic. Let k be the number of these classes. We construct the k direct sums $Q_1, \dots, Q_k$ of the subspaces (4) in one class. Obviously

$$V = Q_1 \dotplus \dots \dotplus Q_k. \tag{5}$$

THEOREM 2 (CLIFFORD'S SECOND THEOREM). *If $k > 1$, then (5) is a decomposition of V into systems of imprimitivity for G.*

PROOF. Let $g \in G$, and consider $g(Q_i)$, $i = 1, \dots, k$. By Lemma 2, $g(Q_i)$ is a direct sum of operator-simple operator-isomorphic subgroups. By Proposition 13.3, every operator-simple subgroup of $V(\Omega)$ is contained in one of the subgroups Q_j, $j = 1, \ldots, k$, and operator-isomorphic operator-simple subgroups are in the same group Q_j. Consequently $g(Q_i) \subset Q_j$. Hence and from the definition of imprimitivity (see §15) it follows that for $k > 1$ (5) is a decomposition of V into systems of imprimitivity for G. ∎

By means of the subspaces (4) we determine the irreducible components of H; that is, the irreducible representations d_j of the form

$$d_j\colon H \to GL(L_j), \quad h \mapsto h \mid L_j, \quad j = 1, \ldots, r. \tag{6}$$

Theorem 2 implies

COROLLARY 1. *The irreducible components of a normal subgroup H of a primitive irreducible group G are pairwise equivalent.*

The number of representations d_j (see (6)) equivalent to a fixed representation d_ν is called the *multiplicity of d_ν*.

COROLLARY 2. *The irreducible components of a normal subgroup H of an irreducible group G have the same multiplicity.*

PROOF. By Lemma 15.1, the subspaces Q_j are of the same dimension, and so are the subspaces L_i. Hence all the irreducible components of H have the same multiplicity $(Q_1 : T)(L_1 : T)^{-1}$. ∎

COROLLARY 3. *If the degree and the multiplicity of an irreducible component of H are equal to 1, then G is monomial.*

LEMMA 4. *Suppose that an irreducible subgroup G of $GL(V)$ can be represented in the form*

$$G = HF, \tag{7}$$

where H and F are elementwise permutable normal subgroups of G. Then the irreducible components of H (F) are pairwise equivalent.

PROOF. Let L_1 be an operator-simple admissible subgroup of $V(\Omega)$, $\Omega = T \cup H$. Then, by (3) and Proposition 13.1, any operator-simple admissible subgroup of $V(\Omega)$ is operator-isomorphic to some group $g(L_1)$, $g \in G$. By (7), $g(L_1) = f(L_1)$, $f \in F$. By Lemma 3, this completes the proof. ∎

We now apply Lemma 4 to the case when T is a field Δ and G an absolutely irreducible subgroup of $GL(n, \Delta)$.

LEMMA 5. *Suppose that an absolutely irreducible subgroup G of $GL(n, \Delta)$ can be represented in the form (7), l is the degree of an irreducible component* *of H, m the degree of an irreducible componet of F, and* $lm \leqslant n$. *Then* $lm = n$, *and the irreducible components of H and F are absolutely irreducible.*

PROOF. By assumption, $[G]_\Delta : \Delta = n^2$. By Lemma 4,

$$[H]_\Delta : \Delta \leqslant l^2, \quad [F]_\Delta : \Delta \leqslant m^2.$$

On the other hand, by (7),

$$n^2 = [G]_\Delta : \Delta \leqslant ([H]_\Delta : \Delta)([F]_\Delta : \Delta) \leqslant l^2 m^2.$$

Hence

$$n \leqslant lm, \quad n = lm, \quad [H]_\Delta : \Delta = l^2, \quad [F]_\Delta : \Delta = m^2.$$

Consequently the irreducible components of H and F are absolutely irreducible. ■

THEOREM 3. *Suppose that an absolutely irreducible subgroup G of* $GL(n, \Delta)$ *can be represented in the form* (7), *that l is the degree of an irreducible component of H, m the degree of an irreducible component of F, and* $lm \leqslant n$. *Then G is conjugate in* $GL(n, \Delta)$ *to a matrix group* $\overline{G} = \overline{H} \dot{\times} \overline{F}$, *where* $\overline{H}$ *is an absolutely irreducible subgroup of* $GL(l, \Delta)$ *and* $\overline{F}$ *an absolutely irreducible subgroup of* $GL(m, \Delta)$, $\overline{H} \cong H$ *and* $\overline{F} \cong F$.

PROOF. By Lemma 5, $lm = n$. By Lemma 4, in some basis B of Δ^n the matrices h of H take the form $h = \overline{h} \dot{\times} E_m$, where $\overline{h}$ ranges over an irreducible subgroup $\overline{H}$ of $GL(l, \Delta)$ isomorphic to H. By Lemma 5, $\overline{H}$ is absolutely irreducible. Hence and by Lemma 15.8, in the basis B the matrices f of F take the form $f = E_l \dot{\times} \overline{f}$, where $\overline{f}$ ranges over an absolutely irreducible subgroup $\overline{F}$ of $GL(m, \Delta)$ isomorphic to F. Consequently the matrices of G have the form $\overline{h} \dot{\times} \overline{f}$ in the basis B, and every matrix $\overline{h} \dot{\times} \overline{f}$ is in G. ■

3. *On the normalizer of a completely reducible group.* Again, let V be a finite-dimensional linear space over a division ring T, and H a completely reducible subgroup of $GL(V)$. As before, we regard V as an additive group $V(\Omega)$ with the operator domain $\Omega = T \cup H$. Let

$$V = L_1 \dotplus \ldots \dotplus L_m \tag{8}$$

be a decomposition of $V(\Omega)$ into a direct sum of operator-simple admissible subgroups

$$L_j, \quad j = 1, \ldots, m. \tag{9}$$

We split the subgroups (9) into classes, putting L_i and L_j into the same class if and only if they are operator-isomorphic. Let k be the number of these classes and $Q_1, \ldots, Q_k$ the direct sums of subgroups L_j in one and the same class.

Obviously

$$V = Q_1 \dotplus \dots \dotplus Q_k. \tag{10}$$

We now construct another decomposition of V into a direct sum of admissible subgroups, by dividing the subgroups (9) into classes in another way: two subgroups L_i and L_j are put in the same class if and only if the groups $H|L_i$ and $H|L_j$ are similar. Let l be the number of these new classes and $W_1, \dots, W_l$ the direct sums of subgroups L_j in one and the same class. Then

$$V = W_1 \dotplus \dots \dotplus W_l. \tag{11}$$

By Proposition 13.3, the decomposition (10) is independent of (8) and is determined by $V(\Omega)$ in a natural manner, uniquely to within the order of the summands, and any two operator-isomorphic operator-simple subgroups of $V(\Omega)$ lie in the same group Q_j.

Similar statements are true for (11):

LEMMA 6. *The decomposition* (11) *is unique to within the order of the summands. Let P and P_1 be operator-simple subgroups of $V(\Omega)$ such that $H|P$ and $H|P_1$ are similar. Then for some μ*

$$P \subset W_\mu, \quad P_1 \subset W_\mu. \tag{12}$$

PROOF. We first prove (12). By Proposition 13.1, P is operator-isomorphic to some group L_i in (8), and P_1 to some group L_j. Since the restrictions $H|P$ and $H|P_1$ are similar, it follows that $H|L_i$ and $H|L_j$ are also similar. Let $L_i \subset Q_\lambda$ and $L_j \subset Q_\nu$. Then, by the properties of (10), $P \subset Q_\lambda$ and $P_1 \subset Q_\nu$. Since $H|L_i$ and $H|L_j$ are similar, there exists a μ, $1 \leqslant \mu \leqslant l$, such that $Q_\lambda, Q_\mu \subset W_\nu$. Hence $P, P_1 \subset W_\mu$. This proves (12). From the construction of (11) and from Proposition 13.1 it follows that $H|U_1$ and $H|U_2$ are similar for any operator-simple subgroups U_1 and U_2 of W_μ. Hence and from (12) the uniqueness of the decomposition (11) follows. ∎

Now let $N(H)$ be the normalizer, and $C(H)$ the centralizer, of H in $GL(V)$.

LEMMA 7. *Each summand W_μ in* (11) *is an $N(H)$-invariant subspace, and each summand Q_j in* (10) *is a $C(H)$-invariant subspace.*

PROOF. Let $g \in N(H)$, and let P be an operator-simple admissible subgroup of W_μ. By Lemma 1, $H|P$ and $H|g(P)$ are similar. By Lemma 6, $g(P) \subset W_\mu$. Hence $g(W_\mu) = W_\mu$. Similarly, the second part of the lemma follows from Lemma 3. ∎

LEMMA 8. *Let $g \in N(H)$. Then*

$$g(Q_i) = Q_j \tag{13}$$

for some j.

PROOF. By construction, Q_i is a direct sum of operator-isomorphic operator-simple subgroups of $V(\Omega)$. By Lemma 2, $g(Q_i)$ is also a direct sum of operator-isomorphic operator-simple subgroups of $V(\Omega)$. Hence $g(Q_i) \subset Q_j$ for some j. It is clear that $Q_i \subset g^{-1}(Q_j) \subset Q_\lambda$. Since Q_i and Q_λ are direct summands, we see that

$$\lambda = i, \quad g^{-1}(Q_j) = Q_i, \quad g(Q_i) = Q_j. \quad \blacksquare$$

Obviously, from the construction of (10) and (11) it follows that each summand W_μ in (11) is a direct sum of certain summands Q_j in (10).

Lemmas 7 and 8 imply

LEMMA 9. *Let $Q_j \subset W_\mu$, and let U be the direct sum of all Q_ν contained in W_μ and having the same dimension as Q_j. Then U is an $N(H)$-invariant subspace of V.*

To end the section, we consider the case when $T = \Delta$ is an algebraically closed field and the irreducible components of H are pairwise equivalent. Under these assumptions, for a suitable choice of a basis of $V = \Delta^n$ the matrices $h \in H$ take the form

$$h = h_1 \dot{\times} E_l, \tag{14}$$

where h_1 ranges over an irreducible subgroup H_1 of $GL(m, \Delta)$ isomorphic to H, and $ml = n$.

LEMMA 10. *Let H be a subgroup of $GL(n, \Delta)$ consisting of matrices of the form* (14). *Then its normalizer $N(H)$ in $GL(n, \Delta)$ is $N_1 \dot{\times} GL(l, \Delta)$, where N_1 is the normalizer of H_1 in $GL(m, \Delta)$.*

PROOF. A direct check shows that

$$N_1 \dot{\times} GL(l, \Delta) \subset N(H). \tag{15}$$

Now let $g \in N(H)$ and $h \in H$. Then

$$ghg^{-1} = g(h_1 \dot{\times} E_l)\, g^{-1} = \bar{h}_1 \dot{\times} E_l, \qquad \bar{h}_1 \in H_1.$$

Obviously the mapping $\gamma_1 \colon h_1 \mapsto \bar{h}_1$ is an automorphism of H_1; it can be extended to an automorphism $\gamma\colon [H_1]_\Delta \longrightarrow [H_1]_\Delta$. Since H_1 is irreducible and Δ algebraically closed, we have $[H_1]_\Delta = \Delta_m$. Therefore γ is an inner automorphism of Δ_m, and so for a given $g \in N(H)$ there is an element g_1 in $GL(m, \Delta)$ such that for any $h_1 \in H_1$

$$g(h_1 \times E_l)\, g^{-1} = g_1 h_1 g_1^{-1} \dot{\times} E_l.$$

Hence it follows that $g_1 \in N_1$. Clearly, for any $h \in H$

$$ghg^{-1} = (g_1 \dot{\times} E_l)\, h\, (g_1 \dot{\times} E_l)^{-1}.$$

Thus, $(g_1 \dot{\times} E_l)^{-1} g$ belongs to the centralizer $C(H)$ of H in $GL(n, \Delta)$. By Lemma 15.8,

$$C(H) = E_m \dot{\times} GL(l, \Delta).$$

Hence

$$g \in (g_1 \dot{\times} E_l)(E_m \dot{\times} GL(l, \Delta)) \subset N_1 \dot{\times} GL(l, \Delta).$$

Together with (15), this proves the lemma. ∎

§17. Some conditions for complete reducibility of a linear group over a field

1. *Semisimple matrices and d-matrices; d-groups.* Let Δ be an arbitrary field, V a linear space of dimension n over Δ, and $a \in \operatorname{End} V$. Clearly, *$a$ is irreducible* (that is, V contains no nontrivial a-invariant subspaces) *if and only if its characteristic polynomial is irreducible in* $\Delta[x]$. It is standard knowledge that *a is completely reducible* (that is, V is a direct sum of irreducible a-invariant subspaces) *if and only if the minimal polynomial $m(x)$ of a has no multiple factors in* $\Delta[x]$. Completely reducible endomorphisms in End V are called *semisimple*, and their matrices are also called semisimple. In other words, a matrix a in Δ_n is *semisimple* over Δ if there is a matrix t in $GL(n, \Delta)$ such that

$$t^{-1}at = \operatorname{diag}[a_1, \ldots, a_k],$$

where a_j are matrices whose characteristic polynomials are irreducible in $\Delta[x]$, $j = 1, \ldots, k$.

We need another definition. A matrix $a \in \Delta_n$ is called a *d-matrix* if there exist an extension Σ of Δ and a matrix $t \in GL(n, \Sigma)$ such that

$$t^{-1}at = \operatorname{diag}[\lambda_1, \ldots, \lambda_n], \quad \lambda_j \in \Sigma; \quad j = 1, \ldots, n.$$

It is clear that for Σ we can take the splitting field of the characteristic polynomial of a.

It follows from the definition of a d-matrix and the properties of minimal polynomials that *a matrix a in Δ_n is a d-matrix if and only if its minimal polynomial $m(x)$ is prime to its derivative $m'(x)$. Hence every d-matrix is semisimple. If Δ is a perfect field, then d-matrices and semisimple matrices are the same thing. If Δ is not perfect, there are semisimple matrices that are not d-matrices.*

A subgroup G of $GL(n, \Delta)$ is called a *d-group* if every element of G is a d-matrix.

LEMMA 1. *Let A be a set of pairwise commuting d-matrices in Δ_n. Then there exist a finite extension Σ of Δ and a matrix $t \in GL(n, \Sigma)$ such that all the matrices of the set $t^{-1}At$ are diagonal.*

PROOF. It is obviously sufficient to consider the case of a finite set A, because A contains a finite maximal system of linearly independent matrices. Let A be finite and let Σ be the splitting field of $f(x) \in \Delta[x]$, where $f(x)$ is the product of the characteristic polynomials of the matrices in A. We proceed by induction on the degree n. For $n = 1$ the lemma is obvious. Let $n > 1$ and suppose that the lemma is proved for matrices of degree less than n. If all the matrices a in A are scalar, we are done. Let a be a nonscalar matrix in A. Since a is a d-matrix, there is a matrix u in $GL(n, \Sigma)$ such that

$$u^{-1}au = \operatorname{diag}[\lambda_1 E_{n_1}, \ldots, \lambda_k E_{n_k}], \qquad k > 1,$$
$$\lambda_i \neq \lambda_j \qquad \text{for} \qquad i \neq j.$$

Consequently the matrices b of the set $u^{-1}Au$ have the form

$$b = \operatorname{diag}[b_1, \ldots, b_k], \qquad b_j \in \Sigma_{n_j}; \qquad j = 1, \ldots, k.$$

Obviously b_j is a d-matrix, and b_j ranges over a certain set B_j of pairwise commuting matrices in Σ_{n_j} as b ranges over $B = u^{-1}Au$. Since $n_j < n$, by the inductive hypothesis there is a matrix t_j in $GL(n_j, \Sigma)$ such that all the matrices of $t_j^{-1}B_jt_j$ are diagonal. Thus all the matrices of the set $t^{-1}At$, where $t = u \cdot \operatorname{diag}[t_1, \ldots, t_k]$, are diagonal. ∎

LEMMA 2. *A set A of pairwise commuting d-matrices in Δ_n is completely reducible.*

PROOF. Let Σ be the field whose existence is asserted in Lemma 1, A_1 the subalgebra of Δ_n generated by E_n and A, and A_2 the subalgebra of Σ_n generated by E_n and A. By Lemma 1, A_2 contains no nonzero nilpotent elements. Consequently the same is true of $A_1 \subset A_2$, so that A_1 is semisimple and A completely reducible. ∎

According to Lemma 1, if $a, b \in \Delta_n$ are any two commuting matrices and $\lambda, \mu \in \Delta$, then ab and $\lambda a + \mu b$ are also d-matrices. The following simple example shows that neither this remark nor Lemma 2 carries over to semisimple matrices. Let p be a prime, and $x^p - \delta$ an irreducible polynomial in $\Delta[x]$, where Δ is a field of characteristic p. Consider the set A consisting of two matrices a and b of degree $n = 2p$:

$$a = \begin{bmatrix} c & c \\ 0 & c \end{bmatrix}, \qquad b = \begin{bmatrix} c & 0 \\ 0 & c \end{bmatrix},$$

where c is a matrix in Δ_p whose characteristic polynomial is $x^p - \delta$. Clearly

$ab = ba$. Moreover,

$$a^p = \left(b + \begin{bmatrix} 0 & c \\ 0 & 0 \end{bmatrix}\right)^p = b^p = \delta E_n.$$

Consequently $x^p - \delta$ is the minimal polynomial of a; that is, a and b are semisimple. It is obvious that A is a reducible set of matrices. A is indecomposable. For if A were decomposable, it would also be completely reducible. However, the group G generated by a and b contains the normal subgroup (u), where

$$u = \begin{bmatrix} E_p & E_p \\ 0 & E_p \end{bmatrix},$$

and by Clifford's theorem it cannot be completely reducible. Thus A is an indecomposable reducible set of commuting semisimple matrices. The matrices ab^{-1} and $a - b$ are not semisimple, though a, b, and b^{-1} are.

LEMMA 3. *Let G be a group of semisimple matrices in $GL(n, \Delta)$, and $\gamma_1, \ldots, \gamma_k$ the irreducible components of G. If all the groups* Im γ_j *are abelian, so is G.*

PROOF. If all the Im γ_j are abelain, then all the matrices in the derived group G' of G are unipotent. Since every matrix in G is semisimple, $G' = (E)$. ∎

LEMMA 4. *A periodic group G of matrices over a field Δ is a d-group if and only if* char $\Delta \notin \Pi(G)$.

PROOF. Let G be a d-group, and $g \in G$ an element of order k. Then, for any prime factor q of k and some extension Σ of Δ, the group Σ^* contains an element of order q. Consequently char $\Delta \notin \Pi(G)$. Conversely, let char $\Delta \notin \Pi(G)$. Then, for any g in G, char Δ does not divide the order k of g. The minimal polynomial of g divides $\psi(x) = x^k - 1$, $(\psi(x), \psi'(x)) = 1$. Thus g is a d-matrix and G a d-group. ∎

COROLLARY 1. *A finite group of matrices over a field Δ is a d-group if and only if its order is not divisible by* char Δ.

LEMMA 5. *Suppose that a group G of matrices over a field Δ has a normal subgroup H such that H is a d-group, G/H a periodic group, and* char $\Delta \notin \Pi(G/H)$. *Then G is a d-group.*

PROOF. For any $g \in G$ there exists an integer $k > 0$ such that $g^k = h \in H$ and char $\Delta \notin \Pi(k)$. Since H is a d-group, the minimal polynomial $m(x)$ of h is prime to $m'(x)$. Obviously $m(x^k)$ annihilates g.

Since

$$(m(x^k),\ km'(x^k)\, x^{k-1}) = 1,$$

$m(x^k)$ does not have multiple roots in any extension of Δ. Thus g is a d-matrix. ∎

2. *On conditions for complete reducibility.*

THEOREM 1 (DIXON [1]). *Let G be a subgroup of $GL(n, \Delta)$ and H a completely reducible subgroup of finite index j in G, where* char $\Delta = 0$ *or* (char $\Delta, j) = 1$. *Then G is completely reducible.*

PROOF. Suppose that G acts on the space Δ^n. Let $Q \neq (0)$ be a G-invariant subspace of Δ^n. We claim that Q has a G-invariant complement. This will prove the theorem. Since H is a completely reducible subgroup of G, we have $\Delta^n = Q \dotplus L$, where L is an H-invariant subspace of Δ^n. We define an endomorphism π of Δ^n by $\pi|Q = 0$ and $\pi|L = e$, where e is the unit element of $\mathrm{Hom}(L, L)$. Let B be a left transversal to H in G. We construct an endomorphism σ_B of Δ^n by setting

$$\sigma_B = \frac{1}{j} \sum_{x \in B} x\pi x^{-1}. \tag{1}$$

The subspace $P = \mathrm{Im}\ \sigma_B = \sigma_B \Delta^n$ is independent of the choice of B. Indeed,

$$x\pi x^{-1}\Delta^n = x\pi\Delta^n = xL.$$

Consequently, if $y = xh$, $h \in H$, then

$$y\pi y^{-1}\Delta^n = y\pi\Delta^n = yL = xhL = xL.$$

It is easy to see that for $g \in G$ the set gB is also a left transversal to H in G, and that

$$g\sigma_B = \frac{1}{j}\left(\sum_{x \in B} (gx)\,\pi\,(gx)^{-1}\right) g = \sigma_{gB}\, g. \tag{2}$$

The subspace P is G-invariant. For it follows from (2) that

$$gP = g\sigma_B\Delta^n = \sigma_{gB}\, g\Delta^n = \sigma_{gB}\Delta^n = \sigma_B\Delta^n = P.$$

The endomorphism σ_B has the following two properties:

$$\mathrm{Ker}\ \sigma_B = Q, \qquad P \cap Q = \mathrm{Im}\ \sigma_B \cap \mathrm{Ker}\ \sigma_B = (0). \tag{3}$$

For if $q \in Q$, then $x\pi x^{-1}q = 0$. Therefore

$$Q \subset \mathrm{Ker}\ \sigma_B. \tag{4}$$

Now let $u = q + l$, $q \in Q$, $l \in L$. Then

$$x^{-1}(q + l) = q_1 + l_1, \qquad q_1 \in Q, \qquad l_1 \in L.$$

Hence $x(q_1 + l_1) = q_1' + x(l_1) = q + l$, $q_1' \in Q$. Thus

$$x(l_1) = q_2 + l, \qquad q_2 \in Q,$$
$$x\pi x^{-1}(q + l) = x\pi(q_1 + l_1) = x(l_1) = q_2 + l.$$

Therefore

$$\sigma_B(q + l) = q_3 + l, \qquad q_3 \in Q. \tag{5}$$

If $\sigma_B(q + l) = 0$, then $l = 0$ by (5). Hence and from (4) we obtain Ker $\sigma_B = Q$. If $\sigma_B(q + l) \in Q$, then $l = 0$ by (5). Consequently $P \cap Q = (0)$, and the proof of (3) is complete. From (3) we derive that $\Delta^n = Q \dotplus P$, where P is a G-invariant subspace. ∎

COROLLARY 1. *If the order of a finite group of matrices over a field* Δ *is not divisible by* char Δ, *the group is completely reducible.*

COROLLARY 2. *A finite d-group* G *of matrices over an arbitrary field* Δ *is completely reducible.*

PROOF. By Lemma 4, char Δ does not divide the order of G, and Corollary 2 follows from this and Corollary 1. ∎

THEOREM 2 (PLATONOV [7]). *Let* Δ *be an arbitrary field, and let* G *be a subgroup of* $GL(n, \Delta)$ *having a completely reducible normal subgroup* H *such that* G/H *is locally finite and* char $\Delta \notin \Pi(G/H)$. *Then* G *is completely reducible.*

PROOF. Let $g_1, \ldots, g_r$ be a maximal linearly independent system of matrices in G, and G_1 the group generated by $g_1, \ldots, g_r$ and H. Since G/H is locally finite, G_1/H is finite. The index $G_1 : H$ is not divisible by char Δ, and so, by Theorem 1, G_1 is completely reducible. Since G_1 and G have the same linear hull, G is also completely reducible. ∎

LEMMA 6. *Let* Δ *be a field of prime characteristic* p *and* b *a d-matrix in* Δ_n *such that* b^p *is diagonal. Then* b *is also diagonal.*

PROOF. We set

$$a = b^p = \operatorname{diag}[\lambda_1 E_{n_1}, \ldots, \lambda_k E_{n_k}], \; \lambda_j \in \Delta, \; \lambda_i \neq \lambda_j \text{ for } i \neq j.$$

Since $ab = ba$, we have

$$b = \operatorname{diag}[b_1, \ldots, b_k], \qquad b_j \in \Delta_{n_j}; \qquad j = 1, \ldots, k.$$

We claim that all the b_j are scalar. For b_j we can write

$$b_j^p = \lambda_j F_{n_j}. \tag{6}$$

Obviously b_j is a d-matrix. Hence for some extension Σ_j of Δ there is a matrix t_j

in $GL(n_j, \Sigma_j)$ such that

$$t_j b_j t_j^{-1} = \operatorname{diag}\left[\rho_1, \ldots, \rho_{n_j}\right].$$

From (6) we find that

$$b_j^p = t_j b_j^p t_j^{-1} = \operatorname{diag}\left[\rho_1^p, \ldots, \rho_{n_j}^p\right] = \lambda_j E_{n_j},$$

where $\rho_1^p = \ldots = \rho_{n_j}^p = \lambda_j$. Since $p = \operatorname{char} \Delta$, we have $\rho_1 = \ldots = \rho_{n_j} = \rho$, and

$$t_j b_j t_j^{-1} = \rho E_{n_j}, \quad b_j = \rho E_{n_j}, \qquad \rho \in \Delta.$$

Thus all the b_j are scalar, and b is diagonal. ■

THEOREM 3. *Let Δ be an arbitrary field and G a d-subgroup of $GL(n, \Delta)$ having an abelian normal subgroup F whose factor group G/F is locally finite. Then G is completely reducible.*

PROOF. If char $\Delta = 0$, the theorem follows from Theorem 2. Let char $\Delta = p$, where p is a prime. It suffices to consider the case when G/F is finite. By Lemma 1 there is then an extension Σ of Δ and a matrix $t \in GL(n, \Sigma)$ such that tFt^{-1} consists of diagonal matrices. We set $F_1 = tFt^{-1}$ and $G_1 = tGt^{-1}$. It is clear that G_1 and F_1 are subgroups of $GL(n, \Sigma)$. Let F_2 be the group of all diagonal matrices in $GL(n, \Sigma)$ that are contained in G_1. Obviously $G_1 \supset F_2 \supset F_1$. By assumption, G_1/F_1 is a finite group. Now let cF_1 be an element of order p^α in G_1/F_1. Then $c \in F_2$. For consider the elements $c, c^p, c^{p^2}, \ldots, c^{p^\alpha}$. By the choice of c, we have $c^{p^\alpha} \in F_1$. By Lemma 6, $c^{p^{\alpha-1}}$ is a diagonal matrix, and so $c^{p^{\alpha-1}} \in F_2$. By repeated application of Lemma 6 we see that $c \in F_2$; that is, $cF \in F_2/F_1$. Thus F_2/F_1 contains a Sylow p-subgroup of G_1/F_1. It follows that the index $G_1 \colon F_2$ is not divisible by p. Now F_2 is a completely reducible subgroup of $GL(n, \Sigma)$. Since

$$t^{-1}F_2(t) \subset G \subset GL(n, \Delta),$$

it follows that $t^{-1}F_2 t$ is a completely reducible subgroup of G, whose index is prime to char Δ. By Theorem 1, G is completely reducible.

We mention without proof another condition for complete reducibility:

THEOREM 4 (SUPRUNENKO [16]). *Let Δ be a perfect field and G a subgroup of $GL(n, \Delta)$ having a completely reducible normal subgroup N. If the factor group G/N is periodic and* char $\Delta \notin \Pi(G/N)$, *then G is completely reducible.*

Examples are known of reducible but indecomposable d-groups (see Platonov and Zalesskiĭ [1], and Merzljakov [2]).

FIFTH CHAPTER

SOLVABLE MATRIX GROUPS

§18. Reducible solvable groups

1. *The unitriangular group.* Let Δ be an associative ring with unit element 1. The group of all matrices of the form

$$\begin{bmatrix} 1 & \alpha_{12} & \alpha_{13} & \dots & \alpha_{1n} \\ 0 & 1 & \alpha_{23} & \dots & \alpha_{2n} \\ \cdot & \cdot & \cdot & \cdot & \cdot \\ 0 & 0 & 0 & \dots & 1 \end{bmatrix},$$

where α_{ij} are arbitrary elements of Δ, $n > 1$, is called the *unitriangular group*, and is denoted by $T(n, \Delta)$.

We shall show that $T(n, \Delta)$ is solvable and that the length of its derived series depends only on n. The following lemma is obvious.

LEMMA 1. *Any two matrices*

$$\begin{bmatrix} E_m & B \\ 0 & E_{n-m} \end{bmatrix},$$

where B is an $m \times (n - m)$ matrix over Δ, commute.

We now define a number $\mu = \mu(n)$ by

$$\mu(n) = [\log_2 (n - 1)] + 1.$$

It is clear that $2^\mu > n - 1 \geqslant 2^{\mu-1}$.

THEOREM 1. *The group $T(n, \Delta)$ is solvable. The length of its derived series is $\mu(n) = [\log_2(n - 1)] + 1$.*

PROOF. Setting $T(n, \Delta) = G_n$, we first show by induction on μ that

$$G_n^{(\mu)} = (E_n) \tag{1}$$

where $G_n^{(\mu)}$ is the μth derived group of G_n. If $\mu(n) \leqslant 2$, then $n \leqslant 4$, and (1) is easily checked for $n \leqslant 4$. Let $\mu(n) > 2$ and suppose the equality is proved for $\mu - 1$. We set $m = 2^{\mu-1}$. Then

$$2^{\mu-1} > m - 1 > 2^{\mu-2}, \qquad \mu(m) = \mu(n) - 1 = \mu - 1.$$

Therefore, by the inductive hypothesis, $G_m^{(\mu-1)} = (E_m)$. This implies that the matrices a in $G_n^{(\mu-1)}$ are such that the $m-1$ diagonals above the principal diagonal consist of zeros. Since $n - m \leqslant m$, we can write a in the form

$$a = \begin{bmatrix} E_m & B \\ 0 & E_{n-m} \end{bmatrix} \tag{2}$$

By Lemma 1, it follows from (2) that $G_n^{(\mu-1)}$ is abelian and that (1) holds. We must now show that

$$G_n^{(\mu-1)} \neq (E_n). \tag{3}$$

From the definition of $\mu = \mu(n)$ we find that $n \geqslant 2^{\mu-1} + 1$. We set $2^{\mu-1} + 1 = \alpha$. In G_n there is the elementary matrix

$$t_{1\alpha} = t_{1\alpha}(1) = E_n + e_{1\alpha}.$$

Now (3) can be derived by means of the tableau

$$\begin{array}{cccccc} t_{12} & t_{13} & t_{15} & t_{19} & \dots\ t_{1,\,2^{\mu-2}+1} & t_{1\alpha} \\ t_{23} & t_{24} & t_{26} & t_{2,\,10} & \dots\ t_{2^{\mu-2}+1,\,\alpha} & \\ t_{34} & t_{35} & t_{37} & t_{3,\,11} & \dots & \\ \cdot & \cdot & & \cdot & & \\ t_{\alpha-1,\,\alpha} & t_{\alpha-2,\,\alpha} & t_{\alpha-4,\,\alpha} & t_{\alpha-8,\,\alpha} & & \end{array}$$

where $t_{ij} = t_{ij}(1)$. For by Lemma 8.3 the elements of the second column in this tableau are commutators of elements of the first column, the elements of the third column are commutators of those of the second, and so on. Therefore

$$t_{13} \in G_n^{(1)}, \quad t_{15} \in G_n^{(2)}, \quad \dots, \quad t_{1,\,2^\beta+1} \in G_n^{(\beta)}.$$

In particular, $t_{1a} = t_{1,2^{\mu-1}+1} \in G_n^{(\mu-1)}$, $G_n^{(\mu-1)} \neq (E_n)$. ∎

We now consider the group $T(n_1, \dots, n_k, \Delta)$, consisting of all matrices of the form

$$a = \begin{bmatrix} E_{n_1} & a_{12} & \dots & a_{1k} \\ 0 & E_{n_2} & \dots & a_{2k} \\ \cdot & \cdot & \cdot & \cdot \\ 0 & 0 & \dots & E_{n_k} \end{bmatrix} \tag{4}$$

where $n_1 + \cdots + n_k = n$ and the a_{ij} are arbitrary $n_i \times n_j$ matrices over Δ.

THEOREM 2. *The length of the derived series of the group* $T(n_1, \dots, n_k, \Delta)$ *is* $\mu(k) = [\log_2(k-1)] + 1$.

PROOF. We set $H_k = T(n_1, \dots, n_k, \Delta)$, and prove first by induction on μ that

$$H_k^{(\mu)} = (E_n), \tag{5}$$

where $\mu = \mu(k)$. For $\mu = 1$ and 2 this is obvious. Let $\mu > 2$ and suppose (5) holds for $\mu-1$. Now k satisfies the inequalities

$$2^{\mu-1}+1 \leqslant k < 2^{\mu}+1.$$

We set $l = 2^{\mu-1}$. Then $2^{\mu-2}+1 < l < 2^{\mu-1}+1$ and, by the inductive hypothesis, $H_l^{(\mu-1)} = (E_n)$. Hence the matrices of $H_k^{(\mu-1)}$ have the form

$$b = \begin{bmatrix} E_{n_1} & 0 & \dots & 0 & b_{1,l+1} & \dots\dots\dots\dots & b_{1k} \\ 0 & E_{n_2} & 0 & \dots & 0 & b_{2,l+2} \dots\dots & b_{2k} \\ \cdot & \cdot & \cdot & \cdot & \cdot & \cdot & \cdot \\ 0 & \dots\dots & 0 & E_{n_{k-l}} & 0 & \dots\ 0 & b_{k-l,k} \\ 0 & \dots\dots & & 0 & E_{n_{k-l+1}} & 0\ \dots & 0 \\ \cdot & \cdot & \cdot & \cdot & \cdot & \cdot & \cdot \\ 0 & \dots\dots & & & \dots & 0 & E_{n_k} \end{bmatrix}. \qquad (6)$$

By the choice of l we have $2l \geqslant k$, i.e. $k-l \leqslant l$. It follows from the last inequality that

$$\begin{aligned} &n_1+n_2+\dots+n_{k-l}+n_{l+1}+\dots+n_k \\ &\leqslant n_1+n_2+\dots+n_{k-l}+n_{k-l+1}+n_{k-l+2}+\dots+n_k = n. \end{aligned}$$

Thus we can write (6) in the form

$$b = \begin{bmatrix} E_s & B_{st} \\ 0 & E_t \end{bmatrix},$$

where $s = n_1 + \dots + n_{k-l}$ and $t = n-s$. Hence and from Lemma 1 it follows that $H_k^{(\mu-1)}$ is an abelian group; that is, $H_k^{(\mu)} = (E_n)$. It remains to show that

$$H_k^{(\mu-1)} \neq (E_n). \qquad (7)$$

We now denote by T_{ij} a matrix of the form (4) for which

$$a_{ij} = \begin{bmatrix} 1 & 0 & \dots & 0 \\ 0 & 0 & \dots & 0 \\ \cdot & \cdot & \cdot & \cdot \\ 0 & 0 & \dots & 0 \end{bmatrix} = \varepsilon_{ij},$$

and the remaining $a_{\lambda\nu}$ all vanish. We set $\alpha = 2^{\mu-1}+1$ and consider the tableau

$$\left.\begin{matrix} T_{12} & T_{13} & T_{15} & \dots & T_{1,2^{\mu-2}+1} & T_{1,\alpha} \\ T_{23} & T_{24} & T_{26} & \dots & T_{2^{\mu-2}+1,\alpha} & \\ \cdot & \cdot & \cdot & & & \\ T_{\alpha-1,\alpha} & T_{\alpha-2,\alpha} & T_{\alpha-4,\alpha} & & & \end{matrix}\right\} \qquad (8)$$

The matrices in the second column of this tableau are commutators of those in the first, the matrices in the third are commutators of those of the second, and so on. For, as is readily seen, $\epsilon_{ij}\epsilon_{jl} = \epsilon_{il}$. If we now set $T_{ij} - E_n = \hat{E}_{ij}$, then

$$\hat{E}_{ij}\hat{E}_{jl} = \hat{E}_{il},$$

and $\hat{E}_{ij}\hat{E}_{\lambda l} = 0$ for $\lambda \neq j$. Hence

$$\left.\begin{aligned} T_{ik}T_{kj} &= E_n + \hat{E}_{ik} + \hat{E}_{kj} + \hat{E}_{ij}, \\ T_{ik}^{-1}T_{kj}^{-1} &= E_n - \hat{E}_{ik} - \hat{E}_{kj} + \hat{E}_{ij}, \end{aligned}\right\} \tag{9}$$

where i, j, k are pairwise distinct indices.

Multiplying the equalities of (9), we find that

$$T_{ik}T_{kj}T_{ik}^{-1}T_{kj}^{-1} = E_n + \hat{E}_{ij} = T_{ij}. \tag{10}$$

The required property of (8) follows from (10). Thus

$$T_{13} \in H_k^{(1)}, \quad T_{15} \in H_k^{(2)}, \ \dots, \ T_{1,\,2^\beta+1} \in H_k^{(\beta)},$$
$$T_{1,\,2^{\mu-1}+1} \in H_k^{(\mu-1)}.$$

This proves (7). ∎

2. *Reduction to primitive groups.* Now let Δ be a field and G an arbitrary subgroup of $GL(n, \Delta)$. In a suitable basis of Δ^n, the matrices g of G take the form

$$g = \begin{bmatrix} a_1(g) & a_{12}(g) & \dots & a_{1k}(g) \\ 0 & a_2(g) & \dots & a_{2k}(g) \\ \cdot & \cdot & \cdot & \cdot \\ 0 & 0 & \dots & a_k(g) \end{bmatrix}, \tag{11}$$

where the mapping $\gamma_i : G \longrightarrow GL(n_i, \Delta)$, $g \longmapsto a_i(g)$ is an irreducible representation of G of degree n_i, and $a_{ij}(g)$ is an $n_i \times n_j$ matrix over Δ, $n_1 + \cdots + n_k = n$ (see §14).

THEOREM 3. *A group G of matrices* (11) *is solvable if and only if all the groups*

$$G_i = \operatorname{Im} \gamma_i, \quad i = 1, \dots, k, \tag{12}$$

are solvable.

PROOF. NECESSITY. If G is solvable, so are all its homomorphic images. Thus all the groups (12) are solvable.

SUFFICIENCY. Suppose that all the groups (12) are solvable. Consider the homomorphism

$$\begin{gathered} f\colon G \to GL(n, \Delta) \\ f(g) = \operatorname{diag}[a_1(g), \dots, a_k(g)], \quad g \in G. \end{gathered} \tag{13}$$

It is clear that the group $\operatorname{Im} f$ is solvable. On the other hand, the matrices of $\operatorname{Ker} f$ are all of the form (4). Hence

$$\operatorname{Ker} f \subset T(n_1, \ldots, n_k, \Delta).$$

This implies that $\operatorname{Ker} f$ is solvable, and hence so is G. ∎

If a group G of matrices of the form (11) is a maximal solvable subgroup of $GL(n, \Delta)$, then the groups (12) are maximal solvable subgroups of $GL(n_i, \Delta)$, $\operatorname{Im} f$ is isomorphic to the direct product of the groups (12), and the elements $a_{ij}(g)$ are arbitrary $n_i \times n_j$ matrices over Δ. Thus a maximal solvable subgroup of $GL(n, \Delta)$ is determined by a sequence of groups $G_1, \ldots, G_k$, where G_i is a maximal solvable irreducible subgroup of $GL(n_i, \Delta)$, and $n_1 + \cdots + n_k = n$.

THEOREM 4. *Let $\mathfrak{G}$ be a reducible maximal solvable subgroup of $GL(n, \Delta)$. Then $\mathfrak{G}$ is conjugate in $GL(n, \Delta)$ to a semidirect product*

$$HT(n_1, \ldots, n_k, \Delta) \tag{14}$$

of groups H and $T(n_1, \ldots, n_k, \Delta)$, where H is the group of all matrices of the form $h = \operatorname{diag}[h_1, \ldots, h_k]$, where $h_j \in G_j$, G_j is a maximal solvable irreducible subgroup of $GL(n_j, \Delta)$, $j = 1, \ldots, k$, and H is the direct product of $G_1, \ldots, G_k$.

PROOF. By Theorem 3, the group (14) is solvable. $\mathfrak{G}$ is conjugate in $GL(n, \Delta)$ to a group G of matrices of the form (11). It is clear that any element $g \in G$ is expressible as

$$g = f(g)u, \quad u \in T(n_1, \ldots, n_k, \Delta), \quad f(g) \in H.$$

Therefore $G \subset HT(n_1, \ldots, n_k, \Delta)$. Since G is a maximal solvable subgroup of $GL(n, \Delta)$,

$$G = HT(n_1, \ldots, n_k, \Delta). \blacksquare$$

THEOREM 5. *Let G be an irreducible solvable subgroup of $GL(n, \Delta)$. Then G is either primitive or conjugate in $GL(n, \Delta)$ to a subgroup of the wreath product*

$$\bar{G} \sim \Gamma, \tag{15}$$

where $\bar{G}$ is a primitive solvable subgroup of $GL(m, \Delta)$, Γ a transitive solvable subgroup of the symmetric group S_k, and $km = n$. In particular, a maximal solvable irreducible subgroup of $GL(n, \Delta)$ is either primitive or conjugate in $GL(n, \Delta)$ to a group (15), *where $\bar{G}$ is a maximal primitive solvable subgroup of $GL(m, \Delta)$ and Γ a maximal transitive solvable subgroup of S_k.*

This follows from Theorem 15.4.

§19. Primitive solvable groups. Boundedness of the length of the derived series of a solvable linear group

1. *Abelian normal subgroups of primitive solvable groups.* Let Δ be an arbitrary field, G a primitive solvable subgroup of $GL(n, \Delta)$, and H an abelian normal subgroup of G.

LEMMA 1. *$GL(n, \Delta)$ contains a solvable subgroup $\mathfrak{C}$ having a maximal* *abelian normal subgroup F such that $\mathfrak{G} \supset G$, $F \supset H$, F is the multiplicative group of an extension field K of ΔE_n, $\Delta E_n \subset K \subset \Delta_n$, and $K:\Delta$ divides n.*

PROOF. By Clifford's theorems (§16), H is completely reducible and its irreducible components are pairwise equivalent. If m_1 is the degree of an irreducible component of H, then m_1 divides n, and the linear Δ-hull $[H] = \Sigma_1$ is a field of degree m_1 over Δ.

We set $H_1 = \Sigma_1^*$. Then $gH_1g^{-1} = H_1$ for $g \in G$. Consequently $G_1 = H_1G$ is a solvable group and H_1 an abelian normal subgroup of G_1. If H_1 is a maximal abelian normal subgroup of G_1, we are done, because we can take $\mathfrak{G}$ and F to be G_1 and H_1, respectively. But if H_1 is contained in another abelian normal subgroup H_1' of G_1, then we consider the solvable group $G_2 = H_2G_1$, where

$$H_2 = \Sigma_2^*, \quad \Sigma_2 = [H_1'], \quad \Sigma_2 \supset \Sigma_1, \quad \Sigma_2 \neq \Sigma_1.$$

If H_2 is a maximal abelian subgroup of G_2, the proof is complete.. But if $H_2 \subset H_2'$, where H_2' is another abelian normal subgroup of G_2, we consider the group

$$G_3 = H_3G_2, \quad H_3 = \Sigma_3^*, \quad \Sigma_3 = [H_2'], \quad \Sigma_3 \supset \Sigma_2, \quad \Sigma_3 \neq \Sigma_2$$

and so on. Since the degrees of the fields $\Sigma_1, \Sigma_2, \Sigma_3, \ldots$ are divisors of n and the inclusions $\Sigma_1 \subset \Sigma_2 \subset \Sigma_3 \subset \cdots$ are strict, the procedure must end, and for some j the group $H_j = \Sigma_j^*$ is a maximal abelian normal subgroup of $G_j = H_jG_{j-1}$. Setting $\mathfrak{G} = G_j$ and $F = H_j$, we obtain what is required. ■

Now let G be a primitive solvable subgroup of $GL(n, \Delta)$ containing a maximal abelian normal subgroup F for which

$$F = K^*, \quad \Delta E_n \subset K \subset \Delta_n, \tag{1}$$

K is an extension of the field ΔE_n, $K:\Delta = m$ and $m \mid n$. Let V be the centralizer of F in G and $\mathcal{G}(K/\Delta)$ the group of all automorphisms of K leaving each element of ΔE_m fixed. We then have

THEOREM 1. *The factor group G/V is isomorphic to a solvable subgroup of $\mathcal{G}(K/\Delta)$.*

PROOF. We consider the mapping

$$f\colon G \to \mathcal{G}(K/\Delta), \quad g \mapsto \sigma_g, \tag{2}$$

where $\sigma_g : K \to K$, $\sigma_g(x) = gxg^{-1}$, $x \in K$. Since $F = K^*$, it follows that $\operatorname{Ker} f = V$. Hence

$$G/V \cong \operatorname{Im} f \subset \mathcal{G}(K/\Delta). \ ■$$

COROLLARY 1. $G:V \leqslant K:\Delta = m$.

In discussing the group V, it is convenient to treat the space Δ^n on which it acts as a space of dimension $r = n/m$ over K. Since every element v of V commutes with every element λ of K, we have $V \subset GL(r, K)$. It is clear that $GL(r, K)$ is the centralizer of F in $GL(n, \Delta)$. In particular, if $K:\Delta = n$, then $F = K^* = GL(1, K) = V$. Together with Theorem 1, this implies

COROLLARY 2. *If* $K:\Delta = n$, *then* $V = F$ *and* G/F *is isomorphic to a solvable subgroup of* $\mathcal{G}(K/\Delta)$, $G:F \leqslant n$.

2. *Abelian normal subgroups of the factor group* G/F. We retain the notation of the preceding subsection, and assume that

$$V \neq F. \tag{3}$$

Then the factor group G/F has a nontrivial abelian normal subgroup A/F contained in V/F (for example, the last term but one of the derived series of V/F).

Let A/F be a nontrivial abelian normal subgroup of G/F contained in V/F.

LEMMA 2. *The center of* A *is* F.

PROOF. $F \subset A \subset V$. Hence F lies in the center $Z_1(A)$ of A. On the other hand, $Z_1(A)$ is an abelian normal subgroup of G. Since F is maximal, $Z_1(A) = F$. ∎

LEMMA 3. (i) *The order of each element of* A/F *is finite and divides* $r = n/m$.

(ii) *Let* ν *be the order of an element* aF *of* A/F. *Then the conjugacy class of* a *in* A *is* $a, \gamma a, \gamma^2 a, \ldots, \gamma^{\nu-1}a$, *where* γ *is an element of order* ν *in* K^*. *There is an element* b *in* A *such that the order of the commutator* $(a, b) = aba^{-1}b^{-1}$ *is* ν.

(iii) *The derived group* A' *of* A *is a finite cyclic group whose order is the same as the exponent of* A/F.

PROOF. For any a and b in A,

$$ab = \lambda ba, \qquad \lambda \in K^*, \qquad a^t b = \lambda^t b a^t.$$

Now, $ab \in GL(r, K)$ and $\det(ab) = \det(\lambda ba) = \lambda^r \det(ab)$. Hence

$$\lambda^r = 1, \qquad a^r b = ba^r. \tag{4}$$

By Lemma 2 we now derive $a^r \in F$ and $(aF)^r = F$. Thus the order ν of aF divides r, and (i) is proved.

Now let $c, d \in A$; then, obviously,

$$(a, c)(a, d) = (a, cd), \qquad (a, c)^r = 1, \qquad (a, c) \in K^*.$$

Thus the set of all commutators (a, c), where c ranges over A and a is fixed, is a finite cyclic group (γ), of order τ, say. Then $(a^\tau, c) = (a, c)^\tau = 1$ for any $c \in A$. Therefore

$$a^\tau \in F, \qquad (aF)^\tau = F, \qquad \nu \mid \tau.$$

On the other hand, $(a, c)^\nu = (a^\nu, c) = 1$, so that $\tau \mid \nu$, and so $\nu = \tau$; that is, the order of γ is ν. This gives (ii). For, $cac^{-1} = (c, a)a = \gamma^\tau a$. On the other hand, $(a, b) = \gamma$ for some b.

It remains to prove (iii). By (4), A' is a finite cyclic subgroup of K^*. Now let ν_1 be the exponent of A/F. Then there is an element a_1 in A such that the order of a_1F in A/F is ν_1. For any $a \in A$, the order ν of $aF \in A/F$ divides ν_1. By (ii), for any $c \in C$ the order of $(a, c) = \gamma^\tau$ divides ν. Consequently $(u, v)^\nu = (u, v)^{\nu_1} = 1$ for any pair u, v in A. On the other hand, A contains an element b_1 such that the order of $\gamma_1 = (a_1, b_1)$ is ν_1. Thus $\gamma_1 \in A'$, $(u, v) \in (\gamma_1)$, and $A' = (\gamma_1)$; that is, A' is of order ν_1. ∎

We now consider the linear K-hull $[A]$ of A.

LEMMA 4. $A : F = [A] : K$.

PROOF. If $c_1, \dots, c_k$ are elements of A, linearly independent over K, they obviously belong to different cosets of F in A. Conversely, let

$$d_1, \dots, d_\mu \tag{5}$$

be representatives of μ different cosets of F in A. We claim that they are linearly independent over K. For any element a of A,

$$ad_ja^{-1} = f_jd_j, \qquad f_j \in F, \tag{6}$$

and for $i \neq j$ there exists an element $a \in A$ such that $f_i \neq f_j$. For suppose that $f_i = f_j$ for every $a \in A$; that is,

$$ad_ia^{-1}d_i^{-1} = ad_ja^{-1}d_j^{-1}, \qquad (a, d_j^{-1}d_i) = 1.$$

It follows that $d_j^{-1}d_i \in F$, and so $d_i \in d_jF$, because F is the center of A. But this contradicts the choice of the elements (5). Now suppose that the system (5) is linearly dependent over K, and that

$$\lambda_1d_1 + \dots + \lambda_\mu d_\mu = 0, \qquad \lambda_i \in K, \tag{7}$$

is a nontrivial dependence relation with the smallest possible number of nonzero coefficients. We may assume that $\lambda_1, \lambda_2 \neq 0$. Let a be an element of A such that

$$ad_1a^{-1}d_1^{-1} = f_1 \neq f_2 = ad_2a^{-1}d_2^{-1}.$$

Then, by (7),

$$f_1(\lambda_1d_1 + \dots + \lambda_\mu d_\mu) - a(\lambda_1d_1 + \dots + \lambda_\mu d_\mu)a^{-1} = 0,$$
$$\lambda_2(f_1 - f_2)d_2 + \dots + \lambda_\mu(f_1 - f_\mu)d_\mu = 0, \qquad \lambda_2(f_1 - f_2) \neq 0.$$

This contradicts the minimality of (7). It is now easy to prove the lemma. Let

$$u_1, \dots, u_k \tag{8}$$

be a maximal system of elements of A linearly independent over K. Then (8) is a K-basis of $[A]$, and all the elements of (8) are in different cosets of F in A. On the other hand, by the preceding remark, (8) is a transversal to F in A. ∎

Since $A \subset GL(r, K)$ and $[A] \subset K_r$, it follows from Lemma 4 that

$$A : F = [A] : K \leqslant r^2 = \frac{n^2}{m^2}. \tag{9}$$

THEOREM 2. *A contains pairs of elements $a_1, b_1, a_2, b_2, \ldots, a_t, b_t$ with the following properties:*

(i) *$(a_i, b_i) = \epsilon_i$ is an element of F of order ν_i; $\nu_{i+1} \mid \nu_i$.*

(ii) *Elements belonging to distinct pairs commute.*

(iii) *Every element a of A has a unique representation in the form*

$$a = f a_1^{\alpha_1} b_1^{\beta_1} \ldots a_t^{\alpha_t} b_t^{\beta_t}, \; f \in F, \quad 0 \leqslant \alpha_i < \nu_i, \quad 0 \leqslant \beta_i < \nu_i.$$

PROOF. Let ν_1 be the maximum order of elements of A/F. Then, by Lemma 3, there is a pair of elements a_1, b_1 in A such that $(a_1, b_1) = \epsilon_1$ is an element of order ν_1. We can express A in the form

$$A = (a_1)(b_1) A_1, \tag{10}$$

where A_1 is the centralizer of the set $\{a_1, b_1\}$ in A. For the number of elements conjugate to a_1 in A is ν_1. Consequently $A : A_1 \leqslant \nu_1^2$. On the other hand, the ν_1^2 products $a_1^{\alpha_1} b_1^{\beta_1}$, $0 \leqslant \alpha_1 < \nu_1$, $0 \leqslant \beta_1 < \nu_1$, belong to distinct cosets of A_1 in A, since

$$(a_1^{\alpha_1} b_1^{\beta_1}, b_1) = \varepsilon_1^{\alpha_1}, \qquad (a_1, a_1^{\alpha_1} b_1^{\beta_1}) = \varepsilon_1^{\beta_1}.$$

Hence $A : A_1 = \nu_1^2$, and (10) holds.

If A_1 is abelian, then $A_1 = F$, $A = (a_1)(b_1)F$, and the theorem is proved. Let $A_1 \neq F$ and let ν_2 be the maximum order of elements of A_1/F. By the choice of ν_1, we know that $\nu_2 \mid \nu_1$. By Lemma 3 and (10) there is a pair of elements a_2, b_2 in A_1 such that $(a_2, b_2) = \epsilon_2$ is an element of order ν_2 in F. Then

$$A_1 = (a_2)(b_2) A_2,$$

where A_2 is the centralizer of the set $\{a_1, b_1, a_2, b_2\}$ in A. Hence

$$A = (a_1)(b_1)(a_2)(b_2) A_2,$$

and the elements $a_1^{\alpha_1} b_1^{\beta_1} a_2^{\alpha_2} b_2^{\beta_2}$, $0 \leqslant \alpha_i < \nu_i$, $0 \leqslant \beta_i < \nu_i$, $i = 1, 2$, belong to distinct cosets of A_2 in A. Since A/F is a finite group, after finitely many steps we arrive at the equality

$$A = (a_1)(b_1) \ldots (a_t)(b_t) F,$$

where the elements $a_1^{\alpha_1} b_1^{\beta_1} \cdots a_t^{\alpha_t} b_t^{\beta_t}$, $0 \leqslant \alpha_i < \nu_i$, $0 \leqslant \beta_i < \nu_i$, $i = 1, \ldots, t$, lie in distinct cosets of F in A. ∎

COROLLARY 1. $A : F = \nu_1^2 \cdots \nu_t^2$.

3. *Finiteness of the index of a maximal abelian normal subgroup.* So far we have assumed that A/F is an arbitrary nontrivial abelian normal subgroup of G/F contained in V/F. Now let A/F be a subgroup of G/F satisfying the following three conditions:

1) A/F is an abelian normal subgroup of G/F.

2) $A/F \subset V/F$.

3) A/F is maximal among the subgroups of G/F satisfying 1) and 2).

THEOREM 3. *A/F is its own centralizer in V/F.*

PROOF. Let B/F be the centralizer of A/F in V/F, and C the centralizer of A in V. Since A/F is normal in G/F, it follows that B is normal in G. Clearly $B \supset C \supset F$. We first show that

$$B = AC. \tag{11}$$

It is obvious that $(a, b) \in F$ for $a \in A$ and $b \in B$. In particular, $(a_j, b) \in F$ and $(b, b_j) \in F$, where a_j and b_j are elements of A having the properties (i), (ii) and (iii) of Theorem 2. Now, since $a_j^{\nu_j}, b_j^{\nu_j} \in F$,

$$(a_j^{\nu_j}, b) = (a_j, b)^{\nu_j} = 1, \quad (b, b_j^{\nu_j}) = (b, b_j)^{\nu_j} = 1.$$

Thus $(a_j, b) = \epsilon_j^{\beta_j}$ and $(b, b_j) = \epsilon_l^{\alpha_j}$, where $j = 1, \dots, t$, $0 \leqslant \beta_j < \nu_j$ and $0 \leqslant \alpha_j < \nu_j$. On the other hand, the element $a = a_1^{\alpha_1} b_1^{\beta_1} \cdots a_t^{\alpha_t} b_t^{\beta_t}$ satisfies the conditions $(a_j, a) = \epsilon_j^{\beta_j}$ and $(a, b_j) = \epsilon_j^{\alpha_j}$, $j = 1, \dots, t$. Hence $(a_j, a^{-1}b) = (a^{-1}b, b_j) = 1$, where $j = 1, \dots, t$; therefore $a^{-1}b \in C$, and so $b \in aC$, whence $B \subset AC$, which implies (11).

Now suppose that $B/F \neq A/F$. Then the next-to-last term D/A of the derived series of B/A is nontrivial, D/A is an abelian normal subgroup of G/A, but D/F cannot be abelian because of the maximality of A/F. Thus D contains elements d and d_1 such that

$$(d, d_1) \notin F. \tag{12}$$

By (11), $d = ac$ and $d_1 = a'c'$, where $a, a' \in A$ and $c, c' \in C$. Consequently, $(d, d_1) = (a, a')(c, c') \in C$. Since $(d, d_1) \in A$, we conclude that

$$(d, d_1) \in A \cap C = F,$$

and this contradicts (12). ∎

COROLLARY 1. *The centralizer of A in V is F.*

COROLLARY 2. *The centralizer of A in G is F.*

THEOREM 4.

$$G : F \leqslant \rho(n) = n^2(n^2 - 1)(n^2 - 2) \dots (n^2 - 2^{l-1}), \tag{13}$$

where $l = [2 \log_2 n]$.

PROOF. There are two possibilities: $V = F$ or $V \neq F$. In the first case it follows from Theorem 1 that $G : F \leqslant m \leqslant n < \rho(n)$. In the second case,

$$G : F = (G : V)(V : A)(A : F). \tag{14}$$

Let us estimate the index $V : A = (V : F) : (A : F)$. By Theorem 3, A/F is its own centralizer in V/F. Consequently the factor group $(V/F)/(A/F)$ is isomorphic to a solvable subgroup of $\mathrm{Aut}(A/F)$. Since $\mathrm{card}(A/F) \leqslant r^2 = n^2/m^2$, we have

$$\mathrm{card}\ \mathrm{Aut}(A/F) \leqslant (r^2 - 1)(r^2 - 2) \ldots (r^2 - 2^{k-1}),$$

where $k = [2\log_2 r]$. Hence, by (13),

$$G : F \leqslant m(r^2 - 1)(r^2 - 2) \ldots (r^2 - 2^{k-1}) \frac{n^2}{m^2}$$
$$= \frac{n^2}{m}(r^2 - 1) \ldots (r^2 - 2^{k-1}) \leqslant \rho(n). \blacksquare$$

THEOREM 5. *Every primitive solvable subgroup* $\mathfrak{G}$ *of* $GL(n, \Delta)$ *has an abelian normal subgroup whose index does not exceed* $\rho(n)$ *(see (13)). The index of a maximal abelian normal subgroup of* $\mathfrak{G}$ *does not exceed* $\rho(n)$.

PROOF.* Let $\mathfrak{H}$ be a maximal abelian normal subgroup of $\mathfrak{G}$. By Lemma 1, $\mathfrak{G}$ is contained in a solvable group G having a maximal abelian normal subgroup F that is the multiplicative group of a field K, with $\mathfrak{H} \subset F$. Since

$$\mathfrak{G}F : F \leqslant G : F \leqslant \rho(n), \qquad \mathfrak{G}F : F = \mathfrak{G} : (\mathfrak{G} \cap F),$$

it follows that $\mathfrak{G} : (\mathfrak{G} \cap F) \leqslant \rho(n)$. Obviously $\mathfrak{G} \cap F$ is an abelian normal subgroup of $\mathfrak{G}$. Since $\mathfrak{H} \subset F$, we have $\mathfrak{G} \cap F \supset \mathfrak{H}$. The maximality of $\mathfrak{H}$ then implies $\mathfrak{G} \cap F = \mathfrak{H}$, and so $\mathfrak{G} : \mathfrak{H} \leqslant \rho(n)$. ■

THEOREM 6. *An irreducible solvable subgroup of* $GL(n, \Delta)$ *has an abelian normal subgroup whose index does not exceed a number* $\rho_1(n)$ *depending only on* n.

PROOF. By Theorem 5, we need consider only imprimitive irreducible solvable subgroups of $GL(n, \Delta)$. Let $\mathfrak{G}$ be such a subgroup. Then, by Theorem 18.5, we may assume that $\mathfrak{G}$ is a subgroup of $\bar{G} \sim \Gamma$, where $\bar{G}$ is a primitive solvable subgroup of $GL(m_1, \Delta)$ and Γ a transitive solvable subgroup of S_{k_1}; $k_1 m_1 = n$. By Theorem 5, $\bar{G}$ has an abelian normal subgroup of index at most $\rho(m_1)$. Thus the direct product of k_1 copies of $\bar{G}$ has an abelian normal subgroup of index at most $(\rho(m_1))^{k_1}$. Hence $\bar{G} \sim \Gamma$ has an abelian normal subgroup of index at most

$$k_1!\,(\rho(m_1))^{k_1} < n!\,\rho(n)^n = \rho_1(n).$$

Since $\mathfrak{G}$ is a subgroup of $G \sim \Gamma$, $\mathfrak{G}$ also has an abelian normal subgroup of index at most $\rho_1(n)$.

THEOREM 7 (MAL'CEV [3], KOLCHIN [1]). *Let* Σ *be an algebraically closed field. Then a solvable subgroup of* $GL(n, \Sigma)$ *has a normal subgroup* N *whose index does not exceed a certain number* $\rho_2(n)$ *depending only on* n. *The matrices of* N *can be reduced simultaneously to triangular form.*

**Editor's note.* Here the roles of G and $\mathfrak{G}$ are reversed in comparison with Lemma 1.

PROOF. By Theorem 6, the fact that Σ is algebraically closed, and Theorem 16.2, an irreducible solvable subgroup of $GL(n, \Sigma)$ has a normal subgroup of index at most $\rho_1(n)$, whose matrices can be simultaneously diagonalized. This implies Theorem 7.

4. *Boundedness of the length of the derived series.*

THEOREM 8 (ZASSENHAUS [1]). *Let Δ be an arbitrary field. Then the length of the derived series of any solvable subgroup $\mathfrak{G}$ of $GL(n, \Delta)$ does not exceed a certain number $l(n)$, depending only on n.*

PROOF. Let $\lambda(\mathfrak{G})$ be the length of the derived series of $\mathfrak{G}$. If $\mathfrak{G}$ is primitive, by Theorem 5 it contains an abelian normal subgroup $\mathfrak{H}$ such that $\mathfrak{G} : \mathfrak{H} \leqslant \rho(n)$. Consequently

$$\lambda(\mathfrak{G}) \leqslant [\log_2 \rho(n)] + 1 = l_1(n). \tag{15}$$

Now let $\mathfrak{G}$ be an irreducible imprimitive subgroup of $GL(n, \Delta)$. Then, by Theorem 18.5, $\mathfrak{G}$ is a subgroup of $\bar{G} \sim \Gamma$, where $\bar{G}$ is a primitive solvable subgroup of $GL(nk_1^{-1}, \Delta)$ and Γ a transitive solvable subgroup of the symmetric group S_{k_1}. Consequently

$$\lambda(\mathfrak{G}) \leqslant \lambda(\bar{G} \sim \Gamma) \leqslant \lambda(\bar{G}) + \lambda(\Gamma) \leqslant [\log_2 k_1!] + \left[\log_2 \rho\left(\frac{n}{k_1}\right)\right] + 1.$$

Hence

$$\lambda(\mathfrak{G}) \leqslant [\log_2 n!] + [\log_2 \rho(n)] + 1 = l_2(n). \tag{16}$$

Finally, let $\mathfrak{G}$ be a reducible solvable subgroup of $GL(n, \Delta)$. Then the matrices of $\mathfrak{G}$ can be reduced simultaneously to the form (11) of §18. By Theorem 18.2 we can write

$$\lambda(\mathfrak{G}) \leqslant [\log_2 (k-1)] + 1 + \lambda(\operatorname{Im} \gamma_j),$$

where γ_j is an irreducible component of $\mathfrak{G}$, of degree $n_j < n$. By (16),

$$\lambda(\operatorname{Im} \gamma_j) \leqslant l_2(n_j) \leqslant l_2(n).$$

Consequently

$$\begin{aligned} \lambda(\mathfrak{G}) &\leqslant [\log_2 (k-1)] + 1 + l_2(n) \leqslant [\log_2 (n-1)] + 1 + l_2(n) \\ &= [\log_2 (n-1)] + [\log_2 n!] + [\log_2 \rho(n)] + 2 = l(n). \end{aligned} \tag{17}$$

Now (15)–(17) imply that $\lambda(\mathfrak{G}) \leqslant l(n)$ for any solvable subgroup $\mathfrak{G}$ of $GL(n, \Delta)$. ■

Note that (17) provides a very coarse estimate of $\lambda(\mathfrak{G})$ for a solvable subgroup $\mathfrak{G}$ of $GL(n, \Delta)$. Dixon [2] has proved that

$$\lambda(\mathfrak{G}) \leqslant n + 2. \tag{18}$$

A group G is said to be *locally solvable* if any finitely generated subgroup of G is solvable.

THEOREM 9 (ZASSENHAUS [1]). *Let Δ be an arbitrary field. Then any locally solvable subgroup $\mathfrak{G}$ of $GL(n, \Delta)$ is solvable.*

PROOF. Given any group G, we define inductively commutators D^j by

$$D^1(g_1, g_2) = g_1 g_2 g_1^{-1} g_2^{-1},$$

$$\begin{aligned} D^j &= D^j(g_1, \ldots, g_{2^j}) \\ &= D^1\left(D^{j-1}(g_1, \ldots, g_{2^{j-1}}), D^{j-1}(g_{2^{j-1}+1}, \ldots, g_{2^j})\right), \end{aligned}$$

$j > 1, g_i \in G$. Let k be a natural number. We claim that *G is solvable and $\lambda(G) \leqslant k$ if and only if*

$$D^k = 1 \tag{19}$$

for any 2^k elements of G. For if $\lambda(G) \leqslant k$, (19) is obviously true. Suppose that (19) holds for any 2^k elements of G. If $k = 1$, then G is abelian and $\lambda(G) = 1$. Let $k > 1$ and suppose the assertion is proved for $k - 1$. Consider the subgroup N of G generated by all the D^{k-1}. Obviously N is an abelian normal subgroup of G. For any 2^{k-1} elements of G/N we have $D^{k-1} = 1$. Consequently $\lambda(G/N) \leqslant k - 1$, and so $\lambda(G) \leqslant k$.

It is now easy to prove the theorem.

By assumption, any finitely generated subgroup $\mathfrak{H}$ of $\mathfrak{G}$ is solvable. By Theorem 8, $\lambda(\mathfrak{H}) \leqslant l(n) = l$. Consequently $D^l = 1$ for any 2^l elements of $\mathfrak{G}$. Hence $\lambda(\mathfrak{G}) \leqslant l$. ■

THEOREM 10. *Any solvable subgroup $\mathfrak{G}$ of $GL(n, \Delta)$ is contained in a maximal solvable subgroup of $GL(n, \Delta)$.*

PROOF. $\mathfrak{G}$ is contained in a maximal locally solvable subgroup $\mathfrak{G}_1$ of $GL(n, \Delta)$. By Theorem 9, $\mathfrak{G}_1$ is solvable. ■

The maximal solvable normal subgroup of a group G, if it exists, is called the *solvable radical* of G.

Since the product of two solvable normal subgroups is a solvable normal subgroup, a group has either no radical at all or a unique one.

THEOREM 11. *A subgroup $\mathfrak{G}$ of $GL(n, \Delta)$ has a solvable radical.*

PROOF. Let $\mathfrak{M}$ be the class of all solvable normal subgroups of $\mathfrak{G}$. The set $\mathfrak{N}$ of all products of the form $a_1 \ldots a_r$, $a_j \in H_j \in \mathfrak{M}$, is clearly a normal subgroup of $\mathfrak{G}$. Every finite set of elements of $\mathfrak{N}$ is contained in the product of finitely many groups $H \in \mathfrak{M}$. Thus $\mathfrak{N}$ is a locally solvable subgroup of $GL(n, \Delta)$. By Theorem 9, $\mathfrak{N}$ is solvable. Obviously, if $H \in \mathfrak{M}$, then $H \subset \mathfrak{N}$. Hence $\mathfrak{N}$ is the solvable radical of $\mathfrak{G}$. ■

§20. Maximal primitive solvable subgroups of the general linear group

1. *The factor group G/V. Irreducibility of V.* Let Δ be an arbitrary field, G a maximal primitive solvable subgroup of $GL(n, \Delta)$, F a maximal abelian

normal subgroup of G, and V the centralizer of F in G. Then, by Lemma 19.1,

$$F = K^*, \quad \Delta E_n \subset K \subset \Delta_n, \tag{1}$$

where K is an extension of Δ, $K:\Delta = m$, and $m \mid n$. There is another field associated with G. By Theorem 14.7, there is an extension Σ of Δ such that G is a subgroup of $GL(l, \Sigma)$.

$$[G]_\Delta = \Sigma_l, \quad (\Sigma:\Delta)\, l = n, \tag{2}$$

and $\Sigma^* E_l$ is the center of G. Since F is a maximal abelian normal subgroup of G, it follows that $\Sigma^* E_l \subset F$ and $\Sigma E_l \subset K$. Hence ΣE_l is simply the set of all elements of K that commute with every element of G. Thus the homomorphism f defined by (2) in §19 satisfies the relation

$$\operatorname{Im} f \subset \mathcal{G}(K/\Sigma) \subset \mathcal{G}(K/\Delta), \tag{3}$$

where $\mathcal{G}(K/\Sigma)$ is the group of all automorphisms of K leaving every element of ΣE_l fixed. Since $\operatorname{Ker} f = V$, we have

$$G/V \cong \operatorname{Im} f \subset \mathcal{G}(K/\Sigma), \quad G:V \leqslant K:\Sigma. \tag{4}$$

LEMMA 1. $[G]_\Delta : \Delta = (G:V)([V]_\Delta : \Delta)$.

PROOF. Let

$$g_1, \ldots, g_\tau \tag{5}$$

be a transversal to V in G. Then the elements (5) are linearly independent over $[V] = [V]_\Delta$. For let

$$g_1 v_1 + g_2 v_2 + \ldots + g_\tau v_\tau = 0, \quad v_j \in [V], \tag{6}$$

be a nontrivial dependence relation with the smallest possible number of nonzero terms. We may assume that $v_1 \neq 0$ and $v_2 \neq 0$. Since g_1 and g_2 are in two distinct cosets of V in G and since V is the centralizer of F in G, there is an element λ in F such that $\lambda_1 \neq \lambda_2$, where $\lambda_j = g_j^{-1} \lambda g_j$, $j = 1, \ldots, \tau$. Evidently $\lambda g_j = g_j \lambda_j$, $j = 1, \ldots, \tau$. Hence and from (6) we obtain

$$\lambda (g_1 v_1 + g_2 v_2 + \ldots + g_\tau v_\tau) - (g_1 v_1 + g_2 v_2 + \ldots + g_\tau v_\tau)\lambda_1 = 0,$$
$$g_2 v_2 (\lambda_2 - \lambda_1) + \ldots + g_\tau v_\tau (\lambda_\tau - \lambda_1) = 0.$$

This contradicts the choice of the relation (6). Thus (5) is linearly independent over $[V]$. Since any element of $[G] = [G]_\Delta$ has the form $g_1 v_1 + \cdots + g_\tau v_\tau$, $v_j \in [V]$, we have

$$[G]:\Delta = (G:V)([V]:\Delta). \ \blacksquare$$

THEOREM 1. (i) $G/V \cong \mathcal{G}(K/\Sigma)$.

(ii) *K is a normal separable extension of Σ with solvable Galois group* $\mathcal{G}(K/\Sigma)$.

(iii) *V is an absolutely irreducible subgroup of $GL(r, K)$.*

PROOF. It follows from (2) that

$$[G]:\Delta = l^2(\Sigma:\Delta) = \frac{n^2}{\Sigma:\Delta}. \tag{7}$$

By (4),

$$G:V \leqslant K:\Sigma, \tag{8}$$

and $G:V = K:\Sigma$ only if (i) and (ii) are true. Since $V \subset GL(r, K)$,

$$[V]:\Delta \leqslant r^2(K:\Delta) = \frac{n^2}{m}. \tag{9}$$

It is obvious that $[V]:\Delta = n^2/m$ only if V is an absolutely irreducible subgroup of $GL(r, K)$. By Lemma 1, (8) and (9),

$$[G]:\Delta = (G:V)([V]:\Delta) \leqslant (K:\Sigma)\frac{n^2}{m} = \frac{(K:\Sigma)\,n^2}{(K:\Sigma)(\Sigma:\Delta)} = \frac{n^2}{\Sigma:\Delta}.$$

By (7), it follows from this and from (8) and (9) that

$$G:V = K:\Sigma, \qquad [V]:\Delta = \frac{n^2}{m}. \tag{10}$$

Now (i), (ii) and (iii) follow from (10). ■

THEOREM 2. $V = F \Longleftrightarrow K:\Delta = n$.

PROOF. By Corollary 19.1.2, $K:\Delta = n \Longleftrightarrow V = F$. Conversely, let $V = F$. Then, replacing V in (10) by F, we obtain

$$K:\Delta = \frac{n^2}{K:\Delta}, \qquad K:\Delta = n. \; ■$$

2. *The factor group A/F.* We now assume that $V \neq F$ and consider the factor group A/F subject to the following conditions:

(i) A/F is an abelian normal subgroup of G/F.

(ii) $A/F \subset V/F$.

(iii) A/F is maximal among the subgroups of G/F satisfying (i) and (ii).

We study G with the help of the invariant series

$$G \supset V \supset A \supset F \supset (E_n). \tag{α}$$

LEMMA 2. *The linear Δ-hull $[A]$ of A is a simple algebra over K. The center of $[A]$ is K.*

PROOF. Obviously, $[A]_\Delta = [A]_K$, because $A \supset F$ and $[A]_\Delta \supset K$. Since A is a normal subgroup of the primitive group G, by Clifford's theorems (§16.2) A is completely reducible, and its irreducible components are equivalent. Consequently, by Theorem 14.4, $[A]_\Delta$ is a simple algebra over Δ, and since $[A]_\Delta = [A]_K$, also over K.

We now show that K is the center of $[A]$. Any nonzero element c of $[A]$ is of the form $c = f_1 d_1 + \cdots + f_\nu d_\nu$, $f_j \in F$, where the d_j are elements of distinct cosets of F in A. If c is in the center of $[A]$, then for any $a \in A$

$$c - aca^{-1} = 0, \quad ad_ja^{-1} = \lambda_j d_j, \quad \lambda_j \in F, \quad j = 1, \ldots, \nu.$$

Therefore

$$f_1(1-\lambda_1)d_1 + \ldots + f_\nu(1-\lambda_\nu)d_\nu = 0.$$

By Lemma 19.4, $d_1, \ldots, d_\nu$ are linearly independent over K. Hence $\lambda_j = 1$, $j = 1, \ldots, \nu$. Since a is an arbitrary element of A, it follows that $\nu = 1, d_1 \in F$, $c = f_1 d_1 \in F$, and K is the center of $[A]$. ∎

LEMMA 3. $[A]_\Delta = [V]_\Delta$.

PROOF. We set $[A] = [A]_\Delta$. With each element $v \in V$ we associate the mapping

$$\varphi_v\colon [A] \to [A], \quad x \to v^{-1}xv. \tag{11}$$

Since $v^{-1}Av = A$, φ_v is an automorphism of $[A]$. It is clear that $\varphi_v(\lambda) = \lambda$ for $\lambda \in K$. By Lemma 2, $[A]$ is a simple algebra over K, with center K. Now any automorphism of a simple algebra that leaves the elements of its center fixed is inner (van der Waerden [2], p. 199). Thus for any $v \in V$ there is an element u in $[A]^*$ such that

$$u^{-1}xu = v^{-1}xv \tag{12}$$

for every $x \in [A]$. Since K is the center of $[A]$, (12) determines u uniquely for fixed v, to within a factor in $F = K^*$. Therefore (12) gives rise to the mapping

$$\psi\colon V \to [A]^*/F, \quad v \mapsto uF.$$

It is readily seen that ψ is a homomorphism. By Corollary 19.3.1 the centralizer of A in V is F, and so Ker $\psi = F$. Now let Im $\psi = U/F$. Then $U/F \cong V/F$. Since U and V are subgroups of $GL(r, K)$, for u and v connected by (12) we have

$$v = du, \tag{13}$$

where $d \in GL(r, K)$ commutes with every element of A. Obviously the element d in (13) is uniquely determined by v to within a factor in F. Therefore (12) and (13) define a mapping

$$\psi_1\colon V/F \to GL(r, K)/F, \quad vF \mapsto dF.$$

This ψ_1 is a homomorphism. For if $vF = dFuF$ and $v_1F = d_1Fu_1F$, then $vv_1F = dd_1Fuu_1F$. Therefore the group D for which $D/F = \operatorname{Im}\psi_1$ is solvable.

Now for $g \in G$ we have

$$g^{-1}Dg = D. \tag{14}$$

For by (13)

$$v_1 = g^{-1}vg = g^{-1}dgg^{-1}ug = d_1u_1, \quad d_1 = g^{-1}dg, \quad u_1 = g^{-1}ug,$$

where $v_1 \in V$ and $u_1 \in g^{-1}[A]g = [A]$. The element d_1 commutes with every element of $[A] = g^{-1}[A]g$. Consequently u_1 and v_1 define the same automorphism of $[A]$; that is, $v_1 = d_1u_1$ as in (13), and $\psi_1(v_1F) = d_1F$, $d_1 \in D$. This implies (14).

We now claim that $D = F$. For GD is solvable by (14). Consequently $D \subset G$. But D is contained in the centralizer of A in G. By Corollary 19.3.2, $D = F$. It is now easy to prove the lemma. By (13), we have $v = du \in F[A] = [A]$, $V \subset [A]$ and $[V] = [A]$. ■

THEOREM 3. *$A : F = n^2/m^2$, and $[A] = K_r$.*

PROOF. By Theorem 1, $[V] = K_r$. Hence, by Lemma 3,

$$[A] : K = [V] : K = K_r : K = r^2 = \frac{n^2}{m^2}.$$

On the other hand, by Lemma 19.4, $A : F = [A] : K$. Thus $A : F = n^2/m^2$. ■

COROLLARY 1. *For every prime divisor q of n/m, there is an element of order q in $K^* = F$. In particular,* $\operatorname{char} \Delta \notin \Pi(n/m)$.

PROOF. The derived group A' of A is obviously a subgroup of $K^* = F$. By Lemma 19.3, the order of A' is equal to the exponent of A/F. Consequently every prime divisor of $A : F$ divides the order of A'. Hence and from Theorem 3 we deduce what is required. ■

COROLLARY 2. *Let n be odd, and R the field of reals. Then every irreducible solvable subgroup of $GL(n, R)$ is monomial.*

PROOF. It suffices to show that $GL(n, R)$ for odd $n > 1$ contains no maximal primitive solvable subgroups. Let $n > 1$ be odd and G a maximal primitive solvable subgroup of $GL(n, R)$. Then $F = K^* = R^*E_n^*$. By Corollary 1 there is an element of order q in R for any prime divisor q of n. Since q is odd, this is impossible. ■

COROLLARY 3. *Let n be odd and Q the field of rationals. If G is a maximal primitive solvable subgroup of $GL(n, Q)$, then $V = F$; that is, $K : Q = n$.*

PROOF. Let $K : Q = m < n$. Then by Corollary 1 there is in K^* an element η of order $q > 2$, where q is a prime divisor of n/m. Consequently $K \supset Q(\eta)$. Since $Q(\eta) : Q = q - 1$, it follows that $K : Q$ is even. But this cannot be, because $K : Q$ divides n and n is odd. ■

COROLLARY 4. *Let n be odd, G a maximal solvable primitive subgroup of $GL(n, Q)$, and $F = K^*$ a maximal abelian normal subgroup of G. Then G is the normalizer of F in $GL(n, Q)$, and the factor group G/F is isomorphic to the group of all Q-automorphisms of K.*

PROOF. By Corollary 3, $K : Q = n$. Consequently $K = Q(\theta)$, where θ is a matrix in $GL(n, Q)$ with an irreducible characteristic polynomial. Since $K^* = F$

is a normal subgroup of G, G is contained in the normalizer $N(F)$ of F in $GL(n, Q)$. Let $\mathcal{G}$ be the group of all Q-automorphisms of K. The set of all elements of the field that are fixed under every automorphism in $\mathcal{G}$ is a field Σ, $K \supset \Sigma \supset QE_n$. As is easily seen, K is a normal extension of Σ and $\mathcal{G}$ is the Galois group $\mathcal{G}(K/\Sigma)$. Since n is odd and the order of $\mathcal{G}$ is equal to $K:\Sigma$ and divides n, it follows that $\mathcal{G}$ is solvable. Every element $a \in N(F)$ defines an automorphism σ_a of $K: \sigma_a(x) = axa^{-1}$, $x \in K$. Obviously $\sigma_a \in \mathcal{G}$ and $\sigma_{ab} = \sigma_a\sigma_b$. Consequently we have the homomorphism $\gamma: N(F) \longrightarrow \mathcal{G}$, $a \longmapsto \sigma_a$. Since $F = K^*$ and $K:Q = n$, it follows that $\operatorname{Ker}\gamma = F$. Since $\mathcal{G}$ is solvable, so is $N(F)$. Hence $G = N(F)$. It remains to prove that $G/F \cong \mathcal{G}$. It suffices to show that γ is an epimorphism. Let σ be any element of $\mathcal{G}$. Clearly σ is completely determined by the matrix $\theta_1 = \sigma(\theta)$. The matrices θ and θ_1 have the same normal form, and therefore $GL(n, Q)$ contains a matrix t such that $t\theta t^{-1} = \theta_1$. Obviously $tKt^{-1} = K$ and $tFt^{-1} = F$. Consequently $t \in N(F)$ and γ is an epimorphism. Hence $G/F = N/F \cong \mathcal{G}$. ∎

THEOREM 4. *A/F is its own centralizer in $GL(r, K)/F$.*

PROOF. Let C/F be the centralizer of A/F in $GL(r, K)/F$ and $c \in G$. Then for $a \in A$ we have $(c, a) \in F$ and $(c, a)^r = 1$. In particular, for the elements a_j and b_j we have, by Theorem 19.2,

$$(c, b_j) = \varepsilon_j^{\alpha_j}, \quad (a_j, c) = \varepsilon_j^{\beta_j}, \quad j = 1, \ldots, t.$$

We set $c_1 = a_1^{\alpha_1}b_1^{\beta_1} \cdots a_t^{\alpha_t}b_t^{\beta_t}$. Then for $j = 1, \ldots, t$

$$(c_1, b_j) = (c, b_j); \quad (a_j, c_1) = (a_j, c);$$
$$(cc_1^{-1}, b_j) = (a_j, cc_1^{-1}) = 1.$$

Hence cc_1^{-1} commutes with every element of A. By Theorem 3, $cc_1^{-1} \in F$, $c \in c_1F \subset A$, and so $C/F = A/F$. ∎

THEOREM 5. *Let*

$$r = q_1^{l_1} \cdots q_k^{l_k} \tag{15}$$

be the canonical decomposition of r, and let Q_j/F be a Sylow q_j-subgroup of A/F, $j = 1, \ldots, k$. Then Q_j/F is the direct product of $2l_j$ groups of order q_j.

PROOF. Obviously Q_j/F is an abelian normal subgroup of G/F of order $g_j^{2l_j}$. We must show that the exponent of Q_j/F is q_j. Suppose that this is false and the exponent of Q_j/F is q_j^α, $\alpha > 1$. Then $\beta = q_j^{\alpha-1} > 1$. We consider the group D/F consisting of all elements $(uF)^\beta$, where uF ranges over Q_j/F. It is obvious that $D \neq F$ and D is normal in G. For any u and v in Q_j

$$(u^\beta, v^\beta) = (u, v)^{\beta^2} = (u, v)^{q_j^{2\alpha_j - 2}} = 1,$$

because the order of the derived group Q_j' of Q_j is q_j^{α}. Hence D is an abelian normal subgroup of G. But this contradicts the maximality of F. ∎

By Theorem 19.2, we infer from Theorem 5:

COROLLARY 1. *Q_j can be represented in the form*

$$Q_j = (u_1)(v_1) \cdots (u_{l_j})(v_{l_j}) F,$$

where

$$(u_i, v_i) = \eta, \quad \eta^{q_j} = 1, \quad \eta \neq 1, \quad \eta \in F; \quad u_i^{q_j}, v_i^{q_j} \in F,$$

and elements belonging to distinct pairs commute.

It is clear that Theorems 3 and 5 completely describe the structure of A/F, as a group of order n^2/m^2 that can be expressed as a direct product of groups of prime order.

Henceforth we regard the groups Q_j, $j = 1, \ldots, k$, of Theorem 5 as subgroups of $GL(r, K)$.

LEMMA 4. (i) *For $\nu \neq j$, the groups Q_ν and Q_j commute elementwise.*

(ii) *The irreducible components of Q_j are pairwise equivalent and absolutely irreducible. The degree of an irreducible component of Q_j is $q_j^{l_j}$.*

PROOF. Let $\nu \neq j$, $a \in Q_\nu$ and $b \in Q_j$. Then $(a, b) = \lambda \in F$, $(a, b)^{q_\nu} = (a^{q_\nu}, b) = \lambda^{q_\nu} = 1$, and $(a, b)^{q_j} = (a, b^{q_j}) = \lambda^{q_j} = 1$. Hence $\lambda = 1$ and $ab = ba$, proving (i). Obviously $A = Q_1 \cdots Q_k$. Hence, by (i), $A = Q_j U_j$, where U_j is a subgroup of A that commutes with Q_j elementwise. As a normal subgroup of the irreducible group A, Q_j is completely reducible. By Lemma 16.4, it follows from $A = Q_j U_j$ that the irreducible components of Q_j are pairwise equivalent. Consequently $[Q_j]_K = T_\rho$, where T is a division ring. Let σ be the degree of an irreducible component of Q_j. Then

$$\rho (T : K) = \sigma, \qquad T_\rho : K = \rho^2 (T : K) = \rho\sigma.$$

On the other hand, by Lemma 19.4

$$[Q_j]_K : K = Q_j : F = q_j^{2l_j}.$$

Hence

$$\rho\sigma = q_j^{2l_j}. \tag{16}$$

Since σ divides r, it follows from (16) that

$$\rho \leqslant \sigma \leqslant q_j^{l_j}. \tag{17}$$

By (16) and (17), $\rho = \sigma = q_j^{l_j}$ and $[Q_j]_K = K_{q_j^{l_j}}$; that is, the degree of each irreducible component of Q_j is $q_j^{l_j}$ and the irreducible components of Q_j are absolutely irreducible, which proves (ii). ∎

THEOREM 6. *There is a K-basis for K^r in which the groups Q_j, $j = 1, \ldots, k$, simultaneously take the form*

$$Q_j = E_{q_1^{l_1} \cdots q_{j-1}^{l_{j-1}}} \dot{\times} Q^j \dot{\times} E_{rq_1^{-l_1} \cdots q_j^{-l_j}},$$

where Q^j is an absolutely irreducible subgroup of $GL(q_j^{l_j}, K)$ isomorphic to Q_j, and $\dot{\times}$ denotes the Kronecker product.

PROOF. By Lemma 4, the irreducible components of Q_1 are absolutely irreducible, pairwise equivalent, and of degree $q_1^{l_1}$. Consequently in some K-basis of K^r the group Q_1 takes the form

$$Q_1 = Q^1 \dot{\times} E_{rq_1^{-l_1}}, \tag{18}$$

where Q^1 is an absolutely irreducible subgroup of $GL(q_1^{l_1}, K)$ isomorphic to Q_1. We now consider the group $U_1 = Q_2 \cdots Q_k$. The groups U_1 and Q_1 commute elementwise. By (18), U_1 has the form

$$U_1 = E_{q_1^{l_1}} \dot{\times} U^1, \tag{19}$$

where U^1 is a subgroup of $GL(rq_1^{-l_1}, K)$ isomorphic to U_1. In particular,

$$Q_2 = E_{q_1^{l_1}} \dot{\times} \bar{Q}_2, \quad \bar{Q}_2 \cong Q_2, \quad \bar{Q}_2 \subset GL(rq_1^{-l_1}, K).$$

We now apply to $\bar{Q}_2$ the same arguments employed for Q_1. The irreducible components of $\bar{Q}_2$ are pairwise equivalent and of degree $q_2^{l_2}$. Consequently in some K-basis of K^r we have both (18) and

$$Q_2 = E_{q_1^{l_1}} \dot{\times} Q^2 \dot{\times} E_{rq_1^{-l_1} q_2^{-l_2}}, \tag{20}$$

where Q^2 is an absolutely irreducible subgroup of $GL(q_2^{l_2}, K)$ isomorphic to Q_2. Since $Q^1 \dot{\times} Q^2$ is an absolutely irreducible subgroup of $GL(q_1^{l_1} q_2^{l_2}, K)$, it follows from (18) and (20) that

$$Q_3 = E_{q_1^{l_1} q_2^{l_2}} \dot{\times} \bar{Q}_3.$$

Proceeding in the same way, we finally obtain what is required. ■

COROLLARY 1. *In a suitable K-basis of K^r, A can be written in the form $A = Q^1 \dot{\times} \cdots \dot{\times} Q^k$, where Q^j is an absolutely irreducible subgroup of $GL(q_j^{l_j}, K)$ isomorphic to Q_j.*

THEOREM 7. *Let P/F be a subgroup of Q_j/F and C the centralizer of P in Q_j. Then $Q_j : C = P : F$.*

PROOF. Let $P : F = q_j^{\nu}$. Then $P = (u_1) \cdots (u_\nu)F$, and P/F is the direct product of ν cyclic groups $(u_j F)$ of order q_j. By Lemma 19.3, the index of the centralizer of u_j in Q_j is q_j. Consequently $i_1 = Q_j : C \leqslant q_l^{\nu}$. Now

$$Q_j = PP_1, \quad P_1 : F = q_j^{2l_j - \nu}.$$

If i_2 is the index of the centralizer of P_1 in Q_j, then $i_2 \leqslant q_j^{2l_j - \nu}$. Since the center of Q_j is F and $Q_j : F = q_j^{2l_j}$, it follows that $q_j^{2l_j} \leqslant i_1 i_2$. Hence $i_1 = q_j^{\nu} = Q_j : C$ ■

The next theorem uses the notation of Theorem 19.2.

THEOREM 8. *Suppose that the pairs of elements* $u_1, v_1, \ldots, u_t, v_t$ *of the algebra* $[A]_K = K_r$ *are subject to three conditions*: (i) $(u_j, v_j) = (a_j, b_j)$, $j = 1, \ldots, t$; (ii) *elements belonging to distinct pairs commute*; (iii) $u_j^{\nu_j} = a_j^{\nu_j}$ *and* $v_j^{\nu_j} = b_j^{\nu_j}$, $j = 1, \ldots, t$. *Then there is an element* x *in* $GL(r, K)$ *such that*

$$xa_jx^{-1} = u_j, \quad xb_jx^{-1} = v_j, \quad j = 1, \ldots, t.$$

PROOF. Consider the r^2 elements

$$u_1^{\alpha_1} \ldots v_t^{\beta_t}, \quad 0 \leqslant \alpha_j < \nu_j, \quad 0 \leqslant \beta_j < \nu_j. \tag{21}$$

If $(\alpha_1, \ldots, \beta_t) \neq (\gamma_1, \ldots, \delta_t)$, then for some j, $1 \leqslant j \leqslant t$, either

$$\left(u_j, u_1^{\alpha_1} \ldots v_t^{\beta_t}\right) \neq \left(u_j, u_1^{\gamma_1} \ldots v_t^{\delta_t}\right),$$

or

$$\left(v_j, u_1^{\alpha_1} \ldots v_t^{\beta_t}\right) \neq \left(v_j, u_1^{\gamma_1} \ldots v_t^{\delta_t}\right).$$

As in the proof of Lemma 19.4, having obtained these inequalities we can prove that the elements (21) are linearly independent over K. Thus the mapping θ : $K_r \longrightarrow K_r$,

$$\sum \lambda_{\alpha_1 \ldots \beta_t} a_1^{\alpha_1} \ldots b_t^{\beta_t} \mapsto \sum \lambda_{\alpha_1 \ldots \beta_t} u_1^{\alpha_1} \ldots v_t^{\beta_t},$$

where $\lambda_{\alpha_1 \cdots \beta_t} \in K$, is bijective. By (i)–(iii), θ is an automorphism of K_r. Obviously $\theta(\lambda) = \lambda$ for $\lambda \in K$. Consequently θ is an inner automorphism of K_r. This completes the proof. ∎

3. *Uniqueness of the invariant series* (α).

THEOREM 9. *A maximal primitive solvable subgroup* G *of* $GL(n, \Delta)$ *has a unique maximal abelian normal subgroup.*

PROOF. Let F and $F_1 \neq F$ be maximal abelian normal subgroups of G. Then, by Lemma 19.1, $F = K^*$ and $F_1 = K_1^*$, where K and K_1 are finite extensions of Δ. As we remarked above (see (2)), G is an absolutely irreducible subgroup of $GL(l, \Sigma)$, where $\Sigma^* E_l$ is the center of G. By Theorem 1, K is a normal separable extension of Σ. Consequently $K = \Sigma(\theta)$, where θ is a primitive element of K. Since F and F_1 are maximal and distinct, we have $K \neq \Sigma E_l$ and $K_1 \neq \Sigma E_l$. Let $\varphi(x) \in \Sigma[x]$ be the minimal polynomial of θ. Then $\varphi(x) = 0$ is a normal equation and the splitting field $\bar{K}$ of $\varphi(x)$ is isomorphic to K. We set $K : \Sigma = \mu$ and $l\mu^{-1} = \rho$, and regard G as a subgroup of $GL(l, \bar{K})$. Over $\bar{K}$, θ takes the form

$$\theta = \mathrm{diag}\,[\theta_1 E_\rho, \ldots, \theta_\mu E_\rho],$$

where θ_j are the roots of $\varphi(x)$. Hence any f in F has the form

$$f = \mathrm{diag}\,[\alpha_1 E_\rho, \ldots, \alpha_\mu E_\rho].$$

G is an imprimitive subgroup of $GL(l, \bar{K})$. By Theorem 16.2, the space $\bar{K}^l$ has a decomposition

$$\bar{K}^l = Q_1 \dotplus \dots \dotplus Q_\mu \tag{22}$$

into systems of imprimitivity for G, and the action of θ on the vectors of Q_j is multiplication by θ_j. There is obviously an element f_1 in F_1 that does not commute with θ. Since $F_1 = K_1^*$ and $e + f_1 \neq 0$, it follows that $e + f_1 \in F$. Since $f_1\theta \neq \theta f_1$, for some j we have $f_1(Q_j) = Q_i$, $i \neq j$. Hence $(e + f_1)(Q_j) = Q_j + Q_i$. This contradicts the decomposition (22) of $\bar{K}^l$ into systems of imprimitivity for G. ∎

We add that Theorem 9 remains valid if G is irreducible, rather than primitive (see Theorem 14 in Suprunenko [8]).

THEOREM 10. *Let B/F be a normal subgroup of G/F contained in Q/F, where Q/F is a Sylow q-subgroup of A/F. Then Q/F is a direct product $Q/F = B/F \times U/F$, where U/F is normal in G/F and the commutator subgroup (B, U) is trivial.*

PROOF. Let U be the centralizer of B in Q. Since U is the intersection with Q of the centralizer of B in G, U is normal in G. The center of B is F. Consequently $U \cap B = F$; that is, $B/F \cap U/F = 1$. By Theorem 7, $Q : U = B : F$. Since $Q : F = (Q : U)(U : F)$, it follows that

$$Q : F = (B : F)(U : F).$$

Thus Q/F is the direct product of its subgroups B/F and U/F. ∎

THEOREM 11. *The group A in the invariant series (α) is uniquely determined by G.*

PROOF. Suppose B has the following properties:

(i) B/F is an abelian normal subgroup of G/F.

(ii) $B/F \subset V/F$.

(iii) B/F is maximal among the subgroups of G/F satisfying (i) and (ii).

We must show that $B = A$. Let q be a prime divisor of r, P/F a Sylow q-subgroup of A/F, and Q/F a Sylow q-subgroup of B/F. We set $D/F = P/F \cap Q/F$. The subgroup D/F is clearly normal in G/F. By Theorem 10,

$$P/F = D/F \times U/F, \quad Q/F = D/F \times W/F$$

where the direct factors U/F and W/F are normal in G/F, and $(D, U) = (D, W) = 1$. Obviously $U/F \cap W/F = 1$. Therefore UW/F is an abelian normal subgroup of G/F; now $R/F = D/F \times UW/F$ is also an abelian normal subgroup of G/F, and R/F is a q-subgroup of V/F. Obviously, $R/F \supset P/F$ and $R/F \supset Q/F$. Hence

$$R/F = P/F, \quad R/F = Q/F, \quad P = Q.$$

Since q is an arbitrary prime factor of r, $A = B$. ∎

It follows from Theorems 9 and 11 that *the invariant series (α) of G is unique.*

THEOREM 12. *Let $\mathfrak{N}$ be the normalizer of A in $GL(n, \Delta)$ and G_1 another maximal primitive solvable subgroup of $GL(n, \Delta)$ such that*

$$G_1 \supset V_1 \supset A \supset F \supset (E_n)$$

is its invariant series (α). *If G and G_1 are conjugate in $GL(n, \Delta)$, they are also conjugate in $\mathfrak{N}$.*

PROOF. Let $tGt^{-1} = G_1$, $t \in GL(n, \Delta)$. We consider the groups tFt^{-1}, tVt^{-1} and tAt^{-1}. Now tFt^{-1} is a maximal abelian normal subgroup of $tGt^{-1} = G_1$. By Theorem 9, $tFt^{-1} = F$. Consequently $tVt^{-1} = V_1$, $tAt^{-1} \subset V_1$, and tAt^{-1}/F is maximal among the abelian normal subgroups of G_1/F contained in V_1/F. By Theorem 11, $tAt^{-1} = A$. Thus $t \in \mathfrak{N}$. ■

4. *The factor group V/A.* We first define two new groups. Let P be an arbitrary field and $HSp(2l, P)$ the set of all matrices $h \in P_{2l}$ such that ${}^t h\Phi_l h = \lambda\Phi_l$, where $\lambda \in P^*$, ${}^t h$ is the transpose of h, and

$$\Phi_l = \begin{bmatrix} 0 & E_l \\ -E_l & 0 \end{bmatrix}.$$

If

$${}^t h_1 \Phi_l h_1 = \lambda_1 \Phi_l, \qquad \lambda_1 \in P^*,$$

then

$${}^t(h_1 h)\, \Phi_l\, (h_1 h) = \lambda\lambda_1 \Phi_l.$$

It is clear that $HSp(2l, P)$ is a subgroup of $GL(2l, P)$. The mapping $h \longmapsto \lambda$ is a homomorphism of the groups $HSp(2l, P) \longrightarrow P^*$, whose kernel consists of all those matrices $h \in P_{2l}$ for which

$${}^t h\Phi_l h = \Phi_l. \tag{23}$$

The group $Sp(2l, P)$ of all matrices $h \in P_{2l}$ satisfying (23) is called the *symplectic group* of degree $2l$ over P. $HSp(2l, P)$ is called the *hypersymplectic group.*

We write a matrix $h \in P_{2l}$ in the form

$$h = \begin{bmatrix} \alpha_{11} & \cdots & \alpha_{1l} & \beta_{11} & \cdots & \beta_{1l} \\ \cdot & \cdot & \cdot & \cdot & \cdot & \cdot \\ \alpha_{l1} & \cdots & \alpha_{ll} & \beta_{l1} & \cdots & \beta_{ll} \\ \gamma_{11} & \cdots & \gamma_{1l} & \delta_{11} & \cdots & \delta_{1l} \\ \cdot & \cdot & \cdot & \cdot & \cdot & \cdot \\ \gamma_{l1} & \cdots & \gamma_{ll} & \delta_{l1} & \cdots & \delta_{ll} \end{bmatrix}$$

and define $2l$-vectors u_i, v_i, x_μ and y_μ $(i, \mu = 1, \ldots, l)$ over P by setting

$$\begin{aligned} u_i &= (\alpha_{1i}, \gamma_{1i}, \ldots, \alpha_{li}, \gamma_{li}), & x_\mu &= (\gamma_{1\mu}, -\alpha_{1\mu}, \ldots, \gamma_{l\mu}, -\alpha_{l\mu}), \\ v_i &= (\beta_{1i}, \delta_{1i}, \ldots, \beta_{li}, \delta_{li}), & y_\mu &= (\delta_{1\mu}, -\beta_{1\mu}, \ldots, \delta_{l\mu}, -\beta_{l\mu}). \end{aligned} \tag{24}$$

By the scalar product of two vectors of the form (24) we mean the sum of products of their corresponding components.

LEMMA 5. *A matrix h in P_{2l} belongs to $HSp(2l, P)$ if and only if the scalar products of the vectors* (24) *are subject to the following conditions*:

1) $u_i x_\mu = 0$, 2) $u_i y_l = \lambda$,

3) $\mu \neq i \Rightarrow u_i y_\mu = 0$, 4) $v_i y_\mu = 0$,

where $\lambda = \lambda(h) \in P^*$.

PROOF. We first observe that for any matrix $h \in P_{2l}$

$$v_i x_\mu = - u_\mu y_l. \tag{25}$$

Now we turn to the matrix ${}^t h \Phi_l h$:

$$
\begin{aligned}
{}^t h\Phi_l h &= {}^t h\,(\Phi_l h)\\
&= \begin{bmatrix} \alpha_{11} \cdots \alpha_{l1} & \gamma_{11} \cdots \gamma_{l1} \\ \cdots\cdots & \cdots\cdots \\ \alpha_{1l} \cdots \alpha_{ll} & \gamma_{1l} \cdots \gamma_{ll} \\ \beta_{11} \cdots \beta_{l1} & \delta_{11} \cdots \delta_{l1} \\ \cdots\cdots & \cdots\cdots \\ \beta_{1l} \cdots \beta_{ll} & \delta_{1l} \cdots \delta_{ll} \end{bmatrix}
\begin{bmatrix} \gamma_{11} \cdots \gamma_{1l} & \delta_{11} \cdots \delta_{1l} \\ \cdots\cdots & \cdots\cdots \\ \gamma_{l1} \cdots \gamma_{ll} & \delta_{l1} \cdots \delta_{ll} \\ -\alpha_{11} \cdots -\alpha_{1l} & -\beta_{11} \cdots -\beta_{1l} \\ \cdots\cdots & \cdots\cdots \\ -\alpha_{l1} \cdots -\alpha_{ll} & -\beta_{l1} \cdots -\beta_{ll} \end{bmatrix}
= \begin{bmatrix} R & S \\ T & U \end{bmatrix},
\end{aligned}
$$

where R, S, T and U are $l \times l$ matrices:

$$R = \| \rho_{i\mu} \|, \quad \rho_{i\mu} = u_i x_\mu; \qquad S = \| \sigma_{i\mu} \|, \quad \sigma_{i\mu} = u_i y_\mu;$$
$$T = \| \tau_{i\mu} \|, \quad \tau_{i\mu} = v_i x_\mu, \qquad U = \| \theta_{i\mu} \|, \quad \theta_{i\mu} = v_i y_\mu.$$

A matrix h belongs to $HSp(2l, P)$ if and only if $R = U = 0$ and $S = -T = \lambda E_l$. The lemma now follows from (25). ∎

It is clear that *Lemma* 5 *remains valid when λ is replaced by the unit element and $HSp(2l, P)$ by $Sp(2l, P)$.*

LEMMA 6. *$Sp(2, P) = SL(2, P)$ and $HSp(2, P) = GL(2, P)$.*

PROOF. Let

$$h = \begin{bmatrix} \alpha & \beta \\ \gamma & \delta \end{bmatrix} \in P_2.$$

Then

$${}^t h\Phi_1 h = \begin{bmatrix} \alpha & \gamma \\ \beta & \delta \end{bmatrix}\begin{bmatrix} 0 & 1 \\ -1 & 0 \end{bmatrix}\begin{bmatrix} \alpha & \beta \\ \gamma & \delta \end{bmatrix} = \begin{bmatrix} \alpha & \gamma \\ \beta & \delta \end{bmatrix}\begin{bmatrix} \gamma & \delta \\ -\alpha & -\beta \end{bmatrix} = \begin{bmatrix} 0 & 1 \\ -1 & 0 \end{bmatrix} \det h.$$

This implies the lemma. ∎

In the case $P = GF(q)$ we write $Sp(2l, q)$ instead of $Sp(2l, P)$.

We now proceed to investigate the factor group V/A. In view of Corollary 6.1 we may assume that A is a Kronecker product

$$A = Q^1 \dot{\times} \ldots \dot{\times} Q^k. \tag{26}$$

Clearly V is a subgroup of the normalizer of A in $GL(r, K)$. Let N be the normalizer of A in $GL(r, K)$ and N_j the normalizer of Q_j in $GL(q_j^{l_j}, K)$, $j = 1, \ldots, k$. We have

THEOREM 13. $N = N_1 \dot{\times} \cdots \dot{\times} N_k$.

PROOF. For any $h \in N$ we have $hQ_jh^{-1} = Q_j$ and

$$h\left(E_{q_1^{l_1}} \dot{\times} \cdots \dot{\times} Q^j \dot{\times} \cdots \dot{\times} E_{q_k^{l_k}}\right) h^{-1} = E_{q_1^{l_1}} \dot{\times} \cdots \dot{\times} Q_{q_j^{l_j}} \dot{\times} \cdots \dot{\times} E_{q_k^{l_k}}.$$

Hence, going over to $[Q_j]_K$, we obtain

$$h\left(E_{q_1^{l_1}} \dot{\times} \cdots \dot{\times} x_j \dot{\times} \cdots \dot{\times} E_{q_k^{l_k}}\right) h^{-1} = E_{q_1^{l_1}} \dot{\times} \cdots \dot{\times} x_j' \dot{\times} \cdots \dot{\times} E_{q_k^{l_k}},$$

$$h\left(E_{q_1^{l_1}} \dot{\times} \cdots \dot{\times} K_{q_j^{l_j}} \dot{\times} \cdots \dot{\times} E_{q_k^{l_k}}\right) h^{-1} = E_{q_1^{l_1}} \dot{\times} \cdots \dot{\times} K_{q_j^{l_j}} \dot{\times} \cdots \dot{\times} E_{q_k^{l_k}}, \tag{27}$$

where $x_j, x_j' \in K_{q_j^{l_j}}$. This defines an algebra automorphism σ_j of $K_{q_j^{l_j}}$:

$$\sigma_j \colon K_{q_j^{l_j}} \to K_{q_j^{l_j}}, \qquad x_j \mapsto x_j', \tag{28}$$

such that $\sigma_j(Q^j) = Q^j$ and $\sigma_j(\lambda E) = \lambda E$ for $\lambda \in K$. Consequently there is a matrix h in $GL(q_j^{l_j}, K)$ such that $\sigma_j(x_j) = h_jx_jh_j^{-1}$ for every x_j in $K_{q_j^{l_j}}$. Since $\sigma_j(Q^j) = h_jQ^jh_j^{-1} = Q^j$, it follows that $h_j \in N_j$. We now set $g = h_1 \dot{\times} \cdots \dot{\times} h_k$. Then for any $a \in A$ we have $gag^{-1} = hah^{-1}$. Hence $g^{-1}h = \lambda \in F$, $h = \lambda h_1 \dot{\times} \cdots \dot{\times} h_k \in N_1 \dot{\times} \cdots \dot{\times} N_k$, $N \subset N_1 \dot{\times} \cdots \dot{\times} N_k$. On the other hand, the inclusion $N_1 \dot{\times} \cdots \dot{\times} N_k \subseteq N$ is obvious. ∎

THEOREM 14. *The factor group N/A is isomorphic to the direct product $N_1/Q^1 \times \cdots \times N_k/Q^k$.*

PROOF. We set

$$M_j = E_{q_1^{l_1} \cdots q_{j-1}^{l_{j-1}}} \dot{\times} N_j \dot{\times} E_{rq_1^{-l_1} \cdots q_j^{-l_j}}. \tag{29}$$

Then, by Theorem 13, $N = M_1 \cdots M_k$. For any $j = 1, \ldots, k$,

$$M_j \cap M_1 \ldots M_{j-1}M_{j+1} \ldots M_k = F.$$

Thus we obtain the direct product

$$N/F = M_1/F \times M_2/F \times \ldots \times M_k/F. \tag{30}$$

On the other hand,

$$A/F = Q_1/F \times \ldots \times Q_k/F. \tag{31}$$

It follows from (30) and (31) that N/A is isomorphic to the direct product $M_1/Q_1 \times \cdots \times M_k/Q_k$. By (29), $M_j/Q_j \cong N_j/Q^j$. ∎

We now consider the factor group N_j/Q^j. To simplify the notation, we set $l_j = l$ and $q_j = q$. By Theorem 5, Q^j can be written in the form

$$Q^j = (c_1)(d_1) \ldots (c_l)(d_l) K^*, \tag{32}$$

where $c_\nu^q, d_\nu^q \in K^*$, $(c_\nu, d_\nu) = \eta$, $\eta^q = 1$, $\eta \neq 1$, $\eta \in K^*$, and elements belonging to distinct pairs commute.

Each element x in N_j determines an automorphism σ_x of $Q^j : \sigma_x(u) = xux^{-1}$, $u \in Q^j$. We now introduce a homomorphism

$$\varphi: \quad N_j \to \operatorname{Aut}(Q^j/K^*) \tag{33}$$

by setting

$$\varphi: x \mapsto \bar{\sigma}_x, \quad \bar{\sigma}_x(uK^*) = \sigma_x(u)K^*, \quad x \in N_j, \quad u \in Q^j.$$

By Theorem 4, Q^j/K^* is its own centralizer in N_j/K^*. Therefore

$$\operatorname{Ker}\varphi = Q^j, \quad N_j/Q^j \cong \operatorname{Im}\varphi. \tag{34}$$

We now apply σ_x to the elements c_i and d_i in (32):

$$\begin{aligned} \sigma_x(c_i) &= c_i' = \lambda_i c_1^{\alpha_{1i}} \dots c_l^{\alpha_{li}} d_1^{\gamma_{1i}} \dots d_l^{\gamma_{li}}, \\ \sigma_x(d_i) &= d_i' = \mu_i c_1^{\beta_{1i}} \dots c_l^{\beta_{li}} d_1^{\delta_{1i}} \dots d_l^{\delta_{li}}, \end{aligned} \tag{35}$$

where $\lambda_i, \mu_i \in K^*$, $0 \leqslant \alpha_{\rho t} < q$, $0 \leqslant \beta_{\rho t} < q$, $0 \leqslant \gamma_{\rho t} < q$, $0 \leqslant \delta_{\rho t} < q$, $i, \rho = 1, \dots, l$. Going over to $\bar{\sigma}_x$, we see from (35) that

$$\begin{aligned} \bar{\sigma}_x(c_iK^*) &= c_1^{\alpha_{1i}} \dots c_l^{\alpha_{li}} d_1^{\gamma_{1i}} \dots d_l^{\gamma_{li}} K^*, \\ \bar{\sigma}_x(d_iK^*) &= c_1^{\beta_{1i}} \dots c_l^{\beta_{li}} d_1^{\delta_{1i}} \dots d_l^{\delta_{li}} K^*, \end{aligned} \tag{36}$$

where the exponents $\alpha_{\rho i}, \beta_{\rho i}, \gamma_{\rho i}$ and $\delta_{\rho i}$ can be taken to be elements of $GF(q)$. Since $\bar{\sigma}_x$ is an automorphism, by (32) the elements c_i' and d_i' of (35) are subject to the following conditions:

$$(c_i', d_i') = \eta, \quad (c_i', c_\mu') = 1, \quad (d_i', d_\mu') = 1, \quad \mu \neq i \Rightarrow (c_i', d_\mu') = 1, \tag{37}$$

$$c_i'^q = c_i^q, \quad d_i'^q = d_i^q. \tag{38}$$

From $(c_i, d_i) = \eta$ we find that $(c_i^\alpha, d_i^\delta) = \eta^{\alpha\delta}$ and $(d_i^\gamma, c_i^\beta) = \eta^{-\beta\gamma}$. Hence

$$(c_i^\alpha d_i^\gamma, c_i^\beta d_i^\delta) = \eta^{\alpha\delta - \beta\gamma}. \tag{39}$$

It follows from (39) that (37) is equivalent to the following congruences:

$$\left.\begin{aligned} &1) \quad \sum_{\rho=1}^{l} (\alpha_{\rho i}\delta_{\rho i} - \beta_{\rho i}\gamma_{\rho i}) \equiv 1, \\ &2) \quad \sum_{\rho=1}^{l} (\alpha_{\rho i}\gamma_{\rho\mu} - \alpha_{\rho\mu}\gamma_{\rho i}) \equiv 0, \\ &3) \quad \sum_{\rho=1}^{l} (\beta_{\rho i}\delta_{\rho\mu} - \beta_{\rho\mu}\delta_{\rho i}) \equiv 0, \\ &4) \quad \mu \neq i \Rightarrow \sum_{\rho=1}^{l} (\alpha_{\rho i}\delta_{\rho\mu} - \beta_{\rho\mu}\gamma_{\rho i}) \equiv 0. \end{aligned}\right\} \pmod q. \tag{40}$$

LEMMA 7. *The matrix*

$$h(x) = \begin{bmatrix} \alpha_{11} & \dots & \alpha_{1l} & \beta_{11} & \dots & \beta_{1l} \\ \cdot & \cdot & \cdot & \cdot & \cdot & \cdot \\ \alpha_{l1} & \dots & \alpha_{ll} & \beta_{l1} & \dots & \beta_{ll} \\ \gamma_{11} & \dots & \gamma_{1l} & \delta_{11} & \dots & \delta_{1l} \\ \cdot & \cdot & \cdot & \cdot & \cdot & \cdot \\ \gamma_{l1} & \dots & \gamma_{ll} & \delta_{l1} & \dots & \delta_{ll} \end{bmatrix}, \tag{41}$$

whose elements are the exponents in (36), *belongs to* $Sp(2l, q)$.

PROOF. We have agreed to regard the exponents as elements of $GF(q)$. Using the scalar products of the vectors (24), we can rewrite (40) in the form

$$1)\ u_i y_i = 1, \quad 2)\ u_i x_\mu = 0,$$
$$3)\ v_i y_\mu = 0, \quad 4)\ \mu \neq i \Rightarrow u_i y_\mu = 0.$$

Hence and from Lemma 5, Lemma 7 follows. ■

Finally, we introduce the mapping

$$\psi\colon N_l \to Sp(2l, q), \quad x \mapsto h(x) \tag{42}$$

(see (41)).

LEMMA 8. *ψ is a homomorphism, and* Ker $\psi = Q^j$. *Thus the factor group N_j/Q^j is isomorphic to a subgroup of $Sp(2l, q)$.*

PROOF. A direct computation shows that $h(xy) = h(x)h(y)$ for $x, y \in N_j$. On the other hand, it follows from (36) and (41) that $h(x)$ is the unit matrix if and only if $\bar{\sigma}_x$ is the identity automorphism. Hence and from Theorem 4 it follows that Ker $\psi = Q^j$. ■

Theorem 14 and Lemma 8 imply

THEOREM 15. *N/A is isomorphic to a direct product of k groups H_j, $j = 1, \ldots, k$, where H_j is a subgroup of $Sp(2l_j, q_j)$; and V/A is isomorphic to a solvable subgroup of the direct product of k symplectic groups $Sp(2l_j, q_j)$, $j = 1, \ldots, k$.*

We now return to the homomorphism (42).

THEOREM 16. *Suppose the elements c_ν and d_ν of Q^j* (see (32)) *are chosen so that $c_\nu^q = d_\nu^q = 1$, $\nu = 1, \ldots, l$, and for $q = 2$ let there be an element of order 4 in K^*. Then ψ is an epimorphism.*

PROOF. Let $h \in Sp(2l, q)$. Using (35), we construct elements c'_i and d'_i, where the elements of h play the part of the exponents α, β, γ and δ. There are two possibilities: $q > 2$ or $q = 2$.

If $q > 2$, we set $\lambda_i = \mu_i = 1$. Then c'_i and d'_i satisfy (37) and (38). By Theorem 8 there exists a $g \in GL(q^l, K)$ such that $gc_ig^{-1} = c'_i$ and $gd_ig^{-1} = d'_i$, $i = 1, \ldots, k$. It is obvious that $g \in N_j$ and $\psi(g) = h$. This proves the theorem for $q > 2$.

Now let $q = 2$. Since there is an element of order 4 in K^*, we can choose λ_i and μ_i in such a way that $c_i'^2 = d_i'^2 = 1$. Hence in this case too (37) and (38) are satisfied, and Theorem 8 is again applicable. ■

It is clear that the assumptions of Theorem 16 are necessarily satisfied when K is an algebraically closed field.

THEOREM 17. *Let $F = \Delta^* E_n$, and let $n = q_1^{l_1} \cdots q_k^{l_k}$ be the canonical decomposition of n. Then G is absolutely irreducible and is conjugate in $GL(n, \Delta)$ to a group $G_1 \dot{\times} \cdots \dot{\times} G_k$, where G_j is a maximal primitive solvable absolutely irreducible subgroup of $GL(q_j^{l_j}, \Delta)$, $j = 1, \ldots, k$.*

PROOF. Since $F = \Delta^* E_n$, F is the center of G. Hence $[G]_\Delta = \Delta_n$ by Theorem 14.7. Therefore G is absolutely irreducible. It follows from $F = \Delta^* E_n$ that $G = V$. Hence by Theorem 13 G is conjugate in $GL(n, \Delta)$ to some subgroup $\bar{G}$ of $N = N_1 \dot{\times} \cdots \dot{\times} N_k$, where $N_j \subset GL(q_j^{l_j}, \Delta)$. Since $\bar{G}$ is a maximal solvable subgroup of $GL(n, \Delta)$, it follows that $\bar{G} = G_1 \dot{\times} \cdots \dot{\times} G_k$, where G_j is a maximal solvable subgroup of $GL(q_j^{l_j}, \Delta)$. But G_j is absolutely irreducible by Lemma 15.8, and is primitive by Lemma 15.9. ∎

5. *The case of an algebraically closed field.* Suppose now that Δ is algebraically closed. Then $F = \Delta^* E_n$, $V = G$, and we can replace (α) by the series

$$G \supset A \supset F \supset (E_n). \qquad (\alpha_1)$$

By Theorem 3, $A : F = n^2$. By Corollary 3.1, char $\Delta \notin \Pi(n)$. Theorem 17 is applicable to G, and we may therefore assume that $G = G_1 \dot{\times} \cdots \dot{\times} G_k$, where G_j is a maximal primitive solvable subgroup of $GL(q_j^{l_j}, \Delta)$, $j = 1, \ldots, k$, and $n = q_1^{l_1} \cdots q_k^{l_k}$ is the canonical decomposition of n. We may thus assume that $n = q^l$, where q is a prime, and G is a maximal primitive solvable subgroup of $GL(q^l, \Delta)$. By Corollary 5.1, A can be represented in the form

$$A = (u_1)(v_1) \ldots (u_l)(v_l) F, \qquad (43)$$

where $(u_i, v_i) = \eta E_n$, $\eta^q = 1$, $\eta \neq 1$, $\eta \in \Delta^*$, u_i^q, $v_i^q \in F$, and elements of distinct pairs commute.

LEMMA 9. *In a suitable basis of Δ^n, A has the form*

$$A = (c_1)(d_1) \ldots (c_l)(d_l) F,$$

where

$$\begin{aligned} c_i &= E_{q^{i-1}} \dot{\times} \operatorname{diag}[1, \eta, \ldots, \eta^{q-1}] \dot{\times} E_{q^{l-i}}, \\ d_i &= E_{q^{i-1}} \dot{\times} \begin{bmatrix} 0 & 1 \\ E_{q-1} & 0 \end{bmatrix} \dot{\times} E_{q^{l-i}}. \end{aligned} \qquad (44)$$

PROOF. For the elements u_i and v_i in (43) we can write $u_i, v_i \in \Delta^* E_n$. Since Δ is algebraically closed, there exist elements ρ_i and τ_i in Δ^* such that

$$(\rho_i u_i)^q = (\tau_i v_i)^q = E_n, \qquad i = 1, \ldots, l.$$

Setting $a_i = \rho_i u_i$ and $b_i = \tau_i v_i$, $i = 1, \ldots, l$, we obtain

$$A = (a_1)(b_1) \ldots (a_l)(b_l) F,$$

where $(a_i, b_i) = \eta E_n$, $a_i^q = b_i^q = E_n$, $i = 1, \ldots, l$, and elements in distinct pairs commute.

On the other hand, it is obvious that $(c_i, d_i) = \eta E_n$, $c_i^q = d_i^q = E_n$, $i = 1, \ldots, l$, and elements in distinct pairs commute.

By Theorem 8, there exists a matrix t in $GL(n, \Delta)$ such that $ta_it^{-1} = c_i$ and $tb_it^{-1} = d_i$, $i = 1, \ldots, l$. Lemma 9 follows from this. ∎

For the rest of this section we assume that

$$A = (c_1)(d_1) \ldots (c_l)(d_l) F,$$

where c_i and d_i are defined by (44), and N is the normalizer of A in $GL(q^l, \Delta)$. Every maximal primitive solvable subgroup G of $GL(q^l, \Delta)$ is conjugate in $GL(q^l, \Delta)$ to a maximal solvable subgroup of N. For every element x of N we have

$$xc_ix^{-1} = \lambda_i c_1^{\alpha_{1i}} \ldots c_l^{\alpha_{li}} d_1^{\gamma_{1i}} \ldots d_l^{\gamma_{li}}, \quad xd_ix^{-1} = \mu_i c_1^{\beta_{1i}} \ldots c_l^{\beta_{li}} d_1^{\delta_{1i}} \ldots d_l^{\delta_{li}},$$

where $\lambda_i, \mu_i \in \Delta^*$ and $\alpha_{\rho i}, \beta_{\rho i}, \gamma_{\rho i}, \delta_{\rho i} \in GF(q)$, $i = 1, \ldots, l$. We also set

$$\psi(x) = \begin{bmatrix} \alpha_{11} & \ldots & \alpha_{1l} & \beta_{11} & \ldots & \beta_{1l} \\ \cdot & \cdot & \cdot & \cdot & \cdot & \cdot \\ \alpha_{l1} & \ldots & \alpha_{ll} & \beta_{l1} & \ldots & \beta_{ll} \\ \gamma_{11} & \ldots & \gamma_{1l} & \delta_{11} & \ldots & \delta_{1l} \\ \cdot & \cdot & \cdot & \cdot & \cdot & \cdot \\ \gamma_{l1} & \ldots & \gamma_{ll} & \delta_{l1} & \ldots & \delta_{ll} \end{bmatrix}.$$

Then $\psi(x) \in Sp(2l, q)$ by Lemma 7, and by Theorem 16 the mapping $\psi : N \longrightarrow Sp(2l, q)$ is surjective. By Theorem 4, Ker $\psi = A$, and so

$$N/A \cong Sp(2l, q). \tag{45}$$

LEMMA 10. *Let $\mathfrak{G}_1$ and $\mathfrak{G}_2$ be two maximal primitive solvable subgroups of $GL(q^l, \Delta)$ such that $A \lhd \mathfrak{G}_1$ and $A \lhd \mathfrak{G}_2$. Then $\mathfrak{G}_1$ and $\mathfrak{G}_2$ are conjugate in $GL(q^l, \Delta)$ if and only if $H_1 = \psi(\mathfrak{G}_1)$ and $H_2 = \psi(\mathfrak{G}_2)$ are conjugate in $Sp(2l, q)$.*

PROOF. Suppose that $\mathfrak{G}_1$ and $\mathfrak{G}_2$ are conjugate in $GL(q^l, \Delta)$. Then by Theorem 12 there is a matrix t in N such that $t\mathfrak{G}_1t^{-1} = \mathfrak{G}_2$. Applying ψ to this equality, we get $uH_1u^{-1} = H_2$, where $u = \psi(t)$. Conversely, suppose that H_1 and H_2 are conjugate in $Sp(2l, q)$. Obviously $\psi^{-1}(H_1)$ and $\psi^{-1}(H_2)$ are solvable subgroups of N. Since the $\mathfrak{G}_i$ are maximal solvable subgroups of N, we have $\mathfrak{G}_i = \psi^{-1}(H_i)$, $i = 1, 2$. Since ψ is surjective, the conjugacy of H_1 and H_2 in $Sp(2l, q)$ implies that of $\psi^{-1}(H_1)$ and $\psi^{-1}(H_2)$ in N. Consequently $\mathfrak{G}_1$ and $\mathfrak{G}_2$ are conjugate in $GL(q^l, \Delta)$. ∎

Now we look for conditions under which a subgroup of N is primitive.

LEMMA 11. *Let B be a subgroup of N containing A. Then B is imprimitive if and only if it has a reducible normal subgroup $D \subset A$ with at least two inequivalent irreducible components.*

PROOF. Sufficiency follows from Theorem 16.2. We prove necessity. Let

$$\Delta^n = L_1 \dot{+} \ldots \dot{+} L_\mu, \quad \mu > 1, \quad n = q^l, \tag{46}$$

be a decomposition of Δ^n into systems of imprimitivity for B. (46) determines a homomorphism $\gamma: B \longrightarrow S_\mu$. Since $B \supset A$, (46) is also a decomposition of Δ^n into systems of imprimitivity for A. Let γ_1 be the restriction of γ to A, $\operatorname{Ker} \gamma_1 = D$. Then $D = A \cap \operatorname{Ker} \gamma$ is a normal subgroup of B. Since A is irreducible, $\gamma_1(A)$ is a transitive subgroup of the symmetric group S_μ. On the other hand, $D \supset F$, and so $\gamma_1(A) \cong A/D$ is abelian. Therefore the order of $\gamma_1(A)$ is equal to μ, $A:D = \mu$, and $D:F = q^{2l}\mu^{-1}$. The irreducible components of D have one and the same degree $\rho \leqslant q^l\mu^{-1}$. Let σ be the number of inequivalent irreducible components of D. Then

$$\sigma \leqslant q^l\rho^{-1}, \quad [D]:\Delta = \rho^2\sigma = D:F = q^{2l}\mu^{-1}.$$

Hence

$$q^{2l}\mu^{-1} = \rho^2\sigma \leqslant \rho^2 q^l \rho^{-1} = \rho q^l \leqslant q^l\mu^{-1}q^l = q^{2l}\mu^{-1}.$$

Consequently $\rho = q^l\mu^{-1}$ and $\sigma = q^l\rho^{-1} = \mu$. Thus D has $\mu > 1$ inequivalent irreducible components. ∎

LEMMA 12. *Let B be a subgroup of N containing A and D a reducible normal subgroup of B with the following two properties:*

1) $F \subset D \subset A$.

2) *D has at least two inequivalent irreducible components.*

Then the center $Z(D)$ of D is different from F.

PROOF. If D is abelian, the lemma follows from 2). Suppose that D is not abelian. Then, as in the proof of Theorem 19.2, we have

$$D = (u_1)(v_1) \ldots (u_k)(v_k) D_k,$$

where D_k is the center of D, $(u_j, v_j) = \eta E_n$, $u_j^q = v_j^q = 1$, $j = 1, \ldots, k$, and elements in distinct pairs commute. D_k cannot coincide with F. For suppose that $D_k = F$, and let W be the centralizer of D in A. Then $D:F = q^{2k}$. By Theorem 7, $W:F = q^{2(l-k)}$. Since $D_k = F$, it follows that $D \cap W = F$. Consequently $A = DW$ and $(D, W) = 1$. It follows that the irreducible components of D are pairwise equivalent (see Lemma 16.4). This contradicts 2), and so $Z(D) \neq F$. ∎

LEMMA 13. *Let B be a subgroup of N containing A. Then B is primitive if and only if F is a maximal abelian normal subgroup of B.*

The necessity follows from Theorems 16.1 and 16.2, the sufficiency from Lemmas 11 and 12.

LEMMA 14. *Let C be an abelian subgroup of A such that $C \supset F$, $C:F = q^\nu$ and $C = (u_1) \cdots (u_\nu)F$, $u_j^q = 1$, $j = 1, \ldots, \nu$. Then $\nu \leqslant l$, and A contains* *elements* $u_{\nu+1}, \ldots, u_l$ *and* $v_1, \ldots, v_l$, *such that* $u_j^q = v_j^q = 1$, $(u_j, v_j) = \eta E_n$, $j = 1, \ldots, l$, *and elements belonging to distinct pairs commute.*

PROOF. $C:F = q^\nu \Rightarrow [C]:\Delta = q^\nu$. Since C is abelian, $[C]:\Delta = q^\nu \leqslant q^l = n$. Thus $\nu \leqslant l$. By Theorem 7, the centralizer $Z(u_1, \ldots, u_\nu)$ of the set $\{u_1, \ldots, u_\nu\}$ in A is a proper subgroup of the centralizer $Z(u_2, \ldots, u_\nu)$. Hence there is an element v_1 in $Z(u_2, \ldots, u_\nu)$ such that $(u_1, v_1) = \eta E_n$, $v_1^q = 1$. There is an element v_2 in $Z(u_1, u_3, \ldots, u_\nu, v_1)$ such that $(u_2, v_2) = \eta E_n$, $v_2^q = 1$. Proceeding in this way, we find v elements $v_1, \ldots, v_\nu$ in A such that $(v_i, v_j) = 1$, $v_j^q = 1$, $(u_i, v_i) = \eta E_n$, and $(u_i, v_j) = 1$, $i \neq j$, $j = 1, \ldots, \nu$. If $\nu = l$, we are done. If $\nu < l$, we can find a pair of elements $u_{\nu+1}, v_{\nu+1}$ in $Z(u_1, v_1, \ldots, u_\nu, v_\nu)$ such that

$$(u_{\nu+1}, v_{\nu+1}) = \eta E_n, \quad u_{\nu+1}^q = v_{\nu+1}^q = 1.$$

The lemma now follows. ∎

A subgroup H of $Sp(2l, q)$ is said to be *s-reducible* if there exist a number ν, $1 \leqslant \nu \leqslant l$, and a matrix $t \in Sp(2l, q)$ such that the matrices of $t^{-1}Ht$ have the form

$$\begin{bmatrix} \alpha_{11} & \ldots & \alpha_{1\nu} & \alpha_{1,\nu+1} & \ldots & \alpha_{1,2l} \\ \cdot & \cdot & \cdot & \cdot & \cdot & \cdot \\ \alpha_{\nu 1} & \ldots & \alpha_{\nu\nu} & \alpha_{\nu,\nu+1} & \ldots & \alpha_{\nu,2l} \\ 0 & \ldots & 0 & \alpha_{\nu+1,\nu+1} & \ldots & \alpha_{\nu+1,2l} \\ \cdot & \cdot & \cdot & \cdot & \cdot & \cdot \\ 0 & \ldots & 0 & \alpha_{2l,\nu+1} & \ldots & \alpha_{2l,2l} \end{bmatrix}.$$

Otherwise H is said to be *s-irreducible.*

THEOREM 18. *Let* $\mathfrak{G}$ *be a subgroup of* N *containing* A. *Then* $\mathfrak{G}$ *is primitive if and only if* $\psi(\mathfrak{G})$ *is an s-irreducible subgroup of* $Sp(2l, q)$.

PROOF. We set $\psi(\mathfrak{G}) = H$. Then $\psi^{-1}(H) = \mathfrak{G}A = \mathfrak{G}$. Suppose that H is s-reducible. Then there is a matrix t in $Sp(2l, q)$ such that all the matrices in $H_1 = t^{-1}Ht$ have the form (47). If $t_1 = \psi^{-1}(t)$, then $\mathfrak{G}_1 = t_1^{-1}\mathfrak{G}t_1 = \psi^{-1}(H_1)$. It follows from the definition of ψ and the form of the matrices in H_1 that for $x \in \mathfrak{G}_1$

$$xa_ix^{-1} = \lambda_i a_1^{\alpha_{1i}} \ldots a_\nu^{\alpha_{\nu i}}, \qquad i = 1, \ldots, \nu, \quad \lambda_i \in \Delta^{\cdot}.$$

Hence $C = (a_1) \cdots (a_\nu)F \neq F$ is an abelian normal subgroup of $\mathfrak{G}_1$. By Lemma 13, $\mathfrak{G}_1$ is imprimitive, and so is $\mathfrak{G}$. Conversely, suppose $\mathfrak{G}$ is imprimitive. Then, by Lemmas 11 and 12, $\mathfrak{G}$ has an abelian normal subgroup C_0 such that $F \subset C_0 \subset A$ and $C_0 : F = q^\nu$, $1 \leqslant \nu \leqslant l$. We write C_0 in the form

$$C_0 = (u_1) \ldots (u_\nu) F, \quad u_j^q = 1, \quad j = 1, \ldots, \nu.$$

By Lemma 14, there are elements $u_{\nu+1}, \ldots, u_l$ and $v_1, \ldots, v_l$ in A such that

$$A = (u_1) \ldots (u_l)(v_1) \ldots (v_l),$$
$$u_j^q = v_j^q = 1, \quad (u_j, v_j) = \eta E_n, \quad j = 1, \ldots, l,$$

and elements of distinct pairs commute. By Theorem 8 there is a matrix t_1 in $GL(q^l, \Delta)$ such that

$$t_1^{-1} u_j t_1 = c_j, \quad t_1^{-1} v_j t_1 = d_j, \quad j = 1, \ldots, l.$$

Obviously $t_1 \in N$, and $C = t_1^{-1} C_0 t_1 = (c_1) \cdots (c_\nu) F$ is an abelian normal subgroup of $\mathfrak{G}_1 = t_1^{-1} \mathfrak{G} t_1$. Let $t = \psi(t_1)$. Then $\psi(\mathfrak{G}) = t^{-1} H t$. Since $C \lhd \mathfrak{G}_1$, the matrices in $t^{-1} H t$ have the form (47). ∎

THEOREM 19. *Let Δ be an algebraically closed field and $n = q^l$, where q is a prime, $q \neq$ char Δ. Let $H_1, \ldots, H_\mu$ be a system of maximal s-irreducible solvable subgroups of $Sp(2l, q)$ such that any maximal s-irreducible subgroup of $Sp(2l, q)$ is conjugate to one and only one subgroup H_j. Then any maximal primitive solvable subgroup of $GL(n, \Delta)$ is conjugate in $GL(n, \Delta)$ to one and only one of the following groups*:

$$\mathfrak{G}_1 = \psi^{-1}(H_1), \ldots, \mathfrak{G}_\mu = \psi^{-1}(H_\mu).$$

$\mathfrak{G}_j$ has a normal subgroup $A = (c_1)(d_1) \cdots (c_l)(d_l)F$, where c_i and d_i are defined by (44) *and $\mathfrak{G}_j / A \cong H_j$, $j = 1, \ldots, \mu$.*

PROOF. Since ψ is surjective, $\psi(\mathfrak{G}_j) = H_j$. By the choice of H_j and Lemma 10, $\psi^{-1}(H_\alpha)$ and $\psi^{-1}(H_\beta)$ are not conjugate in $GL(n, \Delta)$ for $\alpha \neq \beta$. Now let $\mathfrak{G}$ be any maximal solvable subgroup of $GL(n, \Delta)$. By Lemma 9, $\mathfrak{G}$ is conjugate in $GL(n, \Delta)$ to some maximal primitive solvable subgroup $\mathfrak{G}^1$ of N such that $\mathfrak{G}^1 \supset A$. Now $\psi(\mathfrak{G}^1) = H$ is a solvable subgroup of $Sp(2l, q)$. By Theorem 8, H is an s-irreducible subgroup of $Sp(2l, q)$. Since $\mathfrak{G}^1$ is maximal among the solvable subgroups of $GL(n, \Delta)$, it follows that H is a maximal s-irreducible solvable subgroup of $Sp(2l, q)$ and $\mathfrak{G}^1 = \psi^{-1}(H)$. By assumption, H is conjugate in $Sp(2l, q)$ to some H_j. Consequently $\mathfrak{G}^1$ is conjugate to $\psi^{-1}(H_j)$ in $GL(n, \Delta)$, and this completes the proof. ∎

Note that some of the assertions in this subsection can be proved without requiring the field to be algebraically closed. In particular, Lemma 11 can be restated as follows.

LEMMA 11a. *Let Δ be an arbitrary field, let $n = q^l$, where q is a prime, and let A_1 be a subgroup of $GL(n, \Delta)$ with the following properties*:

1. *The center of A_1 is $\Delta^* E_n$, and $A_1 / \Delta^* E_n$ is abelian.*
2. *$[A_1] = \Delta_n$ and $A_1 : F = n^2$.*

Let N_1 be the normalizer of A_1 in $GL(n, \Delta)$ and B_1 a subgroup of N_1 containing A_1. Then B_1 is imprimitive if and only if it has a normal subgroup $D_1 \subset A_1$ with at least two inequivalent irreducible components.

The proof of Lemma 11a is almost the same as that of Lemma 11. The only difference is that the equality $[D]:\Delta = \rho^2\sigma$ must be replaced by an inequality $[D]:\Delta \leqslant \rho^2\sigma$.

By virtue of Theorems 17, 18.4 and 18.5, the following theorem follows from Theorem 19.

THEOREM 20. *Let Δ be an algebraically closed field and n a positive integer. Then the set of all maximal solvable subgroups of $GL(n, \Delta)$ splits into finitely many conjugacy classes; the number of these classes is bounded by a function depending only on n.*

For comparison, we add that Platonov [14] has proved that any algebraic group has only finitely many nonconjugate maximal solvable subgroups.

Concerning real solvable linear groups, see Suprunenko [8].

§21. Solvable matrix groups over a finite field

1. *The case $m = n$.* We first assume that Δ is an arbitrary field. Let G be a maximal solvable primitive subgroup of $GL(n, \Delta)$ and $F = K^*$ a maximal abelian normal subgroup of G, with $K:\Delta = n$. Let $\mathcal{G}$ denote the group of all Δ-automorphisms of K, and $N(F)$ the normalizer of F in $GL(n, \Delta)$.

LEMMA 1. *The factor group $N(F)/F$ is isomorphic to a subgroup of $\mathcal{G}$. If K is a simple extension of Δ, then $N(F)/F \cong \mathcal{G}$. G is a maximal solvable subgroup of $N(F)$, and G/F is isomorphic to a solvable subgroup of $\mathcal{G}$. If K is a simple extension of Δ, then G/F is isomorphic to a maximal solvable subgroup of $\mathcal{G}$.*

PROOF. It is clear that G is a subgroup of $N(F)$. Consider the homomorphism $\gamma : N(F) \longrightarrow \mathcal{G}$, $g \longmapsto \sigma_g$, where $\sigma_g : K \longrightarrow K$, $\sigma_g(x) = gxg^{-1}$. By assumption, $K:\Delta = n$; therefore the centralizer of K in $GL(n, \Delta)$ is F; that is $\operatorname{Ker}\gamma = F$, and $N(F)/F$ is isomorphic to a subgroup of $\mathcal{G}$. We now show that if K is a simple extension of Δ, then γ is surjective. Let $K = \Delta(\theta)$ and $\sigma \in \mathcal{G}$. Then the matrices θ and $\theta_1 = \sigma(\theta)$ have the same irreducible characteristic polynomial in $\Delta[x]$. Thus $\sigma(\theta) = a\theta a^{-1}$, where $a \in GL(n, \Delta)$. It is obvious that $a \in N(F)$ and $\sigma(x) = axa^{-1}$ for $x \in K$. This proves that γ is surjective. Consequently, if K is a simple extension of Δ, then $N(F)/F \cong \mathcal{G}$. ∎

We now apply Lemma 1 to the case of a finite field Δ.

THEOREM 1. *Let $\Delta = GF(p^t)$, and let G be a maximal solvable primitive subgroup of $GL(n, \Delta)$ and $F = K^*$ a maximal abelian normal subgroup of G, with $K:\Delta = n$. Then the following assertions are true:*

(i) *G is a cyclic extension of K^*, and*

$$\operatorname{card} G = n\,(p^{nt} - 1). \tag{1}$$

(ii) *If Δ^n is identified with K, the elements g of G can be written in the form*

$$g(x) = \lambda x^{p^{tu}}, \tag{2}$$

where $x \in K$, $\lambda \in K^*$ and $u = 0, 1, \ldots, n-1$.

PROOF. By assumption, $\Delta = GF(p^t)$ and $K = GF(p^{nt})$. Hence $\mathcal{G} = \mathcal{G}(K, \Delta)$ is cyclic of order n. By Lemma 1, G/F is cyclic of order n. This gives (i). To prove (ii), we put $\Delta^n = K$. Then F consists of all mappings f of the form $f: K \longrightarrow K$, $f(x) = \lambda x$, $x \in K$, $\lambda \in K^*$. By Lemma 1, $N(F): F = n$ and $N(F) = G$. On the other hand, the number of elements g of the form (2) is equal to card $N(F)$, and for such elements $gFg^{-1} = F$. Therefore $G = N(F)$ consists of all elements of the form (2). ∎

We denote the subgroup of $GL(n, \Delta)$ consisting of all elements (2) by

$$G_{n(p^{nt}-1)}. \tag{3}$$

LEMMA 2. *Suppose that the pair $\langle n, \Delta\rangle$, where n is an integer and $\Delta = GF(p^t)$, is subject to the following condition:*

(a) *For any divisor d of n, $1 \leqslant d < n$, there exists a prime divisor q of n/d such that $q \nmid p^{dt} - 1$.*

Then any maximal solvable primitive subgroup of $GL(n, \Delta)$ is conjugate in $GL(n, \Delta)$ to $G_{n(p^{nt}-1)}$.

This assertion follows from Corollary 20.3.1 and Theorem 1.

THEOREM 2. *Let $\Delta = GF(p^t)$ and suppose that for every divisor n_1 of n, $1 < n_1 \leqslant n$, the pair $\langle n_1, \Delta\rangle$ satisfies* (a) *in Lemma 2. Then every maximal irreducible solvable subgroup G of $GL(n, \Delta)$ is conjugate in $GL(n, \Delta)$ either to $G_{n(p^{nt}-1)}$ or to $G_{n_1(p^{n_1 t}-1}$ $\wr$ Γ, where Γ is a maximal transitive solvable subgroup of the symmetric group S_k, $k = n/n_1$.*

PROOF. If G is primitive, the theorem follows from Lemma 2. Let G be imprimitive. Then, by Theorem 18.5, G is conjugate in $GL(n, \Delta)$ to $\mathfrak{G} \wr \Gamma$, where $\mathfrak{G}$ is a maximal primitive solvable subgroup of $GL(n_1, \Delta)$ and Γ a maximal transitive solvable subgroup of the symmetric group S_k, $k = n/n_1$. By Lemma 2, $\mathfrak{G}$ is conjugate in $GL(n_1, \Delta)$ to $G_{n_1(p^{n_1 t}-1)}$. Thus G is conjugate to $G_{n_1(p^{n_1 t}-1)} \wr \Gamma$. ∎

The assumptions of Theorem 2 are satisfied, in particular, if $n = q^l$ and $\Delta = GF(p^t)$, where q is a prime such that $q \nmid p^t - 1$. For in this case it is easy to verify that $(q, p^{nt} - 1) = 1$.

2. *Generators of Q.* Let Δ be an *arbitrary* field, G a maximal primitive solvable subgroup of $GL(n, \Delta)$, A and $F = K^*$ groups in the series (α) (see §20.2), $K: \Delta = m < n$, q a prime divisor of n/m, and Q/F a Sylow q-subgroup of A/F.

LEMMA 3. *Let $Q = BC$, where B and C are subgroups of Q such that $(B, C) = 1$ and $C \supset F$. If D is a normal subgroup of C, if C/D is locally cyclic, and if the center of D lies in F, then $(Q, C) = (D, D)$.*

PROOF. Since $(B, C) = 1$, we have $(Q, C) = (C, C)$. There are only two possibilities: $(D, D) = 1$ or $(D, D) = (Q, Q)$. In the second case the lemma is obvious. We consider the case $(D, D) = 1$. Then, by assumption, $Z(D) = D \subset F$. Since C/D is locally cyclic, C is abelian. Consequently $(Q, C) = (C, C) = 1 = (D, D)$. ∎

We prove a further lemma about Q.

LEMMA 4. *Let Δ be an algebraic extension of a finite field, and $Q_0 = Q \cap SL(n, \Delta)$. Then $(Q_0, Q_0) \neq 1$.*

PROOF. We consider the mapping $\delta : Q \longrightarrow \Delta^*, x \longmapsto \det x$, for which, obviously, $\operatorname{Ker} \delta = Q_0$, $\operatorname{Im} \delta \subset \Delta^*$ and $\operatorname{Im} \delta \cong Q/Q_0$. Since Δ^* is locally cyclic, so is Q/Q_0. Clearly the center $Z(Q_0)$ of Q_0 is normal in G. Hence $Z(Q_0) \subset F$. If we now set $B = F$, $C = Q$ and $D = Q_0$, then B, C and D satisfy the conditions of Lemma 3. Therefore $(Q_0, Q_0) = (Q, Q) \neq 1$. ∎

By Corollary 20.5.1, Q can be expressed in the form

$$Q = (c_1)(d_1) \ldots (c_l)(d_l) F, \tag{4}$$

$$(c_j, d_j) = \eta, \quad \eta^q = 1, \quad \eta \neq 1, \quad \eta \in F, \quad c_j^q, d_j^q \in F, \tag{5}$$

where elements of distinct pairs c_j, d_j commute.

THEOREM 3. *Let Δ be an algebraic extension of a finite field. Then the generators c_ν, d_ν of Q can be chosen so that, apart from* (4) *and* (5), *also*

$$c_\nu^q = d_\nu^q = 1, \qquad \nu = 1, \ldots, l, \tag{6}$$

(for $q > 2$), and

$$c_\nu^2 = \pm 1, \quad d_\nu^2 = \pm 1, \qquad \nu = 1, \ldots, l \tag{7}$$

(for $q = 2$).

PROOF. We set

$$\begin{aligned} B &= (c_1)(d_1) \ldots (c_{j-1})(d_{j-1})(\eta), \\ C &= (c_j)(d_j) \ldots (c_l)(d_l) F, \end{aligned} \quad D = \begin{cases} \{x \mid x \in C,\ x^q = 1\}, & \text{for } q > 2, \\ \{x \mid x \in C,\ x^q = \pm 1\}, & \text{for } q = 2. \end{cases} \tag{8}$$

Then $Q = BC$, $(B, C) = 1$ and $C \supset F$. Let $u, v \in D$. For $q > 2$,

$$u^q = v^q = 1, \; vu = \lambda uv, \; \lambda \in F, \; \lambda^q = 1;$$

$$(uv)^q = \lambda^{\frac{q(q-1)}{2}} u^q v^q = 1.$$

Hence $uv \in D$. For $q = 2$,

$$u^2 = \pm 1, \quad v^2 = \pm 1, \quad vu = \pm uv, \quad (uv)^2 = \pm 1,$$

so that again $uv \in D$. It follows that D is normal in C.

The factor group C/D is locally cyclic. For if $q > 2$, we consider the mapping $\gamma: C \longrightarrow F$, $x \longmapsto x^q$. It is easy to compute that

$$\gamma(x, y) = (y, x)^{\frac{q(q-1)}{2}} x^q y^q = x^q y^q = \gamma(x)\gamma(y).$$

Therefore γ is a homomorphism, $\operatorname{Im}\gamma \subset F$ and $\operatorname{Ker}\gamma = D$. Since Δ is an algebraic extension of a finite field, $F = K^*$ is a locally cyclic group. Thus for $q > 2$ the group $C/D \cong \operatorname{Im}\gamma$ is locally cyclic. But if $q = 2$, we consider the mapping

$$\gamma_1\colon C/\{\pm 1\} \to F/\{\pm 1\}, \qquad \{\pm 1\}x \longmapsto \{\pm 1\}x^2.$$

It is clear that γ_1 is a homomorphism, $\operatorname{Ker}\gamma_1 = D/\{\pm 1\}$, and

$$C/D \cong C/\{\pm 1\}/D/\{\pm 1\} \cong \operatorname{Im}\gamma_1 \subset F/\{\pm 1\}.$$

Hence C/D is also locally cyclic for $q = 2$. To make Lemma 3 applicable we must still show that the center $Z(D)$ of D lies in F. First let $j = 1$. Then $B = (\eta)$, $C = Q$, and D is normal in G. Hence $Z(D) \subset F$. Thus for $j = 1$ the groups B, C and D satisfy the conditions of Lemma 3, so that $(D, D) = (Q, C) = (C, C)$. Hence $(D, D) \neq 1$. We may thus assume that one pair c_1, d_1 lies in D; that is,

$$c_1^q = d_1^q = 1 \qquad \text{for} \quad q > 2,$$
$$c_1^2 = \pm 1,\ d_1^2 = \pm 1 \qquad \text{for} \quad q = 2.$$

Now suppose that the $j - 1$ pairs $c_1, d_1, \dots, c_{j-1}, d_{j-1}$ satisfy the conditions

$$c_1^q = d_1^q = \dots = c_{j-1}^q = d_{j-1}^q = 1 \qquad \text{for} \quad q > 2,$$
$$c_1^2 = \pm 1,\ d_1^2 = \pm 1,\ \dots,\ c_{j-1}^2 = \pm 1,\ d_{j-1}^2 = \pm 1 \qquad \text{for} \quad q = 2,$$

and that B, C and D are defined by (8). We define a group $\bar{Q}$ by

$$\bar{Q} = \{x \mid x \in Q,\ x^q = 1\} \qquad \text{for} \quad q > 2,$$
$$\bar{Q} = \{x \mid x \in Q,\ x^2 = \pm 1\} \qquad \text{for} \quad q = 2.$$

$\bar{Q}$ is normal in G, and so its center $Z(\bar{Q})$ lies in F. Clearly, $B \subset \bar{Q}$. Let $a \in \bar{Q}$; then

$$a = bc, \quad b \in B, \quad c \in C. \tag{9}$$

By (9), $c \in D$; that is, $\bar{Q} = BD$. Since $(B, D) = 1$, it follows that $Z(D) \subset Z(\bar{Q}) \subset F$. Thus B, C and D satisfy the conditions of Lemma 3. Therefore c_j and d_j can be chosen in D. This completes the proof. ■

This theorem enables us to put the matrices of A into a canonical form, provided that the ground field Δ is an algebraic extension of a finite field.

According to §20, the group A in the series (α) can be brought to the form $A = Q^1 \dot{\times} \cdots \dot{\times} Q^k$. Hence it is sufficient to consider the groups $Q^j \cong Q_j$.

THEOREM 4. *Let Δ be an algebraic extension of a finite field, and $q_j = q > 2$. Then in a suitable basis of $K^{q\tau}$ the group Q^j can be put in the form*

$$Q^j = (u_1)(v_1)\ldots(u_l)(v_l)F, \tag{10}$$

with

$$\begin{aligned} u_\nu &= E_{q^{\nu-1}} \dot{\times} c \dot{\times} E_{q^{l-\nu}}, \\ v_\nu &= E_{q^{\nu-1}} \dot{\times} d \dot{\times} E_{q^{l-\nu}}, \end{aligned} \tag{11}$$

where

$$c = \operatorname{diag}[1, \eta, \ldots, \eta^{q-1}], \quad d = \begin{bmatrix} 0 & 1 \\ E_{q-1} & 0 \end{bmatrix},$$

$$\eta^q = 1, \quad \eta \neq 1, \quad \eta \in K^*, \quad \nu = 1, \ldots, l.$$

PROOF. Let $\phi: Q_j \longrightarrow Q^j$ be an isomorphism, and let $\bar{c}_\nu = \phi(c_\nu)$ and $\bar{d}_\nu = \phi(d_\nu)$, $\nu = 1, \ldots, l$ (see (4)). By Theorem 3,

$$\bar{c}_\nu^q = \bar{d}_\nu^q = 1, \quad \nu = 1, \ldots, l.$$

On the other hand, as is readily seen,

$$u_\nu^q = v_\nu^q = 1, \quad (u_\nu, v_\nu) = (\bar{c}_\nu, \bar{d}_\nu) = \eta,$$

and matrices belonging to distinct pairs u_ν, v_ν commute. Hence the mapping $\bar{c}_\nu \longmapsto u_\nu$, $\bar{d}_\nu \longmapsto v_\nu$, $\nu = 1, \ldots, l$, can be extended to a central automorphism of the algebra $[Q^j] = K_{q^l}$. Therefore in $GL(q^l, K)$ there is a matrix t such that $t\bar{c}_\nu t^{-1} = u_\nu$ and $t\bar{d}_\nu t^{-1} = v_\nu$, $\nu = 1, \ldots, l$. ■

THEOREM 5. *Let Δ be an algebraic extension of a finite field, and $q_j = q = 2$.*

(i) *If K^* contains an element of order* 4, *then Q^j is defined by* (10) *and* (11).

(ii) *If K^* contains no element of order* 4, *there are two possibilities*:

1) *Q^j is defined by* (10) *and* (11), *with $q = 2$ and $\eta = -1$.*

2) *Q^j is defined by* (10), *and u_ν, v_ν for $\nu \neq l$ are given by* (11), *where $q = 2$, $\eta = -1$, and*

$$u_l = E_{2^{l-1}} \dot{\times} \begin{bmatrix} 0 & -1 \\ 1 & 0 \end{bmatrix}, \quad v_l = E_{2^{l-1}} \dot{\times} \begin{bmatrix} \gamma & \delta \\ \delta & -\gamma \end{bmatrix},$$

$$\gamma^2 + \delta^2 + 1 = 0, \quad \gamma, \delta \in K.$$

PROOF. Suppose that K^* contains an element of order 4. Then in Theorem 3 we can write $c_\nu^2 = d_\nu^2 = 1$, $\nu = 1, \ldots, l$, instead of (7). Thus Theorem 4 is valid in this case for $q = 2$, and (i) is proved. Now suppose that there are no elements of order 4 in K^*. If $c_\nu^2 = d_\nu^2 = 1$, $\nu = 1, \ldots, l$, then Q^j is defined by (10) and (11). It remains to consider the case when for some ν either $c_\nu^2 = -1$ or $d_\nu^2 = -1$. If $c_\nu^2 = -d_\nu^2 = \pm 1$, the pair c_ν, d_ν can be replaced by a pair c'_ν, d'_ν such that $c'^2_\nu = d'^2_\nu = 1$. For suppose, say, that $c_\nu^2 = -d_\nu^2 = -1$. We then replace c_ν, d_ν by the pair $c_\nu d_\nu, d_\nu$; $(c_\nu d_\nu)^2 = -c_\nu^2 d_\nu^2 = 1$. We may therefore assume that $c_\nu^2 = d_\nu^2 = \pm 1$. We now show that c_j, d_j can be chosen so that

$c_j^2 = d_j^2 = 1, j = 1, \ldots, l-1$, and $c_l^2 = d_l^2 = -1$. Suppose we have two pairs

$$c_\nu^2 = d_\nu^2 = -1, \quad c_\mu^2 = d_\mu^2 = -1, \quad \mu \neq \nu.$$

We then replace the pairs c_ν, d_ν and c_μ, d_μ by $c_\nu, d_\nu d_\mu$ and $c_\nu c_\mu, d_\mu$; $(d_\nu d_\mu)^2 = (c_\nu c_\mu)^2 = 1$. By the preceding arguments we can replace the last two pairs by pairs c'_ν, d'_ν and c'_μ, d'_μ so that $c'^2_\nu = d'^2_\nu = c'^2_\mu = d'^2_\mu = 1$. Consequently c_j and d_j can be chosen so that

$$c_1^2 = d_1^2 = \ldots = c_{l-1}^2 = d_{l-1}^2 = 1, \quad c_l^2 = d_l^2 = -1$$

The matrices u_ν and v_ν in Theorem 5 are subject to the same conditions. Hence, as in the proof of Theorem 4, we obtain (ii). ∎

3. *Irreducible solvable groups of prime degree.* In this subsection we construct all the maximal irreducible solvable subgroups of $GL(q, \Delta)$, where q is a prime and $\Delta = GF(p^t)$. We observe forthwith that $GL(q, \Delta)$ is solvable when $q = 2$ and Δ is either $GF(2)$ or $GF(3)$, for in this case the order of $GL(q, \Delta)$ is 6 or 48.

Let G be a maximal irreducible solvable subgroup of $GL(q, \Delta)$. Three cases are possible:

1) G is imprimitive.

2) G is primitive, and its maximal abelian normal subgroup F is equal to K^*, where $K = GF(p^{qt})$.

3) G is primitive, and its maximal abelian normal subgroup F is $\Delta^* E_q$.

In the first case G is monomial, and by Theorem 18.5 G is conjugate in $GL(q, \Delta)$ to $\Delta^* \wr \Gamma$, where Γ is a maximal transitive solvable subgroup of the symmetric group S_q. By Theorem 4.11 the maximal solvable transitive subgroups of S_q are conjugate in S_q and of order $q(q-1)$. In this case, therefore,

$$\operatorname{card} G = q(q-1)(p^t-1)^q. \tag{12}$$

In the second case, by Theorem 1 G is conjugate in $GL(q, \Delta)$ to $G_{q(p^{qt}-1)}$.

The third case is much more complicated. Suppose that G is primitive and its maximal abelian normal subgroup F is $\Delta^* E_q$. Then G has a normal subgroup

$$\begin{gathered} A = (a)(b)F, \quad a^q \in F, \qquad b^q \in F, \\ (a, b) = \eta E_q, \quad \eta^q = 1, \quad \eta \neq 1, \quad \eta \in \Delta. \end{gathered} \tag{13}$$

Now it follows from (13) that there is an element of order q in Δ^*; hence the third case is possible only if $q \mid p^t - 1$. G is obviously a subgroup of the normalizer N of A in $GL(q, \Delta)$. For $x \in N$ we can write $xax^{-1} = \lambda a^\alpha b^\gamma$ and $xbx^{-1} = \mu a^\beta b^\delta$, where $\lambda, \mu \in \Delta$ and $\alpha, \beta, \gamma, \delta \in GF(q)$. We set

$$\psi\colon N \to Sp(2, q) = SL(2, q), \qquad x \mapsto \begin{bmatrix} \alpha & \beta \\ \gamma & \delta \end{bmatrix}. \tag{14}$$

First let $q = 2$ and suppose that Δ^* contains an element i of order 4; that is,

$p^t - 1 \equiv 0 \pmod 4$. Then by Theorem 4 we can choose

$$a = \begin{bmatrix} 1 & 0 \\ 0 & -1 \end{bmatrix}, \quad b = \begin{bmatrix} 0 & 1 \\ 1 & 0 \end{bmatrix}. \tag{15}$$

By Theorem 20.16, the homomorphism $\psi : N \longrightarrow SL(2, 2)$ is surjective. Since $\operatorname{Ker} \psi = A$, we have $N/A \cong SL(2, 2) \cong S_3$. Consequently N is solvable, $G = N$, and

$$\operatorname{card} G = 6 \cdot 4(p^t - 1) = 24(p^t - 1). \tag{16}$$

The group $SL(2, 2)$ can be generated by the two matrices

$$\sigma = \begin{bmatrix} 1 & 1 \\ 0 & 1 \end{bmatrix}, \quad \tau = \begin{bmatrix} 0 & 1 \\ 1 & 1 \end{bmatrix}.$$

We set

$$g = \operatorname{diag}[1, -i], \quad h = \begin{bmatrix} 1 & i \\ 1 & -i \end{bmatrix}. \tag{17}$$

It is readily checked that $g, h \in N$, $\psi(g) = \sigma$ and $\psi(h) = \tau$. Therefore the elements

$$a,\ b,\ g,\ h,\ \rho E_2, \tag{18}$$

where $\rho \in \Delta^*$ is of order $p^t - 1$, a and b are given by (15), and g and h by (17), form a system of generators of G.

Now let $q = 2$, but suppose that Δ^* contains no elements of order 4. By Lemma 4, the elements a and b in (13) may be chosen so that

$$\det a = \det b = 1. \tag{19}$$

From (19) and the fact that $a^2, b^2 \in \Delta^* E_2$ it follows that $z^2 + 1$ is the minimal polynomial of a and b. We can therefore set

$$a = \begin{bmatrix} 0 & -1 \\ 1 & 0 \end{bmatrix}.$$

Since $(a, b) = b^2 = -E_2$, we now find

$$b = \begin{bmatrix} \theta & \omega \\ \omega & -\theta \end{bmatrix}, \quad \theta^2 + \omega^2 = -1, \quad \theta, \omega \in \Delta^*.$$

As generators for $SL(2, 2)$ we can take the matrices σ and

$$\tau_1 = \begin{bmatrix} 0 & 1 \\ 1 & 0 \end{bmatrix}.$$

It is easy to check that the matrices

$$g_1 = \begin{bmatrix} 1 & -1 \\ 1 & 1 \end{bmatrix}, \quad h_1 = \begin{bmatrix} \theta & \omega - 1 \\ \omega + 1 & -\theta \end{bmatrix} \tag{20}$$

satisfy the following conditions:

$$g_1 a g_1^{-1} = a, \qquad h_1 a h_1^{-1} = b, \quad g_1 b g_1^{-1} = ab, \quad h_1 b h_1^{-1} = a.$$

Hence $g_1, h_1 \in N$ and

$$\psi(g_1) = \sigma, \quad \psi(h_1) = \tau_1, \quad N/A \cong SL(2, 2),$$
$$N = G, \qquad \operatorname{card} G = 24(p^t - 1).$$

Now G can be generated by the matrices

$$a, \ b, \ g_1, \ h_1, \ \rho E_2, \tag{21}$$

where

$$a = \begin{bmatrix} 0 & -1 \\ 1 & 0 \end{bmatrix}, \ b = \begin{bmatrix} \theta & \omega \\ \omega & -\theta \end{bmatrix}, \ \theta^2 + \omega^2 = -1, \ \theta, \ \omega \in \Delta^*,$$

g_1 and h_1 are given by (20), and ρ is an element of order $p^t - 1$ in Δ^*.

This completes discussion of the case $q = 2$.

Now let $q > 2$. Then, by Theorem 4, the matrices a and b may be chosen as follows:

$$a = \operatorname{diag}[1, \eta, \ldots, \eta^{q-1}], \qquad b = \begin{bmatrix} 0 & 1 \\ E_{q-1} & 0 \end{bmatrix},$$
$$a^q = b^q = E_q, \qquad \eta^q = 1, \quad \eta \neq 1, \quad \eta \in \Delta^*.$$

By Theorem 20.16, the homomorphism ψ is surjective. Therefore $N/A \cong SL(2, q)$ and $G = \psi^{-1}(H)$, where H is a maximal solvable subgroup of $SL(2, q)$.

LEMMA 5. *Let H be a subgroup of $SL(2, q)$. Then $\psi^{-1}(H)$ is primitive if and only if H is irreducible.*

PROOF. If H is an irreducible subgroup of $SL(2, q)$, then $\psi^{-1}(H)$ is primitive by Lemma 20.11a. Now let H be a reducible subgroup of $SL(2, q)$. Then there is a matrix d in $GL(2, q)$ such that the matrices in dHd^{-1} have the form

$$\begin{bmatrix} \alpha & \beta \\ 0 & \delta \end{bmatrix}, \qquad \alpha, \ \beta, \ \delta \in GF(q). \tag{22}$$

We may assume that $d \in SL(2, q)$. For let $\det d = \delta_1$ and $\delta_2 \delta_1 = 1$. Then

$$d_1 = \begin{bmatrix} 1 & 0 \\ 0 & \delta_2 \end{bmatrix} d \in SL(2, q),$$

and the matrices in $d_1 H d_1^{-1} = H_1$ have the form (22). Now $\psi^{-1}(H_1)$ is imprimitive. For since the matrices of H_1 are of the form (22), for any g in $\psi^{-1}(H_1)$ we have

$$gag^{-1} = \lambda a^{\alpha}, \qquad g(a)Fg^{-1} = (a)F,$$

so that $(a)F$ is an abelian normal subgroup of $\psi^{-1}(H_1)$. By Theorem 16.2, $\psi^{-1}(H_1)$ is imprimitive. Since H_1 and H are conjugate in $SL(2, q)$, it follows that $\psi^{-1}(H_1)$ and $\psi^{-1}(H)$ are conjugate in N. Thus $\psi^{-1}(H)$ is also imprimitive. ∎

This lemma reduces the construction of all maximal solvable primitive subgroups of $GL(q, \Delta)$ to that of the system of all maximal solvable irreducible subgroups of $SL(2, q)$ and the partition of this system into conjugacy classes.

If $q = 3$, $SL(2, q)$ is a solvable group of order 24. Hence for $q = 3$ we have $G = \psi^{-1}(SL(2, 3))$ and

$$\operatorname{card} G = 24 \cdot 9 \cdot (p^t - 1) = 216(p^t - 1). \tag{23}$$

Now let $q > 3$. Let us find the maximal solvable irreducible subgroups of $SL(2, q)$. As proved above, a maximal solvable irreducible subgroup of $GL(2, q)$ is conjugate in $GL(2, q)$ to one of the following groups:(1)

G_1, of order $2(q-1)^2$ (G_1 is imprimitive),
G_2, of order $2(q^2-1)$ (G_2 is primitive, $\operatorname{card} F = q^2 - 1$),
G_3, of order $24(q-1)$ (G_3 is primitive, $\operatorname{card} F = q - 1$).

A maximal solvable irreducible subgroup of $SL(2, q)$ is the intersection of a maximal solvable irreducible subgroup of $GL(2, q)$ with $SL(2, q)$.

We consider the intersections $U_j = G_j \cap SL(2, q)$, $j = 1, 2, 3$. The group G_1 is the set of all matrices of the form

$$\begin{bmatrix} \lambda_1 & 0 \\ 0 & \lambda_2 \end{bmatrix} \begin{bmatrix} 0 & 1 \\ 1 & 0 \end{bmatrix}^{\tau},$$

$$\lambda_1, \lambda_2 \in GF(q), \quad \tau = 0, 1, \quad \lambda_1, \lambda_2 \neq 0.$$

Therefore U_1 consists of all matrices of the form

$$\begin{bmatrix} \lambda & 0 \\ 0 & (-1)^{\tau}\lambda^{-1} \end{bmatrix} \begin{bmatrix} 0 & 1 \\ 1 & 0 \end{bmatrix}^{\tau},$$

$$\lambda \in GF(q), \quad \tau = 0, 1, \quad \lambda \neq 0;$$

$\operatorname{card} U_1 = 2(q-1)$. U_1 can be generated by two matrices u and v:

$$u = \operatorname{diag}[\rho, \rho^{-1}], \qquad v = \begin{bmatrix} 0 & 1 \\ -1 & 0 \end{bmatrix},$$

where ρ is an element of order $q - 1$ in $GF(q)$. The group G_2 of type $G_{n(p^{nt}-1)}$ (see (3)) consists of the elements g of the form

$$g(x) = \lambda x^{q^{\tau}}, \quad x, \lambda \in GF(q^2), \quad \lambda \neq 0, \quad \tau = 0, 1.$$

G_2 can be generated by two elements g_0 and f_0, where

$$g_0(x) = x^q, \qquad f_0(x) = \mu x, \qquad x \in GF(q^2),$$

and μ is an element of order $q^2 - 1$ in $GF(q^2)$. Let us find $\det g_0$ and $\det f_0$.

---

(1) When $q = 5$, G_1 is not a maximal solvable subgroup of $GL(2, 5)$. Indeed $\operatorname{card} GL(2, 5) = 2^5 \cdot 15$, $\operatorname{card} G_1 = 2^5$ and $\operatorname{card} G_3 = 2^5 \cdot 3$. The group G_1 is a Sylow 2-subgroup of $GL(2, 5)$ and is conjugate in $GL(2, 5)$ to a subgroup of G_3.

Let $\omega \in GF(q^2)\backslash GF(q)$. Then $1, \omega$ is a basis of $GF(q^2)$ over $GF(q)$; $\omega + \omega^q = \nu \in GF(q)$. Next, since $g_0(1) = 1$ and $g_0(\omega) = \omega^q = \nu - \omega$, the matrix of g_0 is

$$\begin{bmatrix} 1 & \nu \\ 0 & -1 \end{bmatrix} \qquad \det g_0 = -1.$$

To calculate $\det f_0$ we choose a basis $1, \mu$ of $GF(q^2)$ over $GF(q)$:

$$f_0(1) = \mu, \quad f_0(\mu) = \mu^2 = -\mu^{q+1} + (\mu^q + \mu)\mu;$$

$$\det f_0 = \begin{vmatrix} 0 & -\mu^{q+1} \\ 1 & \mu^q + \mu \end{vmatrix} = \mu^{q+1}.$$

Therefore the homomorphism $G_2 \longrightarrow (GF(q))^*$, $x \longmapsto \det x$, is surjective and its kernel is U_2. Hence card $U_2 = 2(q + 1)$.

Finally, we consider G_3 and U_3. G_3 is generated either by the elements (18) or by the elements (21) (with p^t replaced by q in both (18) and (21)). In the first case, $GF(q)$ contains an element i of order 4. It is easy to see that $\det a$, $\det b$ and $\det \rho E_2$ are squares in $GF(q)$. Now $h^3 = (2 + 2i)E_2$, so that $\det h$ is also a square in $GF(q)$. By (17), $\det g = -i$. Thus $\det g$ is a square in $GF(q)$ if and only if $q \equiv 1 \pmod 8$. Now suppose that G_3 is generated by the elements (21), $\det g_1 = 2$ and $\det h_1 = 2$. Then $\det x$ is a square in $GF(q)$ for all x in G_3 if and only if 2 is a quadratic residue modulo q; that is, $q \equiv \pm 1 \pmod 8$. It follows that in either case the homomorphism

$$G_3 \longmapsto (GF(q))^*, \qquad x \longmapsto \det x$$

is surjective if and only if $q \equiv \pm 3 \pmod 8$.

Thus, if $q \equiv \pm 3 \pmod 8$, then

$$\text{card } U_3 = \frac{24(q-1)}{q-1} = 24.$$

If $q \equiv \pm 1 \pmod 8$, then

$$\text{card } U_3 = \frac{24(q-1)}{(q-1)/2} = 48.$$

We have now found the groups $U_j = G_j \cap SL(2, q)$, $j = 1, 2, 3$. It should be observed, however, that conjugacy of two groups B and B_1 in $GL(n, P)$ does not necessarily imply conjugacy of $B \cap SL(n, P)$ and $B_1 \cap SL(n, P)$ in $SL(n, P)$. If P is an arbitrary field, B a subgroup of $GL(n, P)$, and the homomorphism

$$B \to P^*, \qquad x \longmapsto \det x \tag{24}$$

is surjective, then conjugacy of B and B_1 in $GL(n, P)$ does imply conjugacy of $B \cap SL(n, P)$ and $B_1 \cap SL(n, P)$ in $SL(n, P)$. For let $B_1 = tBt^{-1}$, $t \in GL(n, P)$. Since the homomorphism (24) is surjective, there is a matrix b in B such that $\det b = (\det t)^{-1}$. We set $d = tb$. Then $\det d = 1$, $dBd^{-1} = B_1$, and

$$B_1 \cap SL(n, P) = dBd^{-1} \cap dSL(n, P)d^{-1} = d(B \cap SL(n, P))d^{-1}.$$

We have seen that the mapping $G_j \longrightarrow (GF(q))^*$, $x \longmapsto \det x$, is always surjective for $j = 1, 2$, while for $j = 3$ it is surjective only if $q \equiv \pm 3 \pmod 8$. It remains to consider the group G_3 for $q \equiv \pm 1 \pmod 8$. Let r be a matrix in $GL(2, q)$ such that $\det r$ is not square in $(GF(q))^*$. As is easy to see, any group of the form dU_3d^{-1}, where $d \in GL(2, q)$, is conjugate in $SL(2, q)$ to one of the two groups U_3 or $U_4 = rU_3r^{-1}$. We claim that U_3 and U_4 are not conjugate in $SL(2, q)$. Suppose that $lU_4l^{-1} = U_3$, $l \in SL(2, q)$. Then $lrU_3(lr)^{-1} = U_3$. Since $G_3 = U_3(GF(q))^*$ for $q \equiv \pm 1 \pmod 8$, it follows that $lrG_3(lr)^{-1} = G_3$. Now $d = lr \notin G_3$, because $\det lr$ is not a square. Hence $(d)G$ is a solvable group containing G_3 as a proper subgroup. This contradicts the maximality of G_3. Thus U_3 and U_4 are not conjugate in $SL(2, q)$.

Thus, if $q > 3$, any maximal primitive solvable subgroup of $GL(q, \Delta)$ with maximal abelian normal subgroup $F = \Delta^* E_q$ is conjugate in $GL(q, \Delta)$ to one of the four groups $\psi^{-1}(U_j)$, $j = 1, \ldots, 4$. Hence and from the preceding arguments of this subsection we obtain

THEOREM 6. *Let $\Delta = GF(p^t)$, where p is a prime. Then the following assertions are true*:

1. *Every maximal irreducible solvable subgroup of $GL(2, \Delta)$ is conjugate in $GL(2, \Delta)$ to one of the following three subgroups*:
   $G_{2,1}$ *of order* $2(p^t - 1)^2$,
   $G_{2,2}$ *of order* $2(p^{2t} - 1)$,
   $G_{2,3}$ *of order* $24(p^t - 1)$.
2. *Every maximal irreducible solvable subgroup of $GL(3, \Delta)$ is conjugate in $GL(3, \Delta)$ to one of the following three subgroups*:
   $G_{3,1}$ *of order* $6(p^t - 1)^3$,
   $G_{3,2}$ *of order* $3(p^{3t} - 1)$,
   $G_{3,3}$ *of order* $216(p^t - 1)$.
3. *Every maximal irreducible solvable subgroup of $GL(q, \Delta)$, where $q > 3$ is a prime, is conjugate in $GL(q, \Delta)$ to one of the following six subgroups*:
   $G_{q,1}$ *of order* $q(q - 1)(p^t - 1)^q$,
   $G_{q,2}$ *of order* $q(p^{qt} - 1)$,
   $G_{q,3}$ *of order* $2(q - 1)q^2(p^t - 1)$,
   $G_{q,4}$ *of order* $2(q + 1)q^2(p^t - 1)$,
   $G_{q,5}$ *of order* $48(p^t - 1)q^2$ *if* $q \equiv \pm 1 \pmod 8$
   or of order $24(p^t - 1)q^2$ *if* $q \equiv \pm 3 \pmod 8$.
   $G_{q,6}$ *of order* $48(p^t - 1)q^2$.

The groups $G_{q,3}, \ldots, G_{q,6}$ exist only if $p^t \equiv 1 \pmod q$, and $G_{q,6}$ exists only if, in addition, $q \equiv \pm 1 \pmod 8$. Moreover,

$$G_{q,j+2} = \psi^{-1}(U_j), \quad G_{q,j+2}/A \cong U_j, \quad j = 1, 2, 3, 4.$$

REMARKS TO THEOREM 6. 1) The groups $G_{2,1}$, $G_{3,1}$ and $G_{q,1}$ are monomial.

2) $G_{2,2}$, $G_{3,2}$ and $G_{q,2}$ are groups of the form $G_{n(p^{nt}-1)}$ ($n = 2, 3, q$; see (3)).

3) The group $G_{2,3}$ exists for $p > 2$, $G_{2,1}$ for $p^t > 2$, and $G_{3,3}$ for $3 \mid p^t - 1$.

4) If $p^t = 2$, then for any prime q every maximal irreducible solvable subgroup of $GL(q, 2)$ is conjugate in $GL(q, 2)$ to the group $G_{q,2}$ of order $q(2^q - 1)$.

5) If $p^t = 3$, the groups $G_{2,1}$ and $G_{2,2}$ are not maximal; they are subgroups of $G_{2,3} = GL(2, 3)$.

6) If $p^t = 5$, the group $G_{2,1}$ is not maximal; it is conjugate to a proper subgroup of $G_{2,3}$.

7) $GL(5, \Delta)$ contains either two or four nonconjugate maximal irreducible solvable subgroups. If $5 \nmid p^t - 1$, these are $G_{5,1}$ and $G_{5,2}$. If $5 \mid p^t - 1$, the only additional subgroups are $G_{5,4}$ and $G_{5,5}$, because by 6) $G_{5,3}$ is not maximal, being conjugate to a proper subgroup of $G_{5,5}$, while $G_{5,6}$ does not exist at all, since $5 \not\equiv \pm 1 \pmod 8$.

The list of the maximal irreducible solvable subgroups of $GL(q, \Delta)$ is complete.

§22. Miscellany

1. *Irreducible solvable groups of prime degree over an algebraically closed field.* Let q be a prime and Δ an algebraically closed field. The preceding results yield a complete classification of the maximal irreducible solvable subgroups of $GL(q, \Delta)$. There are two cases: $q = \operatorname{char} \Delta$ or $q \neq \operatorname{char} \Delta$. In the first case, by Corollary 20.3.1, $GL(q, \Delta)$ has no primitive solvable subgroups, and so any irreducible solvable subgroup of $GL(q, \Delta)$ is monomial. Hence:

THEOREM 1. *Let $q = \operatorname{char} \Delta$. Then any maximal irreducible solvable subgroup of $GL(q, \Delta)$ is conjugate in $GL(q, \Delta)$ to a group*

$$\Delta^* \sim \Gamma, \tag{1}$$

where Γ is a maximal transitive solvable permutation group of degree q and order $q(q-1)$.

Now let $q \neq \operatorname{char} \Delta$. We wish to find the maximal irreducible solvable subgroups of $GL(q, \Delta)$. If such a group is imprimitive, it is obviously conjugate in $GL(q, \Delta)$ to a group (1). It remains to find the maximal primitive solvable subgroups of $GL(q, \Delta)$. By Theorem 20.19, any such subgroup is conjugate in $GL(q, \Delta)$ to a group of the form $\psi^{-1}(H)$, where H is a maximal s-irreducible subgroup of $Sp(2, q) = SL(2, q)$, and conjugacy of H and H_1 in $Sp(2, q)$ implies that of $\psi^{-1}(H)$ and $\psi^{-1}(H_1)$ in $GL(q, \Delta)$. It follows from the definition of s-irreducibility that subgroups of $Sp(2, q)$ are s-irreducible if and only if they are irreducible. To construct all the maximal primitive solvable subgroups of $GL(q, \Delta)$, therefore, we must find the maximal irreducible solvable subgroups of $SL(2, q)$ and split them into conjugacy classes. In the proof of Theorem 21.6 we have found maximal irreducible solvable subgroups U_1, U_2, U_3, U_4 of $SL(2, q)$ such that any maximal solvable irreducible sugroupb of $SL(2, q)$ is conjugate in $SL(2, q)$ to one of them. Hence and from the properties of the U_i we obtain

THEOREM 2. *Let $q \neq$ char Δ. Then every maximal primitive solvable subgroup of $GL(q, \Delta)$ is conjugate in $GL(q, \Delta)$ to one of the following $k = k(q)$ groups*:

$$G_{q_1}, \ldots, G_{q_k}. \tag{2}$$

Each of the groups (2) *has a normal subgroup $A = (a)(b)\Delta^*$, where*

$$a = \operatorname{diag}[1, \eta, \ldots, \eta^{q-1}], \ \eta \in \Delta, \ \eta^q = 1, \quad \eta \neq 1, \ b = \begin{bmatrix} 0 & 1 \\ E_{q-1} & 0 \end{bmatrix}.$$

If $q = 2$ or $q = 3$, then $k = 1$, $G = G_{q_1}$ is the normalizer of A in $GL(q, \Delta)$, $G = \psi^{-1}(SL(2, q))$ and $G/A \cong SL(2, q)$. If $q = 5$, then $k = 2$, $G_{q_1} = \psi^{-1}(U_2)$, $G_{q_2} = \psi^{-1}(U_3)$, $G_{q_1} : A = 12$ and $G_{q_2} : A = 24$. If $q > 5$, $q \equiv \pm 3 \pmod 8$, then $k = 3$, $G_{q_j} = \psi^{-1}(U_j)$, $j = 1, 2, 3$; $G_{q_1} : A = 2(q - 1)$, $G_{q_2} : A = 2(q + 1)$ and $G_{q_3} : A = 24$. If $q > 5$, $q \equiv \pm 1 \pmod 8$, then $k = 4$, $G_{q_j} = \psi^{-1}(U_j)$, $j = 1, 2, 3, 4$; $G_{q_1} : A = 2(q - 1)$, $G_{q_2} : A = 2(q + 1)$ and $G_{q_3} : A = G_{q_4} : A = 48$.

EXAMPLE. If char $\Delta \neq 2$, then $GL(2, \Delta)$ contains exactly three nonconjugate maximal solvable subgroups:

1) The group $G^{(1)}$ of all matrices of the form

$$\begin{bmatrix} \alpha & \beta \\ 0 & \gamma \end{bmatrix}, \quad \alpha, \gamma \in \Delta^*, \quad \beta \in \Delta.$$

2) The group $G^{(2)}$ of all matrices of the form

$$\begin{bmatrix} \lambda_1 & 0 \\ 0 & \lambda_2 \end{bmatrix} \begin{bmatrix} 0 & 1 \\ 1 & 0 \end{bmatrix}^t, \quad \lambda_1, \lambda_2 \in \Delta^*, \quad t = 0, 1.$$

3) The group $G^{(3)} = \psi^{-1}(SL(2, 2))$, which is generated by the matrices

$$a = \operatorname{diag}[1, -1], \quad b = \begin{bmatrix} 0 & 1 \\ 1 & 0 \end{bmatrix},$$

$$g = \operatorname{diag}[1, -i], \quad h = \begin{bmatrix} 1 & i \\ 1 & -i \end{bmatrix},$$

where i is an element of order 4 in Δ^*, together with $\Delta^* E$.

2. *On reducible groups.* Now let Δ be an arbitrary field and M a class of irreducible matrix groups over Δ, not necessarily of the same degree. We consider the matrix groups of degree n over Δ for which the images of their irreducible components lie in M. Let $G_1, \ldots, G_k$ be groups in M such that $n_1 + \cdots + n_k = n$, where n_j is the degree of G_j. We construct the group $G = T(G_1, \ldots, G_k)$

consisting of all matrices of the form

$$g=\begin{bmatrix} a_1 & a_{12} & \dots & a_{1k} \\ 0 & a_2 & \dots & a_{2k} \\ \cdot & \cdot & \cdot & \cdot \\ 0 & 0 & \dots & a_k \end{bmatrix}, \tag{3}$$

where $a_j \in G_j$, and a_{ij} is an arbitrary $n_i \times n_j$ matrix over Δ. The class of all groups of the form $T(G_1, \dots, G_k)$, where $G_j \in M$ and $1 \leqslant k \leqslant n$, is denoted by $\langle M, n\rangle$. We fix a basis

$$u_1, \dots, u_n \tag{4}$$

of Δ^n, the same for all $G \in \langle M, n\rangle$. Then every group G in $\langle M, n\rangle$ can be regarded as a group of automorphisms of Δ^n with matrices of the form (3) in the basis (4) (the numbers k and n_j may vary from group to group in $\langle M, n\rangle$).

Let $G = T(G_1, \dots, G_k) \in \langle M, n\rangle$, n_j the degree of G_j, $j = 1, \dots, k$. We set $m_1 = n_1$ and $m_j = m_{j-1} + n_j$ for $j = 2, \dots, k$, and construct $k - 1$ subspaces of Δ^n:

$$L_1, \dots, L_{k-1}, \tag{5}$$

where L_j is the m_j-dimensional subspace with the basis $u_1, \dots, u_{m_j}$, $j = 1, \dots, k - 1$.

LEMMA 1. *$L_1, \dots, L_{k-1}$ are the only nontrivial G-invariant subspaces of Δ^n.*

PROOF. The subspaces (5) are clearly G-invariant. We introduce the k subspaces $Q_1, \dots, Q_k$, of dimensions $n_1, \dots, n_k$, with the bases $u_1, \dots, u_{n_1}$; $\dots$; $u_{m_{k-1}+1}, \dots, u_n$, respectively. Let L be a nontrivial ($L \neq 0$, $L \neq \Delta^n$) G-invariant subspace, and $v \in L$, $v \neq 0$. Then

$$v = q_1 + \dots + q_t, \quad q_j \in Q_j, \quad q_t \neq 0, \quad t \leqslant k. \tag{6}$$

Since G_t is an irreducible subgroup of $GL(n_t, \Delta)$, it follows that $G_t(q_t)$ contains n_t linearly independent vectors. Consequently L contains, together with v, also the vectors

$$v_1 + r_1, \quad v_2 + r_2, \quad \dots, \quad v_{n_t} + r_{n_t}, \tag{7}$$

where $v_1, \dots, v_{n_t} \in L_{t-1}$ and $r_1, \dots, r_{n_t}$ is a basis of Q_t. Obviously there are elements $\sigma_1, \dots, \sigma_{m_{t-1}}$ in G such that $\sigma_1(r_1) = r_1 + l_1, \dots, \sigma_{m_{t-1}}(r_1) = r_1 + l_{m_{t-1}}$, where $l_1, \dots, l_{m_{t-1}}$ is a basis of L_{t-1} and $\sigma_\nu(v_1) = v_1$, $\nu = 1, \dots, m_{t-1}$. Thus the vectors

$$\sigma_1(r_1 + v_1) = r_1 + l_1 + v_1, \quad r_1 + l_2 + v_1, \quad \dots, \quad r_1 + l_{m_{t-1}} + v_1$$

are in L. Hence, since the vectors (7) are also in L, so are $l_1, \dots, l_{m_{t-1}}$. Thus $L_{t-1} \subset L$. Since the vectors (7) lie in L, it follows from this inclusion that $Q_t \subset L$. Consequently

$$L \supset L_{t-1} \dotplus Q_t = L_t. \tag{8}$$

Now we choose the vector v in L so that the number t in (6) is maximal. Then $L \subset L_{t-1} \dotplus Q_t = L_t$. Hence $L = L_t$. Since $L \neq \Delta^n$, it follows that $t < k$. ■

Suppose now that M is subject to the following condition:

(*) If for some $m \leqslant n$ a group A of the class $\langle M, m\rangle$ is conjugate in $GL(m, \Delta)$ to a subgroup of a group $B \in M$, then $A = B$.

THEOREM 3. *Let $G = T(G_1, \ldots, G_k)$ and $H = T(H_1, \ldots, H_j)$ be groups in $\langle M, n\rangle$, where M is a class of irreducible matrix groups over Δ satisfying (*). If $d^{-1}Gd \subset H$ for $d \in GL(n, \Delta)$, then $G = H$.*

PROOF. Let $d^{-1}Gd \subset H$; let (5) be nontrivial G-invariant subspaces, and $U_1 \subset \cdots \subset U_{j-1}$ nontrivial H-invariant subspaces. Then $d(U_1), \ldots, d(U_{j-1})$ are invariant under dHd^{-1}. Since $G \subset dHd^{-1}$, by Lemma 1 we have $d(U_1) = L_{i_1}, \ldots, d(U_{j-1}) = L_{i_{j-1}}$, where $1 \leqslant i_1 < i_2 < \cdots < i_{j-1} \leqslant k$. Therefore $U_\nu : \Delta = L_{i_\nu} : \Delta$, $\nu = 1, \ldots, j-1$. Hence and by the method of constructing the subspaces L_ν and U_ν,

$$U_\nu = L_{i_\nu}, \quad d(U_\nu) = U_\nu, \quad \nu = 1, \ldots, j-1.$$

Consequently in the basis (4) the matrix d takes the form

$$d = \begin{bmatrix} d_1 & d_{12} & \ldots & d_{1j} \\ 0 & d_2 & \ldots & d_{2j} \\ \cdot & \cdot & \cdot & \cdot \\ 0 & 0 & \ldots & d_j \end{bmatrix}.$$

Hence and from $d^{-1}Gd \subset H$ we obtain j inclusions of the form

$$d_1^{-1}T(G_1, \ldots, G_{t_1})d_1 \subset H_1; \ \ldots; \ d_j^{-1}T(G_{k-t_j+1}, \ldots, G_k)d_j \subset H_j,$$

where $t_1 + \cdots + t_j = k$. Hence, by (*),

$$H_1 = T(G_1, \ldots, G_{t_1}), \ \ldots, \ H_j = T(G_{k-t_j+1}, \ldots, G_k).$$

Since the groups H_ν are irreducible, it follows that $j = k$, $H_1 = G_1, \ldots, H_k = G_k$ and $H = G$. ■

3. *On reducible solvable groups.* We recall that an irreducible solvable group of degree n has an abelian normal subgroup whose index does not exceed a certain number $\rho(n)$ depending only on n.

LEMMA 2. *Let $G_1, \ldots, G_k$ be irreducible solvable matrix groups of degrees $n_1, \ldots, n_k$ over a field Δ, $k > 1$, $n_1 + \cdots + n_k = n$, where Δ contains at least $\rho(n) + 1$ elements. Then $T = T(G_1, \ldots, G_k)$ cannot be contained in an irreducible solvable subgroup of $GL(n, \Delta)$.*

PROOF. Suppose, on the contrary, that $G \subset \Phi$, where Φ is an irreducible solvable subgroup of $GL(n, \Delta)$. Now Φ has an abelian normal subgroup F such that $\Phi : F \leqslant \rho(n)$. There are no unipotent elements in F, for F is a completely reducible abelian group. Let H be the group of all matrices of the form

$$h = \begin{bmatrix} E_{n_1} & h_{12} & \dots & h_{1k} \\ 0 & E_{n_2} & \dots & h_{2k} \\ \cdot & \cdot & \cdot & \cdot \\ 0 & 0 & \dots & E_{n_k} \end{bmatrix},$$

where $h_{\mu\nu}$ is an arbitrary $n_\mu \times n_\nu$ matrix over Δ. It is obvious that $H \subset G \subset \Phi$ and $F \cap H = 1$. It is also clear that H contains at least $\rho(n) + 1$ elements. On the other hand, $H \cong H/F \cap H \cong HF/F$ and $HF : F \leqslant \rho(n)$, so that card $H \leqslant \rho(n)$. This contradiction proves the lemma. ∎

THEOREM 4. *Let Δ be a field containing at least $\rho(n) + 1$ elements, and M a class of matrix groups over Δ having the following properties:*

(i) *For any $m \leqslant n$, every maximal irreducible solvable subgroup of $GL(m, \Delta)$ is conjugate in $GL(m, \Delta)$ to exactly one group in M.*

(ii) *Every group of degree m in M is maximal among the solvable irreducible subgroups of $GL(m, \Delta)$.*

Then every group of the class $\langle M, n\rangle$ is a maximal solvable subgroup of $GL(n, \Delta)$, and every maximal solvable subgroup of $GL(n, \Delta)$ is conjugate in $GL(n, \Delta)$ to exactly one group in $\langle M, n\rangle$.

PROOF. It is obvious that every maximal solvable subgroup of $GL(n, \Delta)$ is conjugate in $GL(n, \Delta)$ to some group in $\langle M, n\rangle$. It follows from Lemma 2 and (i) that M satisfies condition (*) of the preceding subsection. Consequently by Theorem 3 every group in $\langle M, n\rangle$ is a maximal solvable subgroup of $GL(n, \Delta)$. It also follows from Theorem 3 that two distinct groups in $\langle M, n\rangle$ cannot be conjugate in $GL(n, \Delta)$. ∎

Theorem 4 implies, in particular, that if Δ contains at least $\rho(n) + 1$ elements, then the group of all nonsingular triangular matrices of degree n over Δ is a maximal solvable subgroup of $GL(n, \Delta)$. In fact, we have a stronger result:

The group of all nonsingular triangular matrices of degree n over a field Δ is a maximal solvable subgroup of $GL(n, \Delta)$, except when $\Delta = GF(2)$ or $\Delta = GF(3)$ (Suprunenko [23]).

THEOREM 5. *Let Δ be an arbitrary field and G a solvable d-group of matrices over Δ. Then G is completely reducible.*

PROOF. As a solvable group, G has a normal subgroup H of finite index, such that, for any irreducible component γ of H, the group Im γ is abelian. By Lemma 17.3, H is abelian. Hence by Theorem 17.3 the group G is completely reducible. ∎

A matrix group over a field Δ is said to be *nonmodular* if either char $\Delta = 0$ or char $\Delta = p$ (where p is a prime) but G has no normal subgroups of finite index $j \equiv 0 \pmod{p}$.

THEOREM 6. *Let Δ be a perfect field. Then any completely reducible solvable nonmodular matrix group G over Δ is a d-group. In particular, a solvable completely reducible matrix group over a field of characteristic zero is a d-group.*

PROOF. G has a completely reducible abelian normal subgroup F of finite index. Every matrix in F is semisimple. Since Δ is a perfect field, F is a d-group. By Lemma 17.5, G is also a d-group. ∎

4. *On solvable integral matrix groups.* Let Δ be a euclidean domain, and Ω the field of fractions of Δ. We treat $GL(n, \Delta)$ as the group of automorphisms of an n-dimensional free module M over Δ, and $GL(n, \Omega)$ as the group of automorphisms of an n-dimensional linear space $V \supset M$ over Ω. Obviously $GL(n, \Delta) \subset GL(n, \Omega)$.

LEMMA 3. *Let G be a subgroup of $GL(n, \Delta)$ and M_1 a G-invariant submodule of M. Then M contains a G-invariant direct summand M_2 of the same dimension over Δ as M_1.*

PROOF. By standard properties of modules over euclidean domains, M_1 has a basis $b_1, \ldots, b_m$ such that $b_j = a_j\delta_j$, $j = 1, \ldots, m$, where $\delta_j \in \Delta$, $\delta_j \neq 0$, and $a_1, \ldots, a_n$ is a basis of M. By assumption, for any $g \in G$ we can write

$$g(b_j) = b_1\lambda_1 + \ldots + b_m\lambda_m, \quad \lambda_t \in \Delta. \tag{9}$$

On the other hand,

$$g(a_j) = a_1\gamma_1 + \ldots + a_n\gamma_n, \quad \gamma_\nu \in \Delta. \tag{10}$$

Hence

$$g(b_j) = g(a_j)\delta_j = (a_1\gamma_1 + \ldots + a_n\gamma_n)\delta_j. \tag{11}$$

Comparing (9) and (11), we see that $\gamma_\nu = 0$ in (10) if $j \leqslant m$, and $\nu > m$. Therefore the submodule M_2 with the basis $a_1, \ldots, a_m$ is a G-invariant direct summand of M. ∎

LEMMA 4. *Let G be a subgroup of $GL(n, \Delta)$ and suppose that V has a G-invariant subspace P. Then M has a G-invariant direct summand M_2 of the same dimension as P.*

PROOF. Let $M_1 = M \cap P$. It is obvious that M_1 is a G-invariant submodule of M and that $M_1 : \Delta = P : \Omega$. By Lemma 3, M contains a G-invariant direct summand M_2 of dimension $M_1 : \Delta$. ∎

Lemma 4 implies

LEMMA 5. *Let G be a subgroup of $GL(n, \Delta)$. Then the Δ-module M has a* *basis in which all the matrices g in G simultaneously take the form*

$$g=\begin{bmatrix} a_1(g) & a_{12}(g) & \dots & a_{1k}(g) \\ 0 & a_2(g) & \dots & a_{2k}(g) \\ \dots & \dots & \dots & \dots \\ 0 & 0 & \dots & a_k(g) \end{bmatrix}, \tag{12}$$

where $a_j(g)$ and $a_{ij}(g)$ are matrices over Δ, with $a_j(g)$ ranging over an irreducible subgroup G_j of $GL(n_j, \Omega)$ when g ranges over G, $j = 1, \dots, k; n_1 + \dots + n_k = n$.

A group is said to be *polycyclic* if it has a normal series all of whose factors are cyclic.

Now let Z be the ring of rational integers and Q the field of rational numbers.

LEMMA 6. *Suppose that the matrices g of a group G have the form* (12), *where $\Delta = Z$ and $\Omega = Q$, and that all the groups G_j, $j = 1, \dots, k$, are abelian. Then G is polycyclic.*

PROOF. First let $k = 1$. Then g is an irreducible abelian subgroup of $GL(n, Q)$ and its linear Q-hull $[G] = \Sigma$ is an extension of Q of degree n. Let K be the ring of algebraic integers of Σ. Since $G \subset GL(n, Z)$, G is a subgroup of K^*. By Dirichlet's theorem, K^* is finitely generated; therefore both K^* and G are polycyclic. It is now easy to prove the lemma by induction on k. Let $k > 1$ and suppose the lemma is true for groups with fewer than k irreducible components. We consider the homomorphisms

$$g \overset{h_1}{\mapsto} \begin{bmatrix} a_1(g) & 0 & \dots & 0 \\ 0 & a_2(g) & \dots & a_{2k}(g) \\ \dots & \dots & \dots & \dots \\ 0 & 0 & \dots & a_k(g) \end{bmatrix} \overset{h_2}{\mapsto} \begin{bmatrix} a_2(g) & a_{23}(g) & \dots & a_{2k}(g) \\ 0 & a_3(g) & \dots & a_{3k}(g) \\ \dots & \dots & \dots & \dots \\ 0 & 0 & \dots & a_k(g) \end{bmatrix}. \tag{13}$$

By the inductive hypothesis, $\operatorname{Im} h_2$ is polycyclic. $\operatorname{Ker} h_2$ is isomorphic to a subgroup of G_1. Therefore $\operatorname{Ker} h_2$ is also polycyclic. Since $\operatorname{Im} h_1/\operatorname{Ker} h_2 \cong \operatorname{Im} h_2$, it follows that $\operatorname{Im} h_1$ is also polycyclic. Now the matrices of $\operatorname{Ker} h_1$ have the form

$$\begin{bmatrix} E_{n_1} & a_{12} \dots a_{1k} \\ 0 & E_{n-n_1} \end{bmatrix}.$$

Consequently $\operatorname{Ker} h_1$ is a finitely generated abelian group, and so is polycyclic. Since $G/\operatorname{Ker} h_1 \cong \operatorname{Im} h_1$, G is also polycyclic. ∎

THEOREM 7 (MAL'CEV [3]). *A solvable subgroup of $GL(n, Z)$ is polycyclic.*

PROOF. Let U be a solvable subgroup of $GL(n, Z)$. Being solvable, U has a normal subgroup G of finite index such that the image of any irreducible component of G is abelian (irreducibility over Q).

By Lemma 5, the free Z-module on which U acts has a basis in which the matrices g of G have the form (12), with abelian groups G_j. By Lemma 6, G is polycyclic. U/G is a finite solvable group and so is also polycyclic. Thus U is polycyclic. ∎

Auslander [1] and Swan [1] have proved that every polycyclic group is isomorphically embeddable in $GL(n, Z)$ for a suitable n. Merzljakov [3] has shown that the holomorph of a polycyclic group also has a faithful integral representation.

To end the chapter we quote a few more results without proof.

A matrix group over an arbitrary field satisfies a nontrivial identity if and only if it is a finite extension of a solvable group (Platonov [13]). Solvable subgroups of the multiplicative group T^* of a division ring of finite rank n over its center Δ have been studied by Suprunenko [17]. A subgroup G of T^* is said to be *irreducible* if its linear Δ-hull is T. A maximal solvable irreducible subgroup G of T^* has a unique maximal abelian normal subgroup F that is the multiplicative group of a commutative subfield $\Sigma \supset \Delta$ of T; Σ is a normal separable extension of Δ with solvable Galois group $\mathfrak{G}(\Sigma/\Delta)$.

There are two possibilities: 1) $\Sigma : \Delta = \sqrt{n}$; 2) $\Sigma : \Delta < \sqrt{n}$. In the first case, $G/F \cong \mathfrak{G}(\Sigma/\Delta)$. In the second case, the invariant series (α) (see §20.2)

$$G \supset V \supset A \supset F \supset (e)$$

has the following properties:

a) A/F is an abelian group of order $(T:\Delta)(\Sigma:\Delta)^{-2}$ whose Sylow subgroups are elementary.

b) V/A is isomorphic to a solvable subgroup of $\mathrm{Aut}(A/F)$.

c) $V/F \cong \mathfrak{G}(\Sigma/\Delta)$.

In particular, let T be the real quaternion ring with the basis $1, i, j, k$ over the field R of real numbers. Then every maximal solvable subgroup of T^* is conjugate in T^* to one of two subgroups G_1 or G_2: G_1 is generated by the elements $i, j, 1+i, 1+k$ and R^*, and $G_2 = (j)C^*$, $C = R(i)$.

Concerning maximal solvable subgroups of $GL(n, T)$, where T is a division ring of finite rank over its center, see Zalesskiĭ [2].

Let $\mathfrak{G}$ be an algebraic algebra over a field Δ and G a locally solvable subgroup of $\mathfrak{G}^*$. Then the smallest subalgebra of $\mathfrak{G}$ containing G is locally finite-dimensional (Zalesskiĭ [1]).*

---

*Editor's note. For recent work in the theory of solvable linear groups the reader should consult the book by Wehrfritz [3].

SIXTH CHAPTER

PERIODIC LINEAR GROUPS

§23. Finiteness conditions for linear groups. Local finiteness of a matrix group over a field

1. *Trace of a matrix*. Let Δ be an arbitrary field. We consider the mapping

$$\mathrm{tr}\colon \Delta_n \to \Delta, \qquad a = \| \alpha_{ij} \| \mapsto \sum_{\nu=1}^{n} \alpha_{\nu\nu}. \tag{1}$$

The element $\mathrm{tr}\, a \in \Delta$ is called the *trace of* a.

As is clear from (1), tr *is a linear mapping; that is, for any* $a, b, \in \Delta_n$ *and* $\lambda, \mu \in \Delta$,

$$\mathrm{tr}\,(\lambda a + \mu b) = \lambda\, \mathrm{tr}\,(a) + \mu\, \mathrm{tr}\,(b).$$

Obviously $\mathrm{tr}(ab) = \mathrm{tr}(ba)$, and so $\mathrm{tr}(cbc^{-1}) = \mathrm{tr}\, b$ for $c \in GL(n, \Delta)$.

We note, in particular, that *in the n^2-dimensional unitary space of matrices C_n, with the scalar product*

$$(a, b) = \sum_{i,\, j=1}^{n} \alpha_{ij}\bar{\beta}_{ij}, \qquad a = \| \alpha_{ij} \|, \qquad b = \| \beta_{ij} \|,$$

we have

$$(a, b) = \mathrm{tr}\,(ab^*), \tag{2}$$

where $b^* = \|\beta^*_{ij}\|$ *and* $\beta^*_{ij} = \bar{\beta}_{ji}$.

If u is a unitary matrix, $a_1 = uau^{-1}$ and $b_1 = ubu^{-1}$, then by (2), $(a_1, b_1) = (a, b)$.

2. *Burnside's criterion for finiteness of a linear group.*

LEMMA 1 (BURNSIDE [2]). *Let Δ be an arbitrary field, and $\mathfrak{G}$ an absolutely irreducible subgroup of $GL(n, \Delta)$ for which $\mathrm{tr}(\mathfrak{G})$ is a finite set with m elements. Then $\mathfrak{G}$ is a finite group, and*

$$\mathrm{card}\, \mathfrak{G} \leqslant m^{n^2}. \tag{3}$$

PROOF. Since $\mathfrak{G}$ is absolutely irreducible, it contains n^2 linearly independent matrices (Corollary 14.4.2)

$$a_\nu = \|\alpha_{ij}^\nu\|, \quad \alpha_{ij}^\nu \in \Delta, \quad \nu = 1, 2, \ldots, n^2. \tag{4}$$

For $x = \|x_{ij}\| \in \mathfrak{G}$ we have

$$\operatorname{tr}(a_\nu x) = \sum_{i,\, j=1}^{n} \alpha_{ij}^\nu x_{ji} = \tau_\nu \in \operatorname{tr}(\mathfrak{G}). \tag{5}$$

Since the matrices (4) are linearly independent, for fixed $\tau_1, \ldots, \tau_{n^2}$ the system (5) has a unique solution $x = \|x_{ij}\|$. Hence (3) holds. ∎

Lemma 1 implies

LEMMA 2. *If the orders of all elements of an absolutely irreducible subgroup $\mathfrak{G}$ of $GL(n, \Delta)$ are bounded by a number μ, then $\mathfrak{G}$ is a finite group whose order is bounded by a number $\nu(\mu, n)$ that depends only on μ and n.*

PROOF. The roots of the characteristic polynomials of matrices in $\mathfrak{G}$ are roots of unity of a fixed order depending on μ. Hence $\operatorname{tr}(\mathfrak{G})$ is a finite set, and the number of its elements is bounded by a function depending on n and μ only. ∎

THEOREM 1 (BURNSIDE'S CRITERION FOR FINITENESS OF A LINEAR GROUP). *Let Δ be an arbitrary field and $\mathfrak{G}$ a subgroup of $GL(n, \Delta)$ such that the orders of its elements are not divisible by* char Δ *and are bounded by a certain number μ. Then $\mathfrak{G}$ is a finite group whose order is less than a certain number $N(\mu, n)$ depending only on n and μ.*

PROOF. Let Ω be the algebraic closure of Δ. We treat $\mathfrak{G}$ as a subgroup of $GL(n, \Omega)$. In a suitable basis of a space V over Ω the matrices of $\mathfrak{G}$ take the form

$$g = \begin{bmatrix} a_1(g) & a_{12}(g) & \ldots & a_{1k}(g) \\ 0 & a_2(g) & \ldots & a_{2k}(g) \\ \cdot & \cdot & \cdot & \cdot \\ 0 & 0 & \ldots & a_k(g) \end{bmatrix} = \begin{bmatrix} a_1(g) & * \\ 0 & A_2(g) \end{bmatrix}, \tag{6}$$

where $\gamma_\nu \colon g \mapsto a_\nu(g)$, $\nu = 1, \ldots, k$, are irreducible representations of $\mathfrak{G}$. We introduce the homomorphism

$$\gamma\colon \mathfrak{G} \to \overline{\mathfrak{G}}, \quad g \mapsto \operatorname{diag}[a_1(g), \ldots, a_k(g)] \tag{7}$$

and consider Ker γ. If a matrix other than E_n has unitriangular form, its order either is infinite or is equal to p^α, where $p = \operatorname{char} \Delta$. Therefore Ker $\gamma = (E_n)$, and γ is an isomorphism. The criterion now follows from this and Lemma 2. ∎

A simple generalization of Burnside's criterion is

THEOREM 2. *Let Δ be an arbitrary field, Γ a subgroup of $GL(n, \Delta)$. Suppose that there is a natural number k such that* char $\Delta \nmid k$ *and $a^k \in \Delta^* E_n$ for* *any element a of* Γ. *Then the order of* $\Gamma/\Gamma \cap \Delta^* E_n$ *does not exceed a certain number* $\mu(k, n)$ *depending only on k and n.*

PROOF. We treat Γ as a subgroup of $GL(n, \Omega)$, where Ω is the algebraic closure of Δ. Then

$$\Omega^* E_n \cap \Gamma = \Delta^* E_n \cap \Gamma. \tag{8}$$

Now let $G = \Gamma\Omega^* E_n$, and let H be the kernel of the homomorphism det: $G \to \Omega^*$. Since Ω is algebraically closed, for any $g \in G$ there is a $\rho \in \Omega^*$ such that $\rho^{-1} g \in H$. Therefore $G = H\Omega^* E_n$. If $h \in H$, then $h^k = \lambda E_n$, $\lambda \in \Omega^*$, $\lambda^n = 1$. We set $\nu = n$ for char $\Delta = 0$, and $\nu = np^{-\alpha}$, $(\nu, p) = 1$, for char $\Delta = p > 0$. Then $h^{k\nu} = E_n$. By Burnside's criterion, card $H < N(k\nu, n)$. Now,

$$G/\Omega^* E_n = H\Omega^* E_n/\Omega^* E_n \cong H/H \cap \Omega^* E_n.$$

Consequently

$$G : \Omega^* E_n \leqslant N(k\nu, n) = \mu(k, n).$$

Since $\Gamma/\Gamma \cap \Omega^* E_n \cong \Gamma\Omega^* E_n/\Omega^* E_n = G/\Omega^* E_n$, by (8) we have

$$\Gamma : (\Gamma \cap \Delta^* E_n) \leqslant \mu(k, n). \ \blacksquare$$

THEOREM 2a. *Let* Δ *be an arbitrary field,* Γ *an irreducible subgroup of* $GL(n, \Delta)$, *and suppose that there is a natural number k such that* char $\Delta \nmid k$ *and that for any element a in* Γ *the element* a^k *lies in the center Z of* Γ. *Then*

$$\Gamma : Z \leqslant \mu(k, n). \tag{9}$$

PROOF. Since Γ is irreducible, $Z \subset \Sigma^*$, where $\Sigma = [Z]_\Delta$ is an extension of Δ, $\Sigma : \Delta = m$, $m \mid n$. Therefore we can treat Γ as a subgroup of $GL(r, \Sigma)$, $r = nm^{-1}$. By Theorem 2,

$$\Gamma : (\Gamma \cap \Sigma^* E_r) \leqslant \mu(k, r) \leqslant \mu(k, n).$$

It is obvious that $\Gamma \cap \Sigma^* E_r = Z$, and this proves (9). $\blacksquare$

THEOREM 3 (BURNSIDE'S SECOND CRITERION). *Let* Δ *be an arbitrary field and* $\mathfrak{G}$ *a subgroup of* $GL(n, \Delta)$ *having only finitely many classes of conjugate elements. Then* $\mathfrak{G}$ *is a finite group.*

PROOF. Again we treat $\mathfrak{G}$ as a subgroup of $GL(n, \Omega)$, where Ω is the algebraic closure of Δ, and we write the matrices $g \in \mathfrak{G}$ in the form (6). Since the number of conjugacy classes of $\gamma_\nu(\mathfrak{G}) = \mathfrak{G}_\nu$ is finite, the set $\operatorname{tr}(\mathfrak{G}_\nu)$ is also finite, and so $\mathfrak{G}_\nu$ is a finite group. We now use induction on the number k of irreducible components of $\mathfrak{G}$. The criterion has already been proved for $k = 1$. Suppose it is true for groups with $k - 1$ irreducible components, $k > 1$. We consider the representation

$$\varphi(g) = \mathrm{diag}\,[a_1(g),\ A_2(g)],$$

where $g \in \mathfrak{G}$ are matrices of the form (6). By the inductive hypothesis, $A_2(g)$ ranges over a finite group when g ranges over $\mathfrak{G}$, and so $\varphi(\mathfrak{G})$ is finite. Ker φ $= \mathfrak{N}$ consists of the matrices of the form

$$\begin{bmatrix} E_s & b \\ 0 & E_t \end{bmatrix}.$$

Thus $\mathfrak{N}$ is an abelian normal subgroup of finite index of $\mathfrak{G}$. It is easy to see that $\mathfrak{N}$ is also finite. For let n range over $\mathfrak{N}$, and $\bar{n}$ over the conjugacy class of n in N. Since $\mathfrak{N}$ lies in the centralizer of n in $\mathfrak{G}$, the class $\bar{n}$ is finite. The set of all the $\bar{n}$ is also finite; hence $\mathfrak{N}$, and therefore $\mathfrak{G}$, are finite groups. ■

Theorem 3 should be compared with the following proposition.

THEOREM 4. *Let Δ be an arbitrary field, and Γ a subgroup of $GL(n, \Delta)$ with the property that each of its conjugacy classes is finite. Then the center of Γ is of finite index in Γ.*

PROOF. Let $g_1, \ldots, g_m$ be a maximal system of elements of Γ linearly independent over Δ, and Z_j the centralizer of g_j in Γ. Then $m \leqslant n^2$ and $\Gamma\colon Z_j = r_j$. The center Z of Γ is the intersection of the groups Z_j, and so $\Gamma\colon Z \leqslant r_1 \cdots r_m$. ■

Evidently the converse is also true.

3. *Schur's first theorem.*

LEMMA 3. *Let Π be the prime field of Δ, and Φ the field of all elements in $K = \Pi(\theta_1, \ldots, \theta_t)$, $\theta_1, \ldots, \theta_t \in \Delta$, that are algebraic over Π. Then $\Phi\colon \Pi$ is finite.*

PROOF. If each θ_i is algebraic over Π, we are done. Suppose, therefore, that some of the θ_i are transcendental over Π and let $\Sigma = \Pi(\theta_1, \ldots, \theta_r)$ be a pure transcendental extension of Δ such that θ_j for $j > r$ is algebraic over Σ. Then, obviously, $K\colon \Sigma = m < \infty$. Hence for any system

$$\alpha_1, \ \ldots, \ \alpha_{m+1} \tag{10}$$

of $m + 1$ elements of K there is a relation of the form

$$\sigma_1\alpha_1 + \ldots + \sigma_{m+1}\alpha_{m+1} = 0, \tag{11}$$

where σ_ν, $\nu = 1, \ldots, m + 1$, is a polynomial in

$$\theta_1, \ \ldots, \ \theta_r \tag{12}$$

with coefficients in Π, and not all the σ_ν vanish. We express the left-hand side of (11) as a polynomial in (12) whose coefficients k_ρ are of the form

$$k_\rho = a_1^\rho \alpha_1 + \ldots + a_{m+1}^\rho \alpha_{m+1}, \quad a_j^\rho \in \Pi.$$

Now if (10) are elements of Φ, the elements (12) are algebraically independent over $\Pi(\alpha_1, \ldots, \alpha_n)$, since they are algebraically independent over Π. Therefore all the coefficients k_ρ vanish. But the a_j^ρ are not all zero, because some of the σ_ν are different from zero. Consequently the elements (10) are linearly dependent over Π; that is, $\Phi: \Pi \leqslant m$. ∎

In the notation of Lemma 3 we have

LEMMA 4. *In Δ^* there are only finitely many elements of finite order that are roots of polynomials in $K[x]$ of a fixed degree n.*

PROOF. Let ϵ be an element of Δ^* of order μ that is a root of an irreducible polynomial $f(x) \in K[x]$ of degree at most n. Then $f(x) \in \Phi[x]$, and $\Pi(\epsilon): \Pi \leqslant mn$ by Lemma 3. On the other hand, if char $\Pi = 0$, then $\Pi(\epsilon): \Pi = \varphi(\mu)$, where φ is Euler's function. Therefore

$$\varphi(\mu) \leqslant mn. \tag{13}$$

Clearly there are only finitely many numbers μ satisfying (13). If char $\Pi = p > 0$, then $\mu \leqslant p^{mn} - 1$. ∎

LEMMA 5. *Let Δ be an arbitrary field and $\mathfrak{G}$ an absolutely irreducible subgroup of $GL(n, \Delta)$. If all the roots of the characteristic polynomial of each matrix in $\mathfrak{G}$ are roots of unity, then $\mathfrak{G}$ is locally finite.*

PROOF. Let $\mathfrak{H}_1$ be a subgroup of $\mathfrak{G}$ generated by a finite set M_1 of matrices. It is evident that M_1 is contained in a finite absolutely irreducible set $M \subset \mathfrak{G}$. Now M generates an absolutely irreducible subgroup $\mathfrak{H}$ of $\mathfrak{G}$. We claim that $\mathfrak{H}$ is a finite group. Let K be the field obtained by adjoining to the prime field Π of Δ the elements of all the matrices in M. Then $\mathfrak{H}$ is a subgroup of $GL(n, K)$. The roots of the characteristic polynomials of the matrices in $\mathfrak{H}$ are elements of finite order and roots of polynomials of degree n over K. By Lemma 4, the set of roots of the characteristic polynomials of matrices in $\mathfrak{H}$ is finite. Then $\operatorname{tr}(\mathfrak{H})$ is a finite set and, by Lemma 1, $\mathfrak{H}$ is a finite group. ∎

THEOREM 5 (SCHUR'S FIRST THEOREM). *A periodic subgroup of $GL(n, \Delta)$ (Δ an arbitrary field) is locally finite.*

PROOF. Let $\mathfrak{G}$ be a periodic subgroup of $GL(n, \Delta)$. We regard it as a subgroup of $GL(n, \Omega)$, where Ω is the algebraic closure of Δ. Let $\mathfrak{H}$ be a finitely generated subgroup of $\mathfrak{G}$. Then obviously every homomorphic image of it is also finitely generated. In particular, if γ is the homomorphism (7) introduced in the proof of the first finiteness criterion, then $\gamma(\mathfrak{H})$ is finitely generated. By Lemma 5, $\gamma(\mathfrak{H})$ is finite. If char $\Delta = 0$, then Ker $\gamma = (E_n)$, and $\mathfrak{H}$ is a finite group. But if char $\Delta = p > 0$, then Ker γ is a nilpotent p-group. Now a nilpotent p-group is locally finite (Kuroš [1], Russian p. 401; English vol. II, p. 230). $\mathfrak{H}$ is a finite extension of a locally finite group, and so it is also locally finite. But $\mathfrak{H}$ is finitely generated, and therefore finite. ■

For a periodic group G we denote by $\Pi(G)$ the set of all primes dividing the orders of its elements.

Schur's theorem and Corollary 17.1.1 imply

THEOREM 6. *Let G be a periodic subgroup of $GL(n, \Delta)$. If* char $\Delta \notin \Pi(G)$, *then G is completely reducible.*

The theorem on local finiteness remains valid for periodic groups having a complete (adequate) system of linear representations of bounded degrees (Platonov [1]).

In connection with Schur's theorem, we quote without proof the following result of Selberg [1]:

Let char $\Delta = 0$. *Then a finitely generated subgroup of $GL(n, \Delta)$ has a torsion-free normal subgroup of finite index.*

It is known that the theorem on local finiteness does not carry over to arbitrary periodic groups. Novikov and Adjan [1] have proved the existence of finitely generated infinite groups in which the orders of the elements are bounded. Golod [1] has constructed a finitely generated infinite p-group $\mathfrak{G}$. The orders of the elements of $\mathfrak{G}$ are not bounded. It is residually finite and therefore has an adequate system of linear representations (of unbounded degrees).

The problem of the local finiteness of a periodic group is completely solved by the following theorem of Procesi [1]:

A periodic group is locally finite if and only if it is isomorphic to a subgroup of the multiplicative group of a ring R satisfying locally a polynomial identity (every finitely generated subring of R satisfies a certain polynomial identity).

4. *Properties of a linear group connected with properties of the characteristic polynomials of its matrices.* Let Δ be an arbitrary field, Γ a subgroup of $GL(n, \Delta)$. Let $F(\Gamma)$ be the set of all the distinct characteristic polynomials of matrices in Γ.

LEMMA 6. *If Δ is algebraically closed, $n > 1$, and the only polynomial in $F(\Gamma)$ is $(x - 1)^n$, then Γ is reducible.*

PROOF. Suppose, on the contrary, that Γ is irreducible. Since $\operatorname{tr} g = n$ for every $g \in \Gamma$, it follows from Lemma 1 that the order of Γ is 1; that is, $\Gamma = (E_n)$. But then Γ cannot be irreducible. ∎

The next result holds for an arbitrary field Δ.

THEOREM 7 (SUPRUNENKO [18]). *Let Γ be a subgroup of $GL(n, \Delta)$, and suppose that $F(\Gamma)$ consists of polynomials $(x - \lambda)^n$, where $\lambda \in \Delta$. Then there is an element t in $GL(n, \Delta)$ such that the matrices of $t^{-1}\Gamma t$ have the form*

$$\begin{bmatrix} \lambda & \lambda_{12} & \dots & \lambda_{1n} \\ 0 & \lambda & \dots & \lambda_{2n} \\ \cdot & \cdot & \cdot & \cdot \\ 0 & 0 & \dots & \lambda \end{bmatrix}. \tag{14}$$

PROOF. Observe, first of all, that the theorem need be proved only for an algebraically closed field Δ. For let Ω be the algebraic closure of Δ and let t_1 be a matrix in $GL(n, \Omega)$ such that the matrices of $t_1^{-1}\Gamma t_1$ have the form (14). Then clearly all the matrices $g - \lambda(g)E_n$, where $g \in \Gamma$ and $\lambda(g)$ is the unique eigenvalue of g, lie in a nilpotent subalgebra A of Δ_n; that is,

$$\Gamma \subset \Delta^{\cdot} E_n + A, \qquad A^n = 0. \tag{15}$$

By Theorem 13.6 there is a matrix in $GL(n, \Delta)$ such that the matrices of $t^{-1}At$ have the form

$$\begin{bmatrix} 0 & \alpha_{12} & \dots & \alpha_{1n} \\ 0 & 0 & \dots & \alpha_{2n} \\ \cdot & \cdot & \cdot & \cdot \\ 0 & 0 & \dots & 0 \end{bmatrix}$$

By (15), the matrices $t^{-1}\Gamma t$ have the form (14).

We now assume that Δ is algebraically closed, and set $G = \Gamma\Delta^*$. The theorem is proved if we can establish that the degree of any irreducible component γ of G is 1. Let $H = \operatorname{Im} \gamma$, and let m be the degree of H and H_0 the kernel of the homomorphism det: $H \longrightarrow \Delta^*, h \longmapsto \det h$. Since $G = \Gamma\Delta^*$ and Δ is algebraically closed, we have $H = H_0\Delta^*$. The characteristic polynomial of $h_0 \in H_0$ is $(x - \epsilon)^m$, where $\epsilon^m = \det h_0 = 1$. Therefore $\operatorname{tr} h_0$ assumes at most m values. Hence, by Lemma 1, H_0 is finite. There are two cases: $\operatorname{char} \Delta = 0$ or $\operatorname{char} \Delta = p > 0$. In the first case, since the characteristic polynomial of every matrix $h_0 \in H_0$ has the form $(x - \epsilon)^m$ and H_0 is finite, it follows that every matrix in H_0 is scalar and $m = 1$. Now let $\operatorname{char} \Delta = p > 0$. Then we write the order of H_0 as dp^τ, where $(d, p) = 1$. If q is a prime divisor of d, then any Sylow q-subgroup of H_0 consists of scalar matrices, because the characteristic polynomial of any

$h_0 \in H_0$ has the form $(x - \epsilon)^m$ and $q \neq p$. Consequently $H_0 = DP$, where D is the subgroup of scalar matrices in H_0 and P is a Sylow p-subgroup of H_0. If $h_0 \in P$, then $\epsilon = 1$ and the characteristic polynomial of h_0 is $(x-1)^m$. Consequently, by Lemma 6, P is reducible if $m > 1$. Since H_0 is irreducible, this is impossible, and so $m = 1$. ∎

COROLLARY 1 (MAL'CEV [3]; KOLCHIN [1]). *Any unipotent subgroup of $GL(n, \Delta)$ is conjugate in $GL(n, \Delta)$ to a subgroup of the group of unitriangular matrices. Any two maximal unipotent subgroups of $GL(n, \Delta)$ are conjugate in $GL(n, \Delta)$.*

COROLLARY 2. *Let Γ be a subgroup of $GL(n, \Delta)$, and suppose that $F(\Gamma)$ consists of polynomials of the form $(x-\lambda)^n$, $\lambda \in \Delta$. Then Γ is nilpotent.*

Now let Δ be an algebraically closed field, Γ a subgroup of $GL(n, \Delta)$, $f \in F(\Gamma)$, $\lambda_1, \ldots, \lambda_r$ the pairwise distinct roots of f, and

$$\Sigma_f = \{\lambda_1\lambda_1^{-1}, \lambda_2\lambda_1^{-1}, \ldots, \lambda_r\lambda_1^{-1}\}, \quad \Sigma = \bigcup \Sigma_f, \quad f \in F(\Gamma).$$

We denote by $\Phi(\Gamma)$ the subgroup of Δ^* generated by the set Σ. For example, for the group Γ in Theorem 7 we have $\Phi(\Gamma) = (1)$. Below we investigate groups Γ such that $\Phi(\Gamma)$ is a periodic group.

LEMMA 7. *Let H be an irreducible subgroup of $GL(n, \Delta)$ and C the center of H. If $\Phi(H)$ is a periodic group, the following assertions are true*:

1) *H/C is locally finite*; *if* char $\Delta = 0$ *then* $\Pi(H/C) = \Pi(\Phi(H))$, *while if* char $\Delta = p > 0$ *then* $\Pi(\Phi(H)) \subset \Pi(H/C) \subset \Pi(\Phi(H)) \cup \{p\}$.

2) *If $\Phi(H)$ is finite, so is the index H: C.*

PROOF. Let $\mathfrak{H} = H\Delta^*E_n$. Then

$$\Phi(\mathfrak{H}) = \Phi(H), \qquad H/C \cong \mathfrak{H}/\Delta^*E_n,$$

where Δ^*E_n is the center of $\mathfrak{H}$. It is therefore sufficient to prove the lemma for $\mathfrak{H}$. Let $\mathfrak{H}_0$ be the kernel of the homomorphism det: $\mathfrak{H} \longrightarrow \Delta^*$. Since Δ is algebraically closed and $\Delta^*E_n \subset \mathfrak{H}$, we have $\mathfrak{H} = \mathfrak{H}_0\Delta^*E_n$. Let $\lambda_1, \ldots, \lambda_n$ be the eigenvalues of a matrix $h \in \mathfrak{H}_0$. Then

$$\begin{aligned} \lambda_j\lambda_1^{-1} &= \varepsilon_j \in \Phi(\mathfrak{H}_0) = \Phi(\mathfrak{H}), \\ \det h &= \varepsilon_1 \ldots \varepsilon_n\lambda_1^n = 1. \end{aligned} \tag{16}$$

By (16), all the eigenvalues of h are roots of unity. By Lemma 5, $\mathfrak{H}_0$ is a locally finite. Now if $\Phi(\mathfrak{H})$ is finite, then $\mathrm{tr}(\mathfrak{H}_0)$ is a finite set and $\mathfrak{H}_0$ is finite, by Lemma 1. But $\mathfrak{H}/\Delta^*E_n$ is also locally finite or finite, respectively, because

$$\mathfrak{H}/\Delta^*E_n \cong \mathfrak{H}_0/\mathfrak{H}_0 \cap \Delta^*E_n.$$

It remains to consider the set $\Pi(\mathfrak{H}/\Delta^* E_n)$. If $h \in \mathfrak{H}_0$, then, since $\Phi(\mathfrak{H})$ is a periodic group, there is a natural number k such that $\Pi(k) \subset \Pi(\Phi(\mathfrak{H}))$ and the characteristic polynomial of the matrix h^k is $(x-\eta)^n$, $\eta^n = 1$. If char $\Delta = 0$, it follows from the periodicity of $\mathfrak{H}_0$ that $h^k = \eta E_n$. But if char $\Delta = p$, then h^k is an element of order dp^τ, $(d, p) = 1$, and

$$h^{kp^\tau} = \eta^{p^\tau} E_n,$$

$$\Pi(\Phi(\mathfrak{H})) \cup \{p\} \supset \Pi(\mathfrak{H}_0/C \cap \mathfrak{H}_0) = \Pi(\mathfrak{H}/C).$$

On the other hand, it is obvious that $\Pi(\mathfrak{H}/C) \supset \Pi(\Phi(\mathfrak{H}))$. ■

THEOREM 8. *If $\Phi(\Gamma)$ is a periodic group, then Γ has a nilpotent normal subgroup N with the following properties:*

1) $\Phi(N) = (1)$.

2) *Γ/N is locally finite; if* char $\Delta = 0$ *then* $\Pi(\Gamma/N) = \Pi(\Phi(\Gamma))$, *while if* char $\Delta = p > 0$ *then* $\Pi(\Phi(\Gamma)) \cup \{p\} \supset \Pi(\Gamma/N) \supset \Pi(\Phi(\Gamma))$.

3) *If $\Phi(\Gamma)$ is finite, so is Γ/N.*

PROOF. As in the proof of the preceding theorem, we may assume that $\Delta^* E_n \subset \Gamma$. Let γ be the homomorphism (7), $K = \operatorname{Ker} \gamma$ and $N = K\Delta^* E_n$. Clearly $N \lhd \Gamma$ and $\Phi(N) = (1)$. We turn to the group $\operatorname{Im} \gamma = \overline{\Gamma} \cong \Gamma/K$. The matrices in the center Z of $\overline{\Gamma}$ have the form diag $[\lambda_1 E_{n_1}, \ldots, \lambda_k E_{n_k}]$. Since $\Phi(\overline{\Gamma}) = \Phi(\Gamma)$, it follows that $\lambda_j \lambda_1^{-1} \in \Phi(\Gamma)$, $j = 1, \ldots, k$. Consequently

$$\Pi(Z/\Delta^* E_n) \subset \Pi(\Phi(\Gamma)). \tag{17}$$

We now prove that Γ/N is locally finite. Since $\gamma^{-1}(\Delta^* E_n) = N$, we have

$$\Gamma/N \cong \gamma(\Gamma)/\Delta^* E_n = \overline{\Gamma}/\Delta^* E_n.$$

If $\overline{\Gamma}_0$ is the kernel of the homomorphism det: $\overline{\Gamma} \longrightarrow \Delta^*$, then

$$\Gamma/N \cong \overline{\Gamma}_0 \Delta^* E_n/\Delta^* E_n \cong \overline{\Gamma}_0/\Delta^* E_n \cap \overline{\Gamma}_0.$$

Therefore it is sufficient to show that $\overline{\Gamma}_0$ is locally finite. Let $g \in \overline{\Gamma}_0$, and let $\lambda_1, \ldots, \lambda_n$ be the eigenvalues of g. Then

$$\lambda_j \lambda_1^{-1} = \varepsilon_j \in \Phi(\Gamma), \qquad \varepsilon_1 \ldots \varepsilon_n \lambda_1^n = 1.$$

Thus all the λ_j are roots of unity. Now $\overline{\Gamma}_0$ is completely reducible, and each of its irreducible blocks is locally finite. Therefore $\overline{\Gamma}_0$ is locally finite. This shows that Γ/N is locally finite.

If now $\Phi(\Gamma)$ is a finite group, then every irreducible block of $\overline{\Gamma}_0$ is finite, and so Γ_0 is a finite group. Then so is Γ/N.

We consider $\Pi(\Gamma/N) = \Pi(\overline{\Gamma}/\Delta^* E_n)$. Let char $\Delta = p > 0$. By Lemma 7,

$$\Pi(\Phi(\Gamma)) \cup \{p\} \supset \Pi(\overline{\Gamma}/Z).$$

Hence and from (17) it follows that

$$\Pi(\Phi(\Gamma)) \cup \{p\} \supset \Pi(\bar{\Gamma}/\Delta^* E_n).$$

On the other hand, it is obvious that $\Pi(\Gamma/N) \supset \Pi(\Phi(\Gamma))$. If char $\Delta = 0$, then, by Lemma 7 and (17),

$$\Pi(\Phi(\Gamma)) = \Pi(\bar{\Gamma}/\Delta^* E_n). \blacksquare$$

COROLLARY 1. *Let Δ be an algebraically closed field of characteristic zero and Γ a subgroup of $GL(n, \Delta)$ such that $\Phi(\Gamma)$ is periodic. Then the set of all unipotent matrices in Γ is a normal subgroup of Γ.*

PROOF. Let $u \in \Gamma$ be a unipotent matrix, K the kernel of the homomorphism γ defined by (7), and $\gamma(u) \in \bar{\Gamma}$. Obviously $\gamma(u)$ is also unipotent. Therefore det $\gamma(u) = 1$ and $\gamma(u) \in \bar{\Gamma}_0$. Since Γ_0 is a periodic group, it follows that $\gamma(u) = E_n$, and $u \in K$. ▪

COROLLARY 2. *Let Δ be an algebraically closed field of characteristic* 0, *and Γ a subgroup of $GL(n, \Delta)$ such that $\Phi(\Gamma)$ is periodic, and either* $2 \notin \Pi(\Phi(\Gamma))$ *or* card $\Pi(\Phi(\Gamma)) \leqslant 2$. *Then Γ is solvable.*

PROOF. By Theorem 6, $\Pi(\Gamma/N) = \Pi(\Phi(\Gamma))$, and Γ/N is locally finite. If $2 \notin \Pi(\Phi(\Gamma))$, then by the theorem of Feit and Thompson [1], Γ/N is locally solvable. If card $\Pi(\Phi(\Gamma)) \leqslant 2$, then Γ/N is solvable by Burnside's theorem. On the other hand, $\Gamma/N \cong \bar{\Gamma}_0/\bar{\Gamma}_0 \cap \Delta^* E_n$. Hence $\bar{\Gamma}_0$ is locally solvable. By Theorem 19.7, a locally solvable matrix group is solvable. Hence Γ is solvable. ▪

§24. Existence of an abelian normal subgroup of finite index in a periodic linear group over the field of complex numbers

1. *Unitariness of periodic subgroups of $GL(n, C)$.* Let C be the field of complex numbers, V_n an n-dimensional vector space over C, and $U(n)$ the group of all unitary $n \times n$ matrices.

LEMMA 1. *Let $\mathfrak{G}$ be a finite subgroup of $GL(V_n)$. Then a scalar product $\tau = \tau_{\mathfrak{G}}$ can be defined in V_n such that the space $\langle V_n, \tau\rangle$ is unitary and every element of $\mathfrak{G}$ is a unitary transformation of $\langle V_n, \tau\rangle$.*

PROOF. Let σ: $V_n \times V_n \longrightarrow C$ be a scalar product for which the space $\langle V_n, \sigma\rangle$ is unitary. We define another scalar product τ by setting

$$\tau(x, y) = \sum_{h \in \mathfrak{G}} \sigma(hx, hy).$$

for $x, y \in V_n$. Obviously τ: $V_n \times V_n \longrightarrow C$ satisfies all the axioms for a scalar product in a unitary space. For $g \in \mathfrak{G}$ we can write

$$\tau(gx, gy) = \sum_{h \in \mathfrak{G}} \sigma(hgx, hgy) = \sum_{h \in \mathfrak{G}} (hx, hy) = \tau(x, y).$$

Therefore every element g of $\mathfrak{G}$ is a unitary transformation of $\langle V_n, \tau\rangle$. ∎

LEMMA 2. *Let $\mathfrak{G}$ be a finite irreducible subgroup of $GL(V_n)$. Then the scalar product $\tau_{\mathfrak{G}}$ of Lemma 1 is determined by the group $\mathfrak{G}$ up to a factor in C^*.*

PROOF. Let $\tau = \tau_{\mathfrak{G}}$. Let B be an orthonormal basis of $\langle V_n, \tau\rangle$. Then the matrix of the bilinear form τ relative to B is E_n, while the matrices of $\mathfrak{G}$ are unitary; that is, $g^* = g^{-1}$ for $g \in \mathfrak{G}$. Now let φ be another scalar product such that $\langle V_n, \varphi\rangle$ is a unitary space and all elements of $\mathfrak{G}$ are unitary transformations of $\langle V_n, \varphi\rangle$. If Φ is the matrix of φ in the basis B, then $g^*\Phi g = \Phi$ for every g in $\mathfrak{G}$. Since $g^{-1} = g^*$, we have $\Phi g = g\Phi$. Since $\mathfrak{G}$ is irreducible, it now follows that $\Phi = \lambda E_n$ and $\varphi = \lambda\tau$, $\lambda \in C^*$. ∎

THEOREM 1. *Every periodic subgroup Γ of $GL(n, C)$ is conjugate in $GL(n, C)$ to a subgroup of $U(n)$.*

PROOF. By Theorem 23.6, periodic subgroups of $GL(n, C)$ are completely reducible. Hence we need only prove the theorem for an irreducible group Γ. Thus let Γ be an irreducible periodic subgroup of $GL(n, C)$. We treat Γ as a subgroup of $GL(V_n)$. The theorem will be proved if we can define a scalar product τ on V_n for which $\langle V_n, \tau\rangle$ is a unitary space and the operators of Γ are unitary transformations. Since Γ is irreducible and locally finite, it contains an irreducible finite subgroup $\mathfrak{G}$. By Lemma 1 there is a scalar product τ in V_n such that the operators in $\mathfrak{G}$ are unitary transformations of $\langle V_n, \tau\rangle$. If $\Gamma = \mathfrak{G}$, the proof is complete.

Now let $\Gamma \neq \mathfrak{G}$ and $h_1 \in \Gamma\backslash\mathfrak{G}$. Then for the finite group $\mathfrak{H}$ generated by h_1 and $\mathfrak{G}$ we can find a scalar product $\tau_1(x, y)$ such that for $h \in \mathfrak{H}$

$$\tau_1(hx, hy) = \tau_1(x, y).$$

In particular, this is true for an arbitrary matrix h in $\mathfrak{G}$. By Lemma 2, $\tau = \lambda\tau_1$. Consequently

$$\tau(h_1x, h_1y) = \tau(x, y),$$

and the elements of Γ are unitary operators of $\langle V_n, \tau\rangle$. ∎

2. *The matrix space C_n*. In the n^2-dimensional matrix space C_n we introduce the scalar product

$$(a, b) = \sum_{i=1}^{n}\sum_{j=1}^{n} \alpha_{ij}\bar{\beta}_{ij} = \operatorname{tr}(ab^*), \quad \|\alpha_{ij}\| = a, \quad \|\beta_{ij}\| = b \tag{1}$$

and we set $v(a) = \operatorname{tr}(aa^*) = (a, a)$. For $a \in C_n$ and $u \in U(n)$,

$$v(ua) = v(au) = v(a). \tag{2}$$

Indeed,

$$v(au) = \operatorname{tr}(au(au)^*) = \operatorname{tr}(a(uu^*)a^*) = v(a)$$

and, similarly, $v(ua) = v(a)$. By (2), for $u \in U(n)$,

$$v(u) = v(uu^{-1}) = v(E_n) = \operatorname{tr}(E_n) = n.$$

Hence if $\|a\| = (a, a)^{1/2}$ is the length of a vector $a \in C_n$, then

$$\|u\| = n^{1/2}. \tag{3}$$

LEMMA 3. *If a and b are unitary matrices, then*

$$v(E_n - aba^{-1}b^{-1}) \leqslant 2v(E_n - a)\,v(E_n - b). \tag{4}$$

PROOF. By (2) and the fact that a is unitary, we may assume that $a = \operatorname{diag}[\alpha_{11}, \ldots, \alpha_{nn}]$. Further,

$$\begin{aligned} v(E_n - aba^{-1}b^{-1}) &= v(ba - ab) \\ &= v((b - E_n)a + a(E_n - b)) \\ &= \sum_{i=j}^{n}\sum_{j=1}^{n} |(\beta_{ij} - \delta_i^j)\alpha_{jj} + \alpha_{ii}(\delta_i^j - \beta_{ij})|^2 \\ &= \sum_{i=1}^{n}\sum_{j=1}^{n} |\alpha_{ii} - \alpha_{jj}|^2 |\delta_i^j - \beta_{ij}|^2; \end{aligned}$$

$$\begin{aligned} |\alpha_{ii} - \alpha_{jj}|^2 &= |(1 - \alpha_{ii}) - (1 - \alpha_{jj})|^2 \\ &\leqslant 2(|1 - \alpha_{ii}|^2 + |1 - \alpha_{jj}|^2) \leqslant 2v(E_n - a). \end{aligned}$$

Hence and from the preceding equality (4) follows. ∎

LEMMA 4. *Let $a, b \in U(n)$, $c = aba^{-1}b^{-1}$, $ac = ca$ and $v(E_n - b) < 4$. Then $ab = ba$.*

PROOF. It follows from the conditions of the lemma that the matrices a and bab^{-1} commute; we may therefore assume that both these matrices are diagonal, say

$$a = \operatorname{diag}[\lambda_1 E_{n_1}, \ldots, \lambda_r E_{n_r}], \quad \lambda_i \neq \lambda_j \quad \text{for} \quad i \neq j.$$

Since the only difference between bab^{-1} and a is in the order of their diagonal elements, it follows that $bab^{-1} = p^{-1}ap$, where p is a matrix representation of a certain permutation of degree n. The matrix $q = pb$ commutes with a and therefore has the form $q = \operatorname{diag}[q_1, \ldots, q_r]$. We write p in the form

$$p = \begin{bmatrix} l_{11} & \dots & l_{1r} \\ \cdot & \cdot & \cdot \\ l_{r1} & \dots & l_{rr} \end{bmatrix},$$

where l_{ij} is an $n_i \times n_j$ matrix. Now suppose that b and a do not commute. Then $l_{\mu\nu} \neq 0$ for some μ and ν, $\mu \neq \nu$, because $b = p^{-1}q$. Since p is a monomial matrix, det $l_{\mu\mu} = 0$. Hence there is a pair $i, j, i \neq j$, such that $l_{ij} \neq 0$. It is clear that the matrix $l_{\mu\mu}$ has a row of zeros and $l_{\nu\nu}$ a column of zeros. It follows that

$$v(p - q) \geqslant 1 + 1 + (u_1, u_1) + (u_2, u_2) = 4,$$

where u_1 is a row of q_μ, u_2 a column of q_ν, and (u_i, u_i) denotes the scalar product. But

$$v(p - q) = v(p^{-1}(p - q)) = v(E_n - b).$$

Therefore $v(E_n - b) \geqslant 4$, contradicting the assumption. ∎

LEMMA 5. *Let a and b be unitary matrices in a periodic subgroup of $GL(n, C)$, such that $v(E_n - a) < ½$ and $v(E_n - b) < 4$. Then $ab = ba$.*

PROOF. Since a periodic linear group is locally finite, the matrices a and b generate a finite group $\mathfrak{G}$. We consider the sequence

$$b_0, b_1, \dots, b_k, \dots, \tag{5}$$

of matrices in $\mathfrak{G}$, where $b_0 = b$ and $b_k = ab_{k-1}a^{-1}b_{k-1}^{-1}$, $k > 0$. We set $\alpha = v(E_n - a)$. Then, by Lemma 3,

$$v(E_n - b_k) \leqslant 2v(E_n - a)\, v(E_n - b_{k-1}) = 2\alpha v(E_n - b_{k-1}).$$

Hence

$$v(E_n - b_k) \leqslant (2\alpha)^k v(E_n - b), \quad k > 0. \tag{6}$$

Since the number of distinct matrices in the sequence (5) cannot exceed the order of $\mathfrak{G}$, while by (6) we can find for any $\epsilon > 0$ a number $k = k(\epsilon)$ such that $v(E_n - b_k) < \epsilon$, it follows that for some natural number l

$$v(E_n - b_l) = 0, \quad b_l = E_n.$$

On the other hand, by (6) we have $v(E_n - b_k) < 4$ for $k > 0$. Hence and from (7), applying Lemma 4, we find that $b_{l-1} = b_{l-2} = \dots = b_1 = E_n$. ∎

3. *The main theorem.*

THEOREM 2 (SCHUR'S SECOND THEOREM). *A periodic subgroup Γ of $GL(n, C)$ has an abelian normal subgroup A whose index does not exceed a number $\rho(n)$ depending only on n.*

PROOF. By Theorem 1 we may assume that $\Gamma \subset U(n)$. We now set

$$D = \left\{ d \mid d \in \Gamma,\ v(E_n - d) < \frac{1}{2} \right\}.$$

By Lemma 5 the elements of D commute pairwise. Hence they generate an abelian subgroup A of Γ. If $d \in D$, then, by (2), $gdg^{-1} \in D$ for $g \in \Gamma$. Therefore A is normal in Γ. Let b and c be representatives of two distinct cosets of A in Γ. Then

$$v(b - c) \geqslant \frac{1}{2}. \tag{8}$$

For let $v(b - c) < \frac{1}{2}$. Then

$$v(E_n - b^{-1}c) = v(b(E_n - b^{-1}c)) = v(b - c) < \frac{1}{2},$$

$b^{-1}c \in D$, $b^{-1}c \in A$. This contradicts the choice of b and c, proving (8).

We now identify the matrix space C_n with the $2n^2$-dimensional euclidean space R^{2n^2}. If $a, b \in C_n$ and

$$a = \| \alpha_{\mu\nu} \|,\quad b = \| \beta_{\mu\nu} \|,\quad \alpha_{\mu\nu} = \lambda_{\mu\nu} + i\rho_{\mu\nu},\quad \beta_{\mu\nu} = \sigma_{\mu\nu} + i\tau_{\mu\nu},$$

we regard a as a vector in R^{2n^2} with the coordinates

$$\lambda_{11}, \rho_{11}, \ldots, \lambda_{\mu\nu}, \rho_{\mu\nu}, \ldots, \lambda_{nn}, \rho_{nn},$$

and the scalar product of vectors a and b is given by

$$\sum_{\mu=1}^{n} \sum_{\nu=1}^{n} (\lambda_{\mu\nu}\sigma_{\mu\nu} + \rho_{\mu\nu}\tau_{\mu\nu}). \tag{9}$$

It follows from (9) that the scalar square of a matrix a in C_n is the same as its scalar square in R^{2n^2}, namely $v(a)$.

Now let B be transversal to A in Γ. Since $\Gamma \subset U_n$, by (3) all the points of B lie on a sphere S of radius $n^{1/2}$ in R^{2n^2}. By (8) we have $\|b - c\| \geqslant \frac{1}{2}$ for two distinct points $b, c \in B$. It follows that card $B < \rho(n)$, where $\rho(n)$ depends only on n, and so $\Gamma : A < \rho(n)$ ■

Schur's theorem implies the following result.

THEOREM 3 (JORDAN'S THEOREM). *A finite subgroup of $GL(n, C)$ has an abelian normal subgroup whose index is less than a certain number $\rho(n)$ depending only on n.*

Another obvious consequence of Schur's theorem is the following:

A periodic simple subgroup of $GL(n, C)$ is finite, or order less than $\rho(n)$.

Schur's theorem remains true for periodic linear groups over an arbitrary field of characteristic zero, but it does not carry over to fields of prime characteristic. There are even simple infinite periodic linear groups. For example, let Δ

be the algebraic closure of $GF(2)$. Then $G = SL(2, \Delta)$ is a periodic infinite simple group. For G is periodic because Δ^* is periodic. G is simple because $Z(G) = (E_2)$, and $G/Z(G) = PSL(2, \Delta)$.

It is just as easy to give an example of a solvable periodic matrix group having no abelian normal subgroup of finite index. Let Δ be the algebraic closure of $GF(p)$, and G the group of all matrices

$$\begin{bmatrix} \alpha & \beta \\ 0 & \gamma \end{bmatrix},$$

where $\alpha, \gamma \in \Delta^*$ and $\beta \in \Delta$. G is obviously a solvable periodic subgroup of $GL(2, \Delta)$. However, as is easy to verify, G has no abelian normal subgroup of finite index.

For fields of prime characteristic, Brauer and Feit [1] have proved the following analog of Jordan's theorem:

Let p be a prime. There exists an integer-valued function $f_p(m, n)$ such that if Δ is a field of characteristic p and G a finite subgroup of $GL(n, \Delta)$ with a Sylow p-subgroup of order at most p^m, then G has an abelian normal subgroup A of index $G: A < f_p(m, n)$.

Later we shall need the following result.

THEOREM 4. *Let G be an irreducible finite subgroup of $GL(n, C)$, where C is the field of complex numbers, and let A be an abelian normal subgroup of G. Then $G: A$ is divisible by n.*

For a proof, see, for example Serre [1], p. II–17.

§25. Sylow subgroups of the general linear group

1. *General remarks. Sylow p-subgroups of $GL(n, \Delta)$ when $p = \text{char } \Delta$.* Let Δ be an arbitrary field, Π a nonempty set of primes, and G any Sylow Π-subgroup of $GL(n, \Delta)$. In particular, when Π is the set of all primes, G is obviously a maximal periodic subgroup of $GL(n, \Delta)$. In a suitable basis of Δ^n, the matrices of G have the form

$$g = \begin{bmatrix} a_1(g) & a_{12}(g) & \dots & a_{1k}(g) \\ 0 & a_2(g) & \dots & a_{2k}(g) \\ \dots & \dots & \dots & \dots \\ 0 & 0 & \dots & a_k(g) \end{bmatrix},$$

where $\gamma_j\colon G \to GL(n_j, \Delta), g \mapsto a_j(g), j = 1, \dots, k$, are the irreducible components of G. It is clear that the $\text{Im}(\gamma_j)$ are Π-groups. If $\text{char } \Delta = 0$, or $\text{char } \Delta = p > 0$ but $p \notin \Pi$, then G is completely reducible (Theorem 23.6), and we may assume that $a_{ij}(g) = 0$ for $i \neq j$.

We define a representation γ by setting

$$\gamma: G \to GL(n, \Delta), \quad g \to \operatorname{diag}[a_1(g), \dots, a_k(g)]. \tag{2}$$

If char $\Delta \notin \Pi$, then γ is faithful, $\operatorname{Im}(\gamma_j)$ is a Sylow Π-subgroup of $GL(n_j, \Delta)$, and G is the direct product of the groups $\operatorname{Im}(\gamma_1), \dots, \operatorname{Im}(\gamma_k)$.

LEMMA 1. *Let* char $\Delta \in \Pi$. *Then* $G = FN$, *where* $F = \operatorname{Im} \gamma$, $N = \operatorname{Ker} \gamma$, *and* $F \cap N = (E_N)$; *F is the direct product of* $\operatorname{Im}(\gamma_1), \dots, \operatorname{Im}(\gamma_k)$, $\operatorname{Im}(\gamma_j)$ *is a Sylow Π-subgroup of $GL(n_j, \Delta)$, and N is the set of all matrices of the form*

$$c = \begin{bmatrix} E_{n_1} & x_{12} & \dots & x_{1k} \\ 0 & E_{n_2} & \dots & x_{2k} \\ \cdot & \cdot & \cdot & \cdot \\ 0 & 0 & \dots & E_{n_k} \end{bmatrix}, \tag{3}$$

where $x_{\alpha\beta}$ is an arbitrary $n_\alpha \times n_\beta$ matrix over Δ.

PROOF. By (1), (2) and (3), we have $g = \gamma(g)c$, $\gamma(g) \in F$, $c \in N$. Hence $g \in FN$. On the other hand, FN is a Π-group. Thus $G = FN$. ∎

We begin by selecting the simple special case of Sylow Π-subgroups of $GL(n, \Delta)$ when $\Pi = \{p\}$ and $p = \text{char } \Delta$.

THEOREM 1. *Let p be a prime and $p =$* char Δ. *Then the Sylow p-subgroups of $GL(n, \Delta)$ are all conjugate in $GL(n, \Delta)$ to the group of all unitriangular matrices.*

PROOF. The set of all unipotent elements of $GL(n, \Delta)$ is precisely the set of all its p-elements. For suppose that $a \in GL(n, \Delta)$ and $a^{p^\alpha} = 1$. Then

$$a^{p^\alpha} - 1 = (a-1)^{p^\alpha} = 0.$$

Therefore a is a unipotent element. Now let u be a unipotent element of $GL(n, \Delta)$. Then $(u-1)^n = 0$. If ν is such that

$$p^{\nu - 1} < n \leqslant p^\nu, \tag{4}$$

then $u^{p^\nu} = 1$, so that u is a p-element. Hence the class of all Sylow p-subgroups of $GL(n, \Delta)$ and that of all its maximal unipotent subgroups coincide. But a maximal unipotent subgroup of $GL(n, \Delta)$ is conjugate in $GL(n, \Delta)$ to the group of all unitriangular matrices. ∎

Note that *if ν is defined by* (4) *and $p =$* char Δ, *then p^ν is the maximum order of a p-element of $GL(n, \Delta)$.*

We now mention a few properties of irreducible and completely reducible Sylow Π-subgroups of $GL(n, \Delta)$.

LEMMA 2. *Let G be an irreducible Sylow Π-subgroup of $GL(n, \Delta)$. Then G is either primitive or expressible as a wreath product $\mathfrak{G} \sim \Gamma$, where $\mathfrak{G}$ is a primitive Sylow Π-subgroup of $GL(m, \Delta)$ and Γ a transitive Sylow Π-subgroup of the symmetric group S_k, $mk = n$.*

This is a special case of Theorem 15.4.

Note that the converse of Lemma 2 is false: the wreath product of a primitive Sylow Π-subgroup of $GL(m, \Delta)$ and a transitive Sylow Π-subgroup of S_k need not be a Sylow Π-subgroup of $GL(mk, \Delta)$. For example, let

$$\Pi = \{2\}, \quad \Delta = GF(q), \quad q = 4t - 1, \quad \mathfrak{G} = \{\pm 1\} \subset \Delta^{\bullet}.$$

Then $\mathfrak{G} \sim S_2$ is a subgroup of $GL(2, \Delta)$, of order 8. But

$$16 \mid \operatorname{card} GL(2, \Delta) = (q^2 - 1)(q^2 - q).$$

Thus $\mathfrak{G} \sim S_2$ is not a Sylow 2-subgroup of $GL(2, \Delta)$.

From Lemma 2 and the general remarks at the beginning of this section it follows that when char $\Delta \notin \Pi$, *a Sylow Π-subgroup of $GL(n, \Delta)$ is conjugate in $GL(n, \Delta)$ to the group G of matrices of the form*

$$g = \operatorname{diag}[g_1, \ldots, g_t], \tag{5}$$

where $g_1, \ldots, g_t$ range independently over $G_1, \ldots, G_t$, respectively, $G_j = \mathfrak{G}_j \sim \Gamma_j$, $\mathfrak{G}_j$ is a primitive Sylow Π-subgroup of $GL(m_j, \Delta)$, and Γ_j is a transitive Sylow Π-subgroup of S_{k_j}, $j = 1, \ldots, t$, $\Sigma_1^t m_j k_j = n$.

The group G of matrices (5) is sometimes denoted by

$$G = \operatorname{diag}[\mathfrak{G}_1 \sim \Gamma_1, \ldots, \mathfrak{G}_t \sim \Gamma_t]. \tag{6}$$

LEMMA 3. *Suppose that a group G of the form* (6) *is a Sylow Π-subgroup of $GL(n, \Delta)$.*

(i) *If Π is the set of all primes, then no two of the groups $\mathfrak{G}_\mu$ and $\mathfrak{G}_\nu$, $\mu \neq \nu$, are conjugate.*

(ii) *If $\Pi = \{p\}$, then every group $\mathfrak{G}_j \sim \Gamma_j$, $1 \leqslant j \leqslant t$, is repeated fewer than p times on the diagonal of* (6).

PROOF. (i) Let Π be the set of all primes. Then G is a maximal periodic subgroup of $GL(n, \Delta)$. If $\mathfrak{G}_1 = \mathfrak{G}_2$, then

$$\operatorname{diag}[\mathfrak{G}_1 \sim S_{k_1}, \mathfrak{G}_2 \sim S_{k_2}]$$

is a proper subgroup of $\mathfrak{G}_1 \sim S_{k_1 + k_2}$. This implies (i).

(ii) Let $\Pi = \{p\}$. Then $k_j = p^{\alpha_j}$. If $\mathfrak{G}_1 \sim \Gamma_1 = \cdots = \mathfrak{G}_p \sim \Gamma_p$, then

$$\operatorname{diag}[\mathfrak{G}_1 \sim \Gamma_1, \ldots, \mathfrak{G}_p \sim \Gamma_p]$$

is a proper subgroup of $\mathfrak{G}_1 \sim \Gamma$, where Γ is a Sylow *p-subgroup of* $S_{p^{\alpha_1+1}}$. ∎

2. *Sylow p-subgroups of $GL(n, \Delta)$ when $p \neq$ char Δ and Δ is algebraically closed.*

LEMMA 4. *Let Δ be an algebraically closed field, p a prime other than* char Δ, *and $n > 1$. Then $GL(n, \Delta)$ has no primitive p-subgroups.*

PROOF. Let Φ be a primitive p-subgroup of $GL(n, \Delta)$. By Theorem 23.5, Φ is locally finite. Since a finite p-group is nilpotent, Φ is locally nilpotent and therefore, by Theorem 19.9, solvable. By Theorem 19.5, Φ has an abelian normal subgroup H of finite index k. By the primitivity of Φ and Theorem 16.2, we have $H \subset \Delta^* E_n$. Clearly $G = \Phi\Delta^*$ is also primitive;

$$G/\Delta^*E_n = \Phi\Delta^*/\Delta^*E_n \cong \Phi/\Delta^*E_n \cap \Phi = \Phi/H.$$

Hence $G: \Delta^*E_n = k$. We consider the mapping det: $G \longrightarrow \Delta^*$;

$$\operatorname{Ker}(\det) = G \cap SL(n, \Delta) = G_0.$$

Let $g \in G$, $\det g = \lambda \in \Delta^*$. Since Δ is algebraically closed, there is an element ρ in Δ^* such that $\rho^n = \lambda$. Obviously

$$\rho^{-1}g \in G_0, \quad g = \rho g_0, \quad g_0 \in G_0;$$
$$G = G_0\Delta^*, \quad G/\Delta^*E_n \cong G_0/\Delta^*E_n \cap G_0.$$

Therefore G_0 is a finite group. Since Φ is locally nilpotent, G_0 is nilpotent. Hence G is also nilpotent. Let Z_1 be the center of G, Z_2/Z_1 the center of G/Z_1, and $c \in Z_2 \backslash Z_1$. Then $(c)Z_1$ is an abelian normal subgroup of G, and $(c)Z_1 \not\subset \Delta^*E_n$. Since G is primitive, this contradicts Clifford's theorem. ∎

Lemma 4 implies

LEMMA 5. *Let Δ be an algebraically closed field and p a prime other than* char Δ. *Then every p-subgroup of $GL(n, \Delta)$ is monomial.*

Lemmas 2 and 4 imply

LEMMA 6. *Let Δ and p be as in Lemma 4, and suppose that $GL(m, \Delta)$ has irreducible p-subgroups. Then $m = p^\alpha$, and the irreducible Sylow p-subgroups of $GL(p^\alpha, \Delta)$ are conjugate in $GL(p^\alpha, \Delta)$ to $G_{p^\alpha} = P \sim N_{p^\alpha}$, where P is a Sylow p-subgroup of Δ^* and N_{p^α} a Sylow p-subgroup of S_{p^α}.*

THEOREM 2. *Let Δ be an algebraically closed field and p a prime other than* char Δ. *Then any two Sylow p-subgroups of $GL(n, \Delta)$ are conjugate. Suppose, further, that*

$$n = a_0 + a_1 p + \ldots + a_\mu p^\mu,$$
$$0 \leqslant a_\nu < p, \quad \nu = 0, \ldots, \mu, \quad a_\mu \neq 0.$$

Then every Sylow p-subgroup of GL(n, Δ) is conjugate in GL(n, Δ) to

$$\mathrm{diag}[\ldots, G_{p^\nu}, \ldots], \tag{7}$$

where $G_{p^\nu} = P \sim N_{p^\nu}$, *P is a Sylow p-subgroup of* Δ^*, N_{p^ν} *is a Sylow p-subgroup of* S_{p^ν}, *and* G_{p^ν} *is repeated in* (7) a_ν *times.*

PROOF. It follows from the complete reducibility of a Sylow p-subgroup in $GL(n, \Delta)$ and from Lemma 6 that a Sylow p-subgroup of $GL(n, \Delta)$ is conjugate in $GL(n, \Delta)$ to a group of the form (7); we need only show, therefore, that G_{p^ν} occurs in (7) a_ν times. Suppose that it occurs in (7) b_ν times. Then $n = b_0 + b_1 p + \cdots + b_\lambda p^\lambda$. It is sufficient to prove that $b_\nu < p$ for $\nu = 0, 1, \ldots, \lambda$. Suppose on the contrary that $b_\nu \geqslant p$ for some ν. Then G contains the direct factor $U = \mathrm{diag}\,[G_{p^\nu}, \ldots, G_{p^\nu}]$, where G_{p^ν} is repeated p times. Obviously $N_{p^{\nu+1}} = N_{p^\nu} \sim N_p$. Hence it is easy to conclude that U is a proper subgroup of $G_{p^{\nu+1}} = P \sim N_{p^{\nu+1}}$. Consequently the group defined by (7) is not a Sylow p-subgroup of $GL(n, \Delta)$. Hence $b_\nu = a_\nu$. ∎

It is obvious that *if p and Δ satisfy the conditions of the theorem, a Sylow p-subgroup P of* Δ^* *is a group of type* p^∞. Hence and from Theorem 2 it follows that *a Sylow p-subgroup of GL(n, Δ) is specified up to isomorphism by the two numbers n and p*; in other words *if* Δ_1 *is another algebraically closed field with* char $\Delta_1 \neq p$, *then the Sylow p-subgroups of* $GL(n, \Delta_1)$ *and* $GL(n, \Delta)$ *are isomorphic.*

3: *Sylow p-subgroups of GL(n, Δ) when Δ is arbitrary.* These groups have been studied by Vol'vačev [1]. We present his results without proof.

THEOREM 3. *The Sylow p-subgroups of GL(n, Δ), where Δ is an arbitrary field, are almost always conjugate in GL(n, Δ). The only exception occurs when* $p = 2$ *and Δ is subject to the following three conditions*: 1) Δ^* *contains no element of order* 4, 2) -1 *is a sum of two squares in* Δ, 3) *if* $\alpha, \beta \in \Delta$, $i^2 = -1$, *and* $\alpha + \beta i$ *is a* 2*-element of* $\Delta(i)$, *then* $(\alpha + \beta i)(\alpha - \beta i) = 1$. *In this exceptional case the set of all Sylow* 2*-subgroups of GL(n, Δ) splits into* $[n/2] + 1$ *conjugacy classes.*

The structure of the Sylow p-subgroups of $GL(n, \Delta)$ depends on certain properties of the pair $\langle \Delta, p \rangle$. A pair $\langle \Delta, 2 \rangle$ for which Δ satisfies the above conditions 1), 2), 3) will be called *exceptional.* An example of an exceptional pair $\langle \Delta, 2 \rangle$ is $\Delta = Q(\sqrt{-5})$, where Q is the field of rationals.

Let P_p denote a Sylow p-subgroup of Δ^*. We split the set of all pairs $\langle \Delta, p \rangle$, where Δ is an arbitrary field and p a prime other than char Δ, into classes $\mathfrak{M}_1, \ldots, \mathfrak{M}_5$ as follows:

1) $\langle\Delta, p\rangle \in \mathfrak{M}_1$ if P_p is a group of type p^∞.

2) $\langle\Delta, p\rangle \in \mathfrak{M}_2$ if the order of P_p is $p^l > 2$.

3) $\langle\Delta, 2\rangle \in \mathfrak{M}_3$ if P_2 is a group of order 2 but $\langle\Delta, 2\rangle$ is not an exceptional pair.

4) $\langle\Delta, 2\rangle \in \mathfrak{M}_4$ if $\langle\Delta, 2\rangle$ is an exceptional pair.

5) $\langle\Delta, p\rangle \in \mathfrak{M}_5$ if $p > 2$ and $P_p = (1)$.

Suppose that $\langle\Delta, p\rangle$ is in $\mathfrak{M}_1$ or $\mathfrak{M}_2$, and that $n = a_0 + a_1 p + \cdots + a_\mu p^\mu$, $a_\mu \neq 0$, and $0 \leqslant a_\nu < p$ for $\nu = 0, \ldots, \mu$. Then every Sylow p-subgroup of $GL(n, \Delta)$ is conjugate in $GL(n, \Delta)$ to $G = \operatorname{diag}[\ldots, G_{p^\nu}, \ldots]$, where the direct factor $G_{p^\nu} = P_p \wr N_{p^\nu}$ is repeated a_ν times. The difference between the first and second case is that when $\langle\Delta, p\rangle \in \mathfrak{M}_1$ a Sylow p-subgroup of $GL(n, \Delta)$ is infinite, while when $\langle\Delta, p\rangle \in \mathfrak{M}_2$ it is finite.

Let $\langle\Delta, 2\rangle \in \mathfrak{M}_3$ and $n = a_0 + a_1 2 + \cdots + a_\mu 2^\mu$, $a_\mu = 1$, and $a_\nu = 0$ or 1 for $\nu = 0, \ldots, \mu - 1$. Then every Sylow 2-subgroup of $GL(n, \Delta)$ is conjugate in $GL(n, \Delta)$ to $G = \operatorname{diag}[\ldots, G_{2^\nu}, \ldots]$, where $G_{2^0} = \{\pm 1\}$ and $G_{2^\nu} = \mathfrak{G} \wr N_{2^{\nu-1}}$ for $\nu > 0$; $\mathfrak{G}$ is a Sylow 2-subgroup of $GL(2, \Delta)$, and the direct factor G_{2^ν} is repeated a_ν times.

Let $\langle\Delta, 2\rangle \in \mathfrak{M}_4$. Then every Sylow 2-subgroup of $GL(2, \Delta)$ is conjugate in $GL(2, \Delta)$ to one of two nonconjugate subgroups $\mathfrak{G}_1$ and $\mathfrak{G}_2$. We express $[n/2]$ in the form

$$\left[\frac{n}{2}\right] = m + l, \tag{8}$$

where m and l are nonnegative integers. Clearly $n = a_0 + 2(m + l)$, $a_0 = 0, 1$. We write down the binary representations of m and l:

$$m = b_0 + b_1 2 + \ldots + b_\lambda 2^\lambda, \qquad l = c_0 + c_1 2 + \ldots + c_\mu 2^\mu.$$

With each representation of $[n/2]$ in the form (8) we associate the group

$$G_{m,l} = \operatorname{diag}\left[\ldots, G^1_{2^\alpha}, \ldots, G^2_{2^\beta}, \ldots\right], \tag{9}$$

where the direct factors $G^1_{2^\alpha} = \mathfrak{G}_1 \wr N_{2^{\alpha-1}}$ and $G^2_{2^\beta} = \mathfrak{G}_2 \wr N_{2^{\beta-1}}$ are repeated b_α and c_β times, respectively, and for $a_0 = 1$ the group $G_{m,l}$ also contains the direct factor $\{\pm 1\}$. Then each of the $[n/2] + 1$ groups $G_{m,l}$ in (9) is a Sylow 2-subgroup of $GL(n, \Delta)$, and every Sylow 2-subgroup of $GL(n, \Delta)$ is conjugate in $GL(n, \Delta)$ to exactly one such group $G_{m,l}$.

Finally, let $\langle\Delta, p\rangle \in \mathfrak{M}_5$, and $\Sigma = \Delta(\epsilon)$, where ϵ is a primitive pth root of unity. We set $m = \Sigma : \Delta$ and write the number n in the form $n = c + mn_0$, $0 \leqslant c < m$. We write n_0, in turn, in the form

$$n_0 = b_0 + b_1 p + \ldots + b_\mu p^\mu, \quad b_\mu \neq 0, \quad 0 \leqslant b_j < p,$$
$$j = 0, \ldots, \mu.$$

Then every Sylow p-subgroup of $GL(n, \Delta)$ is conjugate in $GL(n, \Delta)$ to

$$G = \mathrm{diag}[1, \ldots, 1, \ldots, G_{mp^\alpha}, \ldots],$$

where 1 occurs c times, $G_{mp^\alpha} = P'_p \wr N_{p^\alpha}$, P'_p is a Sylow p-subgroup of Σ^*, Σ^* is regarded as a subgroup of $GL(m, \Delta)$, and G_{mp^α} is repeated b_α times.

4. *On Sylow p-subgroups of $GL(n, T)$ when T is a division ring.* The following result of Zalesskiĭ [3] is quoted without proof.

THEOREM 4. *Let T be a division ring, C its center, and $T: C = p^{2\alpha}m^2$, $(m, p) = 1$. If $p = 2$, C^* is assumed to contain an element of order 4. Then the set of all Sylow p-subgroups of $GL(n, T)$ splits into at most $\alpha + 1$ conjugacy classes. In particular, if $((T: C), p) = 1$, then the Sylow p-subgroups of $GL(n, T)$ are conjugate.*

5. *Sylow Π-subgroups of $GL(n, C)$ over the field C of complex numbers.*

LEMMA 7. *$GL(n, C)$ contains irreducible Π-subgroups if and only if $\Pi(n) \subset \Pi$.*

PROOF. Let G be an irreducible Π-subgroup of $GL(n, C)$. By Theorem 23.5, G is locally finite and therefore contains a finite irreducible subgroup $\mathfrak{G}$. By Theorem 24.4, $n \mid \mathrm{card}\ \mathfrak{G}$, and so $\Pi(n) \subset \Pi(\mathfrak{G}) \subset \Pi$. Conversely, let $\Pi(n) \subset \Pi$. We consider the group G generated by the following two matrices a and b:

$$a = \mathrm{diag}[1, \varepsilon, \ldots, \varepsilon^{n-1}],$$
$$b = e_{21} + e_{32} + \cdots + e_{n, n-1} + e_{1n},$$

where ϵ is an element of order n in C^*. As is easily checked, $ab = \epsilon ba$; card $G = n^3$, so that G is a Π-group. Clearly G is irreducible. ∎

LEMMA 8. *Let G be a primitive Sylow Π-subgroup of $GL(n, C)$. Then the order of the intersection $G_0 = G \cap SL(n, C)$ is smaller than a certain number $\mu(n)$ depending only on n, and G can be expressed in the form*

$$G = G_0 PE_n, \tag{10}$$

where P is a Sylow Π-subgroup of C^* and PE_n the unique maximal abelian normal subgroup of G.

PROOF. Evidently $G \supset PE_n$. On the other hand, by Theorem 16.2 any abelian normal subgroup of G contains only scalar matrices. Therefore PE_n is the unique maximal abelian normal subgroup of G. By Theorem 24.2 $G: PE_n < \rho(n)$. We now prove (10). Since G is a Π-group, $\det a = \alpha \in P$ for $a \in G$. Let γ be a root of the equation $x^n = \alpha$. Then $\gamma \in P$, because $\Pi(n) \subset \Pi$. Obviously $\det(\gamma^{-1}a) = 1$, $\gamma^{-1}a \in G_0$. This gives (10). It remains to estimate card G_0.

Clearly

$$G_0 \cap PE_n = SL(n, C) \cap PE_n.$$

Consequently $\operatorname{card}(G_0 \cap PE_n) = n$. Now,

$$G : PE_n < \rho(n), \qquad G/PE_n = G_0PE_n/PE_n \cong G_0/PE_n \cap G_0.$$

Hence $\operatorname{card} G_0 < \mu(n) = n\rho(n)$. ∎

LEMMA 9. *Let G be an irreducible Sylow* Π*-subgroup of* $GL(n, C)$. *Then either G is primitive or there exists exactly one nonrefinable decomposition of* C^n *into systems of imprimitivity for G.*

PROOF. Suppose that C^n has two nonrefinable decompositions

$$C^n = Q_1 \dotplus \ldots \dotplus Q_\nu \quad \text{and} \quad C^n = L_1 \dotplus \ldots \dotplus L_\mu \tag{11}$$

into systems of imprimitivity for G. By Lemma 15.5, G then has two representations, $G = \mathfrak{G} \sim \Gamma$ and $G = \mathfrak{G}_1 \sim \Gamma_1$, where Γ and Γ_1 are transitive Sylow Π-subgroups of S_ν and S_μ, respectively, $\mathfrak{G}$ is a primitive Sylow Π-subgroup of $GL(n\nu^{-1}, C)$ and $\mathfrak{G}_1$ a primitive Sylow Π-subgroup of $GL(n\mu^{-1}, C)$. Since $G = \mathfrak{G} \sim \Gamma$, it follows that G has an abelian normal subgroup A whose matrices a in a basis corresponding to the first decomposition (11) have the form

$$a = \operatorname{diag}[\lambda_1 E_m, \ldots, \lambda_\nu E_m], \qquad m = n\nu^{-1},$$

where the λ_j range independently over P;

$$G : A \leqslant \nu!\,(\mathfrak{G} : PE_n)^\nu < \nu!\rho(m).$$

Similarly, it follows from the second decomposition in (11) that G has another abelian normal subgroup B of finite index. We express A as $A_1 \cdots A_\nu$, where A_j consists of the matrices

$$\operatorname{diag}[E_m, \ldots, E_m, \lambda_j E_m, E_m, \ldots, E_m].$$

Since $A_jB/B \cong A_j/A_j \cap B$, it follows that $A_j : D_j < \infty$, where $D_j = A_j \cap B$. Consequently $D_j \neq (E_n)$. The product $D_1 \cdots D_\nu = D$ is direct and is contained in B. Hence every subspace Q_j is invariant under B. Now the restriction $B|Q_j$ is a normal subgroup of $\mathfrak{G}$. Since $\mathfrak{G}$ is primitive and $B|Q_j$ abelian, every matrix in $B|Q_j$ is scalar. On the other hand, if we treat C^n as an additive group with the operator domain $C \cup B$, then the one-dimensional subspaces of Q_i and Q_j, $i \neq j$, are not operator-isomorphic. Therefore both decompositions (11) are obtained from a decomposition of C^n into a direct sum of operator-simple groups by combining all operator-isomorphic summands. By Proposition 13.3 this decomposition is unique. Therefore $\nu = \mu$, and after a suitable renumbering $L_j = Q_j$, $j = 1, \ldots, \nu$. ∎

THEOREM 5. *A Sylow* Π*-subgroup of* $GL(n, C)$ *has a unique maximal abelian normal subgroup.*

PROOF. Let G be a Sylow Π-subgroup of $GL(n, C)$. It is easy to see that we need only consider the case when G is irreducible. For primitive groups the theorem has already been proved (see Lemma 8). Suppose now that G is imprimitive, and let

$$C^n = Q_1 \dotplus \ldots \dotplus Q_k \tag{12}$$

be a nonrefinable decomposition of C^n into systems of imprimitivity for G. This decomposition (12) defines an abelian normal subgroup A of G whose matrices, in a basis corresponding to the decomposition, have the form

$$a = \operatorname{diag}[\lambda_1 E_m, \ldots, \lambda_k E_m], \qquad \lambda_j \in P, \qquad m = nk^{-1}.$$

A is the direct product of k copies of P. Now let B be a maximal abelian normal subgroup of G. Then PE_n is a proper subgroup of B. By Theorem 16.2 there exists a decomposition $C^n = R_1 \dotplus \cdots \dotplus R_t$ of C^n into systems of imprimitivity for G such that $B|R_j = PE_{n/t}$. This decomposition can be refined to a nonrefinable decomposition, which, by Lemma 9, must coincide with (12). Thus $B \subset A$. Since B is a maximal abelian normal subgroup of G, we have $B = A$. ∎

LEMMA 10. *Let* $n = km$*; let* $\mathfrak{G}$ *be a primitive Sylow* Π*-subgroup of* $GL(m, C)$ *and* Γ *a transitive Sylow* Π*-subgroup of* S_k*. Then the wreath product*

$$H = \mathfrak{G} \sim \Gamma \tag{13}$$

is an irreducible Sylow Π*-subgroup of* $GL(n, C)$.

PROOF. H is irreducible by Lemma 15.4. It remains to prove that it is not a subgroup of another Π-subgroup of $GL(n, C)$. Suppose that H is contained in a Sylow Π-subgroup G of $GL(n, C)$. H has an abelian normal subgroup A, namely the set of all matrices

$$a = \operatorname{diag}[a_1, \ldots, a_k], \qquad a_j \in PE_m.$$

We express A as a direct product $A = A_1 \times \cdots \times A_k$, where A_j consists of the elements $\operatorname{diag}[E_m, \ldots, a_j, \ldots, E_m]$, $A_j \cong P$. Let B be a maximal abelian normal subgroup of G. Then $G: B < \infty$ and $A_j: (A_j \cap B) < \infty$. Consequently $A_j \cap B = D_j \neq (E_n)$. The product $D_1 \cdots D_k = D$ is direct and is a subgroup of B. Now if

$$C^n = Q_1 \dotplus \ldots \dotplus Q_k \tag{14}$$

is the nonrefinable decomposition of C^n into systems of imprimitivity for H

defined by (13), then, since $D \subset B$, every subspace Q_j is invariant under B. Hence the one-dimensional subspaces of Q_i are not operator-isomorphic to the one-dimensional subspaces of Q_j for $i \neq j$ (the operator domain is $C \cup B$). Therefore the nonrefinable decomposition of C^n into systems of imprimitivity for G is a refinement of (14). Since $G \supset H$, it must coincide with (14). Thus $G = H$. ∎

THEOREM 6. *For any fixed set* Π *of primes, the number of conjugacy classes of Sylow* Π*-subgroups does not exceed a certain bound* $\tau(n)$ *depending only on* n.

PROOF. It is clearly sufficient to prove the theorem for irreducible Sylow Π-subgroups. For primitive Π-subgroups the theorem follows from Lemma 8, because a primitive Sylow Π-subgroup of $GL(n, C)$ is uniquely determined by a finite group of order less than $\mu(n)$ (see (10)). By Lemma 2, an imprimitive Sylow Π-subgroup G of $GL(n, C)$ is conjugate to a group of the form $\mathfrak{G} \sim \Gamma$, where $\mathfrak{G}$ is a primitive Sylow Π-subgroup of $GL(m, C)$, and Γ a transitive Sylow Π-subgroup of S_k, $km = n$. This proves the theorem. ∎

6. *On maximal periodic subgroups of* $GL(n, C)$. The following two theorems are obtained from Theorems 5 and 6 by taking Π to be the set of all primes.

THEOREM 5a. *A maximal periodic subgroup of* $GL(n, C)$ *has a unique maximal abelian normal subgroup.*

THEOREM 6a. *The set of all maximal periodic subgroups of* $GL(n, C)$ *splits into finitely many conjugacy classes. The number of these classes is bounded by a certain number* $\tau(n)$ *depending only on* n.

Lemma 10 implies

LEMMA 10a. *Let* $n = km$, *and let* $\mathfrak{G}$ *be a maximal periodic primitive subgroup of* $GL(m, C)$. *Then* $\mathfrak{G} \sim S_k$ *is a maximal periodic subgroup of* $GL(n, C)$.

COROLLARY 1. *Let* P *be the periodic part of* C^*. *Then the group of all monomial matrices whose nonzero elements lie in* P *is a maximal periodic subgroup of* $GL(n, C)$.

THEOREM 7. (i) *Every maximal periodic subgroup of* $GL(n, C)$ *is conjugate in* $GL(n, C)$ *to a group of the form*

$$G = \mathrm{diag}\,[G_1, \ldots, G_t], \qquad G_j = \mathfrak{G}_j \sim S_{k_j}, \tag{15}$$

where $\mathfrak{G}_j$ *is a primitive maximal periodic subgroup of* $GL(m_j, C)$, $\Sigma_1^t k_j m_j = n$, *and no two of the groups* $\mathfrak{G}_1, \ldots, \mathfrak{G}_t$ *are conjugate.*

(ii) *Every group of the form* (15) *is a maximal periodic subgroup of* $GL(n, C)$.

PROOF. (i) follows from Lemmas 2 and 3. Let us prove (ii). Let G be given by (15) and suppose that it is a proper subgroup of a maximal periodic subgroup F in $GL(n, C)$. Further, let

$$C^n = R_1 \dotplus \ldots \dotplus R_t \tag{16}$$

be a decomposition of C^n into G-invariant G-irreducible subspaces R_j, and

$$C^n = U_1 \dotplus \ldots \dotplus U_s \tag{17}$$

a decomposition into F-invariant F-irreducible subspaces U_j. The subspaces $R_1, \ldots, R_t$ are the only G-invariant G-irreducible subspaces of C^n. Thus each U_j in (17) is the direct sum of certain terms of (16). Since $F \neq G$, it follows from Lemma 10a that some U_j contains more than one term of (16), say $U_1 = R_1 \dotplus \cdots \dotplus R_\gamma$, $\gamma > 1$. Now we consider the two groups $\Phi = F|U_1$ and $\Psi = G|U_1$. Since

$$\Psi = \operatorname{diag}[G_1, \ldots, G_\gamma], \quad G_\alpha = \mathfrak{G}_\alpha \sim S_{k_\alpha}, \quad \alpha = 1, \ldots, \gamma,$$

it follows that Ψ has an abelian normal subgroup A of finite index, which is the direct product of k groups $A_1, \ldots, A_k$, $k = k_1 + \cdots + k_\gamma$, $A_j \cong P$, where P is the periodic part of C^*. If

$$R_\alpha = Q_{\alpha 1} \dotplus \ldots \dotplus Q_{\alpha k_\alpha}$$

is a nonrefinable decomposition of R_α into systems of imprimitivity for G_α, we see that $A|Q_{\alpha\beta} \cong P$. Now Φ has an abelian normal subgroup B of finite index. Also, $D_j = A_j \cap B$ has finite index in A_j. Since A_j is an infinite group, we have $D_j \neq (E)$. The product $D = D_1 \cdots D_k$ is direct and is a subgroup of B. Consequently the subspaces $Q_{\alpha\beta}$ are invariant under B. We claim* that

$$U_1 = \dot{\sum} Q_{\alpha\beta} \tag{18}$$

is a nonrefinable decomposition into systems of imprimitivity for Φ. For let us decompose each $Q_{\alpha\beta}$ into a direct sum of B-invariant one-dimensional subspaces. Since $B \supset D$, two distinct subspaces of the form $Q_{\alpha\beta}$ do not have operator-isomorphic one-dimensional subspaces (the operator domain is $B \cup C^*$). Hence the decomposition of U_1 into systems of imprimitivity for Φ that is determined by B in accordance with Clifford's theorem is a refinement of (18). Therefore the unique nonrefinable decomposition of U_1 into systems of imprimitivity for Φ is a refinement of (18). But $\Phi \supset \Psi$, and $Q_{\alpha\beta}$ is a direct summand of the nonrefinable decomposition of R_α into systems of imprimitivity for G_α. Consequently

---

**Editor's note.* The argument in support of this claim has been changed from that given in the original.

(18) is the nonrefinable decomposition of U_1 into systems of imprimitivity for Φ. Since Φ is a maximal periodic linear group, we have $\Phi = \mathfrak{G} \sim S_k$, where $\mathfrak{G}$ is a primitive group. Since $\Phi \supset \Psi$, the groups $\mathfrak{G}_j$, $j = 1, \ldots, \gamma$, are conjugate to a subgroup of $\mathfrak{G}$. But $\mathfrak{G}_j$ is a maximal periodic subgroup of $GL(m_j, C)$. Therefore, $\mathfrak{G}_j$ is conjugate to $\mathfrak{G}$; that is, $\mathfrak{G}_1, \ldots, \mathfrak{G}_\gamma$ are conjugate to one another. This contradicts the construction of the group (15). ∎

The theorems just proved on maximal periodic subgroups of $GL(n, C)$ remain valid for any algebraically closed field of characteristic zero. If Δ is an algebraically closed field whose characteristic does not belong to Π, then Lemmas 7 and 9, Theorem 5 and Lemma 10 carry over without modification to Sylow Π-subgroups of $GL(n, \Delta)$. Theorem 6, however, requires a slight change on transition from C to Δ: the bound $\tau(n)$ must be replaced by $\tau(n, p)$, where $p = \operatorname{char} \Delta$.

§26. Structure theorems on periodic matrix groups over a field

Taken together, the conditions of periodicity and existence of a faithful linear representation over a field are such strong restrictions that groups satisfying both have many of the properties of finite groups. Here we quote without proof some theorems on periodic matrix groups, due to Platonov [11], which can be regarded as far-reaching generalizations of classical theorems in the theory of finite groups*.

THEOREM 1. *Let G be a periodic matrix group over a field and N a normal subgroup of G such that*

$$\Pi(N) \cap \Pi(G/N) = \varnothing.$$

Then G can be represented as a product $G = DN$, $D \cap N = E$. If also $G = D_1N$, $D_1 \cap N = E$, then the subgroups D and D_1 are conjugate in G.

THEOREM 2. *In a periodic matrix group over a field the Sylow p-subgroups are conjugate.*

A Π-subgroup H of a periodic group G is called a *Hall Π-subgroup* if it contains a Sylow p-subgroup of G for every $p \in \Pi$.

In the next theorem $\Pi' = \Pi \backslash \Pi(G)$.

THEOREM 3. *Let G be a periodic matrix group over a field.*

(i) *If G is solvable, and H and H' are Sylow Π- and Π'-subgroups, respectively, of G, then $G = HH'$.*

**Editor's note.* For recent developments in the theory of periodic linear groups, the reader should consult the book by Wehrfritz [3].

(ii) *If for any $p \in \Pi(G)$, G contains a Sylow p-subgroup that has a Hall complement, then G is solvable.*

Let G be a periodic group. A system $\{P_k\}$ of Sylow p-subgroups P_k of G, one for each $p_k \in \Pi(G)$, is called a *Sylow basis* for G if G is the product of the P_k and if for any finite subsystem

$$P_{k_1}, \ldots, P_{k_m} \tag{1}$$

the group H generated by (1) has $\Pi(H) = \{p_{k_1}, \ldots, p_{k_m}\}$.

THEOREM 4. (i) *A solvable periodic matrix group G over a field has a Sylow basis; any two such bases are conjugate in G.*

(ii) *If a periodic matrix group over a field has a Sylow basis, it is solvable.*

We end the chapter with two theorems of Platonov [11] on algebraic matrix groups.

THEOREM 5. *Sylow p-subgroups of an algebraic matrix group G over an algebraically closed field are conjugate.*

THEOREM 6. *Let Π be an arbitrary set of primes. Sylow Π-subgroups of a solvable algebraic matrix group over an algebraically closed field are conjugate.*

SEVENTH CHAPTER

NILPOTENT AND LOCALLY NILPOTENT MATRIX GROUPS

§27. Irreducible nilpotent matrix groups

1. *Finiteness of the index of the center.* Let G be an arbitrary group. The *upper central series*

$$Z_0 = (e) \subset Z_1(G) \subset Z_2(G) \subset \ldots \subset Z_n(G) \subset \ldots$$

of G is defined inductively: $Z_{j+1}(G)/Z_j(G)$ is the center of $G/Z_j(G)$. The group $Z_j(G)$ is called the *jth hypercenter* of G. G is said to be *nilpotent* if $G = Z_l(G)$ for some l. The smallest l for which $G = Z_l(G)$ is the *nilpotency class of G.*

In particular, a nontrivial abelian group is nilpotent of class 1, and the trivial group is nilpotent of class 0.

If G and H are elementwise commuting subgroups of a group, H abelian and G nilpotent of class $l > 0$, then GH is also nilpotent of class l.

The *lower central series*

$$G_0 = G \supset G_1 \supset G_2 \supset \ldots \supset G_l \supset \ldots$$

of G is also defined inductively: $G_{j+1} = (G_0, G_j)$, $j = 0, 1, \ldots$, where (G_0, G_j) is the commutator subgroup of G_0 and G_j.

A group G is nilpotent of class l if and only if l is the smallest number for which G_l is trivial.

Thus, *the upper and lower central series of a nilpotent group have the same length* (Kuroš [1], Russian p. 388; English vol. II, p. 214).

Now let Δ be an arbitrary field, G an irreducible nilpotent subgroup of $GL(n, \Delta)$ of class l, and

$$Z_0 = (E) \subset Z_1 \subset Z_2 \subset \ldots \subset Z_l = G$$

the upper central series of G.

LEMMA 1. *G is contained in a nilpotent subgroup $\mathfrak{G}$ of $GL(n, \Delta)$ with the following properties:*

(i) *$\mathfrak{G}$ and G have the same nilpotency class l.*

(ii) *$\mathfrak{G}$ is an absolutely irreducible subgroup of $GL(r, \Sigma)$, where Σ is an extension of Δ, $\Sigma : \Delta = n/r$.*

(iii) $Z_1(\mathfrak{G}) = \Sigma^* E_r \supset Z_1 = Z_1(G)$.

PROOF. Let $[G]$ be the linear Δ-hull of G. Since G is irreducible, by Theorem 14.4 we have $[G] = T_\rho$, where T is a division ring, $T\colon \Delta = n/\rho$. There are two possibilities: T either is a field or is noncommutative. In the first case, setting $\Sigma = T$, $r = \rho$ and $\mathfrak{G} = GT^*$, we see that $\mathfrak{G}$ satisfies (i), (ii) and (iii). In the second case, G is contained in another nilpotent subgroup G_1 of $GL(n, \Delta)$, of class l, such that $[G_1]: \Delta > [G]: \Delta$. For if T is noncommutative, there is a matrix h in $GL(n, \Delta)\backslash T_\rho$ that commutes with every element of T_ρ. Now $G_1 = (h)G$ is a nilpotent subgroup of $GL(n, \Delta)$ of class l. Since $h \in GL(n, \Delta)\backslash T_\rho$, we see that $[G_1] \neq [G]$. Hence, $[G_1]: \Delta > [G]: \Delta$. This completes the proof. ■

LEMMA 2. *The order of every element of Z_2/Z_1 is a divisor of n.*

PROOF. By Lemma 1, $G \subset \mathfrak{G} \subset GL(r, \Sigma)$ and $Z_1 \subset \Sigma^* E_r$, $r|n$. For $a \in Z_2$ and $g \in G$ we have $ag = \sigma ga$, $\sigma \in \Sigma^*$. Hence

$$\det(ag) = \sigma^r \det(ga), \qquad \sigma^r = 1.$$

Therefore

$$a^r g = \sigma^r g a^r = g a^r, \qquad a^r \in Z_1, \qquad (aZ_1)^r = Z_1. \ ■$$

LEMMA 3. *Let ν be the order of an element aZ_1 of Z_2/Z_1. Then there is an element g in G such that the order of $(a, g) = aga^{-1}g^{-1}$ is also ν.*

PROOF. By Lemma 1, $G \subset GL(r, \Sigma)$ and $Z_1 \subset \Sigma^* E_r$. For any element g of G,

$$(a, g) \in \Sigma^* E_r, \qquad (a, g)^r = 1.$$

If $h \in G$, then $(a, g)(a, h) = (a, gh)$. Hence all the commutators (a, g), $g \in G$, form a finite cyclic group, of order τ, say. Then for $g \in G$

$$(a^\tau, g) = (a, g)^\tau = 1, \quad (aZ_1)^\tau = Z_1, \qquad \nu \mid \tau.$$

But $(a, g)^\nu = (a^\nu, g) = 1$, $\tau|\nu$, and so $\tau = \nu$. ■

COROLLARY 1. char Δ *cannot divide the order of any element of Z_2/Z_1.*

LEMMA 4. *Let $Z_1 = \Delta^* E_n$. Then $Z_2\colon Z_1 = [Z_2]\colon \Delta$, where $[Z_2]$ is the linear Δ-hull of Z_2.*

PROOF. Let

$$d_1, \ldots, d_t \tag{1}$$

be part of a transversal to Z_1 in Z_2. We claim that the system (1) is linearly independent over Δ. Let

$$\lambda_1 d_1 + \lambda_2 d_2 + \cdots + \lambda_t d_t = 0, \qquad \lambda_j \in \Delta, \tag{2}$$

be a nontrivial dependence relation with the smallest number of nonzero terms. We may assume that $\lambda_1, \lambda_2 \neq 0$. For $g \in G$,

$$gd_jg^{-1}=\gamma_jd_j, \qquad \gamma_j\in\Delta^*,$$

and for $i\neq j$ there is an element $g\in G$ such that

$$\gamma_i\neq\gamma_j, \tag{3}$$

because d_i and d_j lie in distinct cosets of Z_1. Let g be such an element. Then, by (2),

$$\begin{aligned}\gamma_1(\lambda_1d_1+\lambda_2d_2+\dots+\lambda_td_t)\\ -g(\lambda_1d_1+\lambda_2d_2+\dots+\lambda_td_t)g^{-1}\\ =\lambda_2(\gamma_1-\gamma_2)d_2+\dots+\lambda_t(\gamma_1-\gamma_t)d_t=0.\end{aligned}$$

This contradicts the choice of the relation (2). Therefore (1) is a linearly independent system, and $Z_2\colon Z_1\leqslant[Z_2]\colon\Delta$. Now let $u_1,\dots,u_\tau$ be elements of Z_2 which form a Δ-basis of $[Z_2]$. Then they lie in τ distinct cosets of Z_1 in Z_2; that is, $Z_2\colon Z_1\geqslant[Z_2]\colon\Delta$. This completes the proof. ∎

LEMMA 5. *Let k be the exponent of Z_2/Z_1. Then the exponent of Z_{j+1}/Z_j, $j=1,\dots,l-1$, is a divisor of k.*

PROOF. For $j=1$ the lemma is trivial. Let $j>1$ and suppose the lemma is true for Z_j/Z_{j-1}. If $a\in Z_{j+1}$, $g\in G$, then $ga=bag$, where $b\in Z_j$. Therefore

$$gZ_{j-1}aZ_{j-1}=bZ_{j-1}aZ_{j-1}gZ_{j-1}.$$

Since $bZ_{j-1}gZ_{j-1}=gZ_{j-1}bZ_{j-1}$, we have

$$gZ_{j-1}(aZ_{j-1})^k=(bZ_{j-1})^k(aZ_{j-1})^kgZ_{j-1}.$$

By the inductive hypothesis, $(bZ_{j-1})^k=Z_{j-1}$. Hence $(aZ_{j-1})^k\subset Z_j$, $(aZ_j)^k=Z_j$. ∎

COROLLARY 1. *The order of each element of G/Z_1 is a divisor of k^{l-1}, where k is the exponent of Z_2/Z_1.*

LEMMA 6. *Let Δ be an algebraically closed field. Then G is a monomial group; that is, there is a basis $u_1,\dots,u_n$ of Δ^n such that for every $g\in G$*

$$g(u_1)=\lambda_1u_{i_1},\ \dots,\ g(u_n)=\lambda_nu_{i_n},$$

where $\lambda_j\in\Delta^*$ and $i_1,\dots,i_n$ is a permutation of $1,\dots,n$.

PROOF. We claim that for $n>1$ a nilpotent irreducible subgroup of $GL(n,\Delta)$ is imprimitive if Δ is algebraically closed. Let $a\in Z_2\backslash Z_1$. Then $(a)Z_1$ is an abelian normal subgroup of G and $(a)Z_1\not\subset\Delta^*E_n$. By Theorem 16.2, G is imprimitive. Now let $\Delta^n=V_1+\dots+V_t$ be a nonrefinable decomposition of Δ^n into systems of imprimitivity for G, and

$$H_j=\{h_j\mid h_j\in G,\ h_j(V_j)=V_j\}.$$

Then the restriction $H_j|V_j=G_j$ is a nilpotent primitive group of degree $nt^{-1}=n_1$. Hence $n_1=1$ and $t=n$. Thus G is monomial. ∎

THEOREM 1. *Let Δ be an arbitrary field and G an irreducible nilpotent subgroup of $GL(n, \Delta)$ of class l. Then*

$$G : Z_1(G) \leqslant n!\, n^{(l-1)(n-1)} = \nu(n, l),$$

and char $\Delta \nmid (G : Z_1(G))$.

PROOF. By Lemma 1, $G \subset \mathfrak{G}$, where $\mathfrak{G}$ is an absolutely irreducible nilpotent subgroup of $GL(r, \Sigma)$ of class l, $Z_1(\mathfrak{G}) = \Sigma^* E_r \supset Z_1(G)$ and $r|n$. If Ω is the algebraic closure of Σ, then $\mathfrak{G}$ is an irreducible, and by Lemma 6 a monomial, subgroup of $GL(r, \Omega)$, $Z_1(\mathfrak{G}) \subset \Omega^* E_r$. Hence there is a homomorphism $\gamma: \mathfrak{G} \longrightarrow S_r$, where S_r is the symmetric group of degree r, such that Ker $\gamma = \mathfrak{H}$ is the subgroup of all diagonal matrices in $\mathfrak{G} \subset GL(r, \Omega)$. It is clear that $\mathfrak{H} \supset Z_1(\mathfrak{G}) = \Sigma^* E_r$. Now let $h = \operatorname{diag}[\omega_1, \ldots, \omega_r]$, $\omega_j \in \Omega^*$, be any matrix in $\mathfrak{H}$. Then, by Lemma 5, $h^{k^{l-1}} \in Z_1(\mathfrak{G})$. We can write h in the form

$$h = \omega_1 \operatorname{diag}[1, \alpha_2, \ldots, \alpha_r].$$

Since

$$\omega_1^{k^{l-1}} = \omega_2^{k^{l-1}} = \ldots = \omega_r^{k^{l-1}},$$

it follows that $\alpha_j^{k^{l-1}} = 1$ for $j = 2, \ldots, r$. Now, if $h_1 \in \mathfrak{H}$ and $h_1 = \omega_1^1 \operatorname{diag}[1, \alpha_2, \ldots, \alpha_r]$, then $h_1 h^{-1} \in Z_1(\mathfrak{G})$. Therefore

$$\mathfrak{H} : Z_1(\mathfrak{G}) \leqslant (k^{l-1})^{r-1} = k^{(l-1)(r-1)}.$$

Since $\mathfrak{G}$: $\mathfrak{H} \leqslant r!$, we have

$$\mathfrak{G} : Z_1(\mathfrak{G}) \leqslant r!\, k^{(l-1)(r-1)} \leqslant n!\, n^{(l-1)(n-1)} = \nu(n, l).$$

Since $G : Z_1(G) = \mathfrak{G}$: $Z_1(\mathfrak{G}) \leqslant \nu(n, l)$, we have G: $Z_1(G) \leqslant \nu(n, l)$. ■

Using Lemma 17.5, we infer from Theorem 1 the following result.

COROLLARY 1. *If Δ is a perfect field, then an irreducible nilpotent subgroup of $GL(n, \Delta)$ is a d-group.*

2. *Irreducible nilpotent matrix groups over an algebraically closed field.*

THEOREM 2. *Let Ω be an algebraically closed field and G an irreducible nilpotent subgroup of $GL(n, \Omega)$. Then $\Pi(G/Z_1(G)) = \Pi(n)$.*

PROOF. By Lemma 6, G is monomial. Hence there is a homomorphism $\gamma: G \longrightarrow S_n$ such that Ker $\gamma \supset Z_1(G)$. Since G is irreducible, Im γ is transitive. Consequently $n|$card(Im γ). Obviously

$$G : Z_1(G) = (G : \operatorname{Ker} \gamma)(\operatorname{Ker} \gamma : Z_1(G)), \quad n \mid G : Z_1(G),$$
$$\Pi(n) \subset \Pi(G/Z_1(G)).$$

On the other hand, by Lemma 5 the order of every element of $G/Z_1(G)$ divides k^{l-1}, where $k|n$. Therefore $\Pi(G/Z_1(G)) \subset \Pi(n)$. ■

COROLLARY 1. *If the ground field is algebraically closed, then $\Pi(k) =$*

$\Pi(n)$, *where k is the exponent of* $Z_2(G)/Z_1(G)$.

COROLLARY 2. *If* Ω *is algebraically closed and* char $\Omega | n$, *then* $GL(n, \Omega)$ *has no irreducible nilpotent subgroups.*

COROLLARY 3. *Let* Δ *be an arbitrary field and G an absolutely irreducible nilpotent subgroup of* $GL(n, \Delta)$. *Then* $\Pi(G/Z_1(G)) = \Pi(n)$.

THEOREM 3. *Let* Ω *be an algebraically closed field and G an irreducible nilpotent subgroup of* $GL(n, \Omega)$ *such that* $Z_1(G) = \Omega^* E_n$. *Then* $G_0 = G \cap SL(n, \Omega)$ *is a finite group and* $\Pi(G_0) = \Pi(n)$.

PROOF. Let $g \in G$. Since Ω is algebraically closed, there is an element ω in Ω^* such that $\det g = \omega^n$. Therefore

$$\omega^{-1} E_n g \in G_0, \qquad G = G_0 Z_1(G).$$

But

$$G : Z_1(G) = G_0 Z_1(G) : Z_1(G) = G_0 : (G_0 \cap Z_1(G)),$$
$$\operatorname{card}(G_0 \cap Z_1(G)) = n.$$

Hence $G_0 : (E_n) = n(G : Z_1(G))$. Thus G_0 is a finite group, and $\Pi(G_0) = \Pi(n)$. ■

COROLLARY 1. *Let H be an irreducible nilpotent subgroup of* $SL(n, \Omega)$, *where* Ω *is algebraically closed. Then H is finite and* $\Pi(H) = \Pi(n)$.

PROOF. Let $G = H\Omega^*$ and $G_0 = G \cap SL(n, \Omega)$. By Theorem 3, G_0 is finite and $\Pi(G_0) = \Pi(n)$. Obviously $H \subset G_0$; hence $\Pi(H) \subset \Pi(n)$, and H is finite. But

$$G/\Omega^* E_n \cong H/H \cap \Omega^* E_n.$$

Therefore

$$\Pi(H/H \cap \Omega^* E_n) = \Pi(G/\Omega^* E_n) = \Pi(n), \qquad \Pi(H) \supset \Pi(n).$$

Hence $\Pi(H) = \Pi(n)$. ■

Let Ω be an algebraically closed field, and $\mathfrak{M}_l$ the set of all subgroups of $GL(n, \Omega)$ satisfying the following conditions:

(i) Every group in $\mathfrak{M}_l$ is irreducible and nilpotent of class l.

(ii) No group of $\mathfrak{M}_l$ is contained in another nilpotent subgroup of $GL(n, \Omega)$ of class l.

THEOREM 4. $\mathfrak{M}_l$ *splits into finitely many conjugacy classes of subgroups in* $GL(n, \Omega)$.

PROOF. Let $G \in \mathfrak{M}_l$. Then $Z_1 = Z_1(G) = \Omega^* E_n$. By Theorem 3, $G = G_0\Omega^*$. Hence G is uniquely determined by the finite group G_0, whose order is at most $n\nu(n, l)$. There are only finitely many nonisomorphic groups of order at most $n\nu(n, l)$. Every finite group has only finitely many inequivalent irreducible linear representations of degree n over Ω. This proves the theorem. ∎

Our next aim is to classify all maximal irreducible nilpotent subgroups of $GL(n, \Omega)$, of class 2, where Ω is an algebraically closed field.

THEOREM 5. *Let Ω be an algebraically closed field and G an irreducible nilpotent subgroup of $GL(n, \Omega)$. If $Z_2 = Z_2(G)$ is irreducible, then $G = Z_2$, so that G is of class 2.*

PROOF. We write the abelian group Z_2/Z_1, where $Z_1 = Z_1(G)$, as a product of ρ cyclic groups of orders $l_1, \ldots, l_\rho$:

$$Z_2/Z_1 = (c_1Z_1) \ldots (c_\rho Z_1);$$

$$Z_2 = (c_1)(c_2) \ldots (c_\rho) Z_1, \qquad Z_2 : Z_1 = l_1 l_2 \ldots l_\rho.$$

By Lemma 3 there is an element g_j in G such that $(c_j, g_j) = \epsilon_j E_n$, where ϵ_j is an element of order l_j in Ω^*. On the other hand, for $g \in G$ we have $(c_j, g) = \epsilon_j^\alpha E_n$. Hence $G = (g_j)F_j$, where F_j is the centralizer of c_j in G. As is easy to see, $G : F_j = l_j$. Now, if F is the centralizer of Z_2 in G, then $G : F \leqslant l_1 \cdots l_\rho$, because F is the intersection of $F_1, \ldots, F_\rho$. Since Z_2 is irreducible, $F = Z_1$. Therefore $G : Z_1 \leqslant l_1 \cdots l_\rho = Z_2 : Z_1$. Hence $G = Z_2$. ∎

THEOREM 6. *Let G be an irreducible nilpotent subgroup of $GL(n, \Omega)$ of class 2 over an algebraically closed field Ω, and $Z_1(G) = \Omega^* E_n$. Then G is maximal among the nilpotent subgroups of $GL(n, \Omega)$ of class 2.*

PROOF. By Lemma 4, $G : Z_1(G) = n^2$. If $\mathfrak{G}$ is a nilpotent subgroup of $GL(n, \Omega)$ of class 2 and $\mathfrak{G} \supset G$, then $Z_1(\mathfrak{G}) = Z_1(G)$ and $\mathfrak{G} : Z_1(G) = n^2$. Hence $\mathfrak{G} = G$. ∎

THEOREM 7. *Let Ω be an algebraically closed field and let* char $\Omega \nmid n$. *Write n in the form*

$$n = k_1 k_2 \ldots k_t, \tag{4}$$

where $k_i > 1$, $i = 1, \ldots, t$, and $k_j | k_i$ for $i < j$. Then the following assertions are true:

(i) *Every decomposition (4) of n determines a maximal irreducible nilpotent subgroup $\Gamma_{k_1 \cdots k_t}$ of $GL(n, \Omega)$ of class 2, whose elements g have the form*

$$g = \lambda a_1^{\alpha_1} b_1^{\beta_1} \ldots a_t^{\alpha_t} b_t^{\beta_t}, \qquad \lambda \in \Omega^*; \tag{5}$$

where

$$a_j = E_{k_1 \ldots k_{j-1}} \dot{\times} \operatorname{diag}\left[1, \varepsilon_j, \ldots, \varepsilon_j^{k_j - 1}\right] \dot{\times} E_{k_{j+1} \ldots k_t},$$

$$b_j = E_{k_1 \ldots k_{j-1}} \dot{\times} \begin{bmatrix} 0 & 1 \\ E_{k_j - 1} & 0 \end{bmatrix} \dot{\times} E_{k_{j+1} \ldots k_t},$$

ϵ_j being an element of order k_j in Ω^*.

(ii) *Two different decompositions of n in the form* (4) *determine nonisomorphic groups* $\Gamma_{k_1 \cdots k_t}$.

(iii) *Every maximal irreducible nilpotent subgroup of* $GL(n, \Omega)$ *of class* 2 *is conjugate to one and only one subgroup* $\Gamma_{k_1 \cdots k_t}$ *determined by a decomposition of n in the form* (4).

PROOF. We consider $\Gamma_{k_1 \cdots k_t} = (a_1)(b_1) \cdots (a_t)(b_t)\Omega^*$. It is obvious that $\Gamma_{k_1 \cdots k_t} \supset \Omega^* E_n$. It follows from (5) that $(a_j, b_j) = \epsilon_j E_n$, and $(a_i, b_j) = (a_i, a_j) = (b_i, b_j) = E_n$ for $i \neq j$. Hence $\Gamma_{k_1 \cdots k_t}$ is nilpotent of class 2. The same relations imply the linear independence of the n^2 elements

$$a_1^{\alpha_1} b_1^{\beta_1} \ldots a_t^{\alpha_t} b_t^{\beta_t}, \quad 0 \leqslant \alpha_j < k_j, \quad 0 \leqslant \beta_j < k_j.$$

Thus $\Gamma_{k_1 \cdots k_t}$ is an irreducible nilpotent subgroup of $GL(n, \Omega)$ of class 2, and $Z_1(\Gamma_{k_1 \cdots k_t}) = \Omega^* E_n$. Together with Theorem 6, this proves (i).

Now,

$$\bar{\Gamma} = \Gamma_{k_1 \ldots k_t}/Z_1 = (a_1 Z_1)(b_1 Z_1) \ldots (a_t Z_1)(b_t Z_1),$$
$$Z_1 = Z_1(\Gamma_{k_1 \ldots k_t});$$

that is, $\bar{\Gamma}$ is the direct product of $2t$ cyclic groups of orders $k_1, k_1, \ldots, k_t, k_t$. This proves (ii).

Let G be an arbitrary maximal irreducible nilpotent subgroup of $GL(n, \Omega)$ of class 2. Then, as in the proof of Theorem 19.2, we can show that

$$G = (c_1)(d_1) \ldots (c_t)(d_t)\Omega^*, \qquad (c_\nu, d_\nu) = \varepsilon_\nu E_n,$$

where ϵ_ν is an element of order k_ν in Ω^*, elements of distinct pairs commute, $k_1 \cdots k_t = n$ and $k_{j+1} | k_j$, $j = 1, \ldots, t-1$. It is clear that $c_\nu^{k_\nu}, d_\nu^{k_\nu} \in \Omega^* E_n$. Since Ω is algebraically closed, we may assume that $c_\nu^{k_\nu} = d_\nu^{k_\nu} = E_n$. Therefore $G \cong \Gamma_{k_1 \cdots k_t}$, and the isomorphism

$$\lambda c_1^{\alpha_1} d_1^{\beta_1} \ldots c_t^{\alpha_t} d_t^{\beta_t} \mapsto \lambda a_1^{\alpha_1} b_1^{\beta_1} \ldots a_t^{\alpha_t} b_t^{\beta_t}, \quad \lambda \in \Omega^*,$$

can be extended to an automorphism

$$\sum \lambda_{\alpha_1 \ldots \beta_t} c_1^{\alpha_1} d_1^{\beta_1} \ldots c_t^{\alpha_t} d_t^{\beta_t} \mapsto \sum \lambda_{\alpha_1 \ldots \beta_t} a_1^{\alpha_1} b_1^{\beta_1} \ldots a_t^{\alpha_t} b_t^{\beta_t}$$

of $\Omega_n = [G] = [\Gamma_{k_1 \cdots k_t}]$. It follows that G and $\Gamma_{k_1 \cdots k_t}$ are conjugate in $GL(n, \Omega)$. ∎

COROLLARY 1. *If* char $\Omega \nmid n$, *then the number of conjugacy classes of maximal irreducible nilpotent subgroups of* $GL(n, \Omega)$ *of class* 2 *is the same as the number of nonisomorphic abelian groups of order n.*

COROLLARY 2. *Two maximal irreducible nilpotent subgroups of* $GL(n, \Omega)$ *of class* 2 *are conjugate in* $GL(n, \Omega)$ *if and only if their factor groups modulo the center are isomorphic.*

3. *An infinite chain of irreducible nilpotent subgroups of $GL(n, \Omega)$.* The following lemma holds for any ground field Δ.

LEMMA 7. *Let*

$$h = \begin{bmatrix} 0 & 1 \\ E_{n-1} & 0 \end{bmatrix} \in GL(n, \Delta), \tag{6}$$

and let d be an arbitrary diagonal matrix in $SL(n, \Delta)$. Then there is a diagonal matrix c in $GL(n, \Delta)$ such that

$$(c, h) = chc^{-1}h^{-1} = d. \tag{7}$$

If c and f are diagonal matrices in $GL(n, \Delta)$ such that $(c, h) = (f, h)$, then $f = \lambda c$, $\lambda \in \Delta^*$.

PROOF. Let

$$d = \operatorname{diag}[\delta_0, \delta_1, \ldots, \delta_{n-1}] \in SL(n, \Delta).$$

As is easy to verify, the matrix

$$c = \operatorname{diag}[1, \gamma_1, \ldots, \gamma_{n-1}],$$
$$\gamma_j = \delta_1 \cdots \delta_j, \qquad j = 1, \ldots, n-1,$$

satisfies (7).

Now let c and f be diagonal matrices in $GL(n, \Delta)$ such that $(c, h) = (f, h)$. Then $(c^{-1}f, h) = E_n$. But a simple computation shows that a diagonal matrix commuting with h must be scalar. Hence $c^{-1}f = \lambda E_n$, $\lambda \in \Delta^*$, and so $f = \lambda c$. ∎

Now let Ω be algebraically closed, and char $\Omega \nmid n$. By means of Lemma 7 we construct an ascending chain of irreducible nilpotent subgroups of $GL(n, \Omega)$, containing groups of any nilpotency class $l > 1$.

We first construct an infinite chain $A_1, A_2, \ldots$ of subgroups of the group of nonsingular diagonal $n \times n$ matrices over Ω, where $A_1 = \Omega^* E_n$, and A_j for $j > 1$ is the set of all diagonal matrices f in $GL(n, \Omega)$ such that $(f, h) = fhf^{-1}h^{-1} \in A_{j-1}$, and h is a matrix of the form (6) in $GL(n, \Omega)$.

LEMMA 8. *For any $j \geqslant 1$, the factor group A_{j+1}/A_j is cyclic of order n.*

PROOF. We first consider A_2. Let $c_2 = \operatorname{diag}[1, \epsilon, \ldots, \epsilon^{n-1}]$, where ϵ is an element of order n in Ω^*. It is easily checked that $c_2 h c_2^{-1} h^{-1} = \epsilon E_n$. Now, if

$$uhu^{-1}h^{-1} = \lambda E_n \in A_1 = \Omega^* E_n,$$

then $\lambda^n = 1$. Therefore $\lambda = \epsilon^\alpha$, $0 \leqslant \alpha < n$. Hence, by Lemma 7,

$$A_2 = (c_2)\, \Omega^* E_n = (c_2)\, A_1.$$

Thus for $j = 1$ the lemma is proved.

Let $j > 1$ and suppose that $A_j = (c_j)A_{j-1}$; let $1, c_j, \ldots, c_j^{n-1}$ be a transversal to A_{j-1} in A_j. We want to show that A_{j+1}/A_j is cyclic of order n. Two matrices u and v in A_{j+1} lie in the same coset of A_j in A_{j+1} if and only if the matrices (u, h) and (v, h) lie in the same coset of A_{j-1} in A_j. For let

$$uhu^{-1}h^{-1} = d_{j-1}vhv^{-1}h^{-1}, \quad d_{j-1} \in A_{j-1}.$$

Then $v^{-1}uh = d_{j-1}hv^{-1}u$, whence

$$(v^{-1}u, h) = d_{j-1} \in A_{j-1}, \quad v^{-1}u \in A_j, \quad u \in vA_j.$$

Conversely, let $u = d_jv$, $d_j \in A_j$. Then

$$(u, h) = d_jvhv^{-1}d_j^{-1}h^{-1} = d_j(vhv^{-1}h^{-1})hd_j^{-1}h^{-1}$$
$$= (v, h)(d_j, h) = d_{j-1}(v, h),$$
$$(u, h) \in (v, h)A_{j-1}.$$

By Lemma 7, there is a matrix c_{j+1} in A_{j+1} such that $(c_{j+1}, h) = \lambda c_j$, where $\det(\lambda c_j) = 1$, $\lambda \in \Omega^*$. It is obvious that $(c_{j+1}^\alpha, h) = \lambda^\alpha c_j^\alpha$. Hence and from the preceding remark it follows that $1, c_{j+1}, \ldots, c_{j+1}^{n-1}$ is a transversal to A_j in A_{j+1}. ∎

We now introduce groups Γ_n^l by setting $\Gamma_n^l = (h)A_l$, $l > 1$.

THEOREM 8. (i) *Γ_n^l is an irreducible subgroup of $GL(n, \Omega)$, nilpotent of class l.*

(ii) *A_j is the jth hypercenter of Γ_n^l for $j < l$; that is, $A_j = Z_j(\Gamma_n^l)$.*

(iii) *Γ_n^l is maximal among the nilpotent subgroups of $GL(n, \Omega)$ of class l.*

(iv) *$\Gamma_n^l : Z_1(\Gamma_n^l) = n^l$.*

PROOF. The group Γ_n^l is irreducible, because it contains the n^2 linearly independent matrices $c_2^\alpha h^\beta$, $0 \leqslant \alpha < n$, $0 \leqslant \beta < n$. From the irreducibility of Γ_n^l and the fact that $A_1 = \Omega^* E_n \subset \Gamma_n^l$ it follows that A_1 is the center of Γ_n^l. Let $1 < j < l$ and suppose we have proved that $A_{j-1} = Z_{j-1}(\Gamma_n^l)$. We want to prove that

$$A_j = Z_j(\Gamma_n^l). \tag{8}$$

With this aim we consider the commutator group (A_j, Γ_n^l):

$$(A_j, \Gamma_n^l) = (A_j, (h)A_l) = (A_j, (h)) \subset A_{j-1}, \quad A_j \subset Z_j(\Gamma_n^l).$$

On the other hand, every diagonal matrix of $Z_j(\Gamma_n^l)$ is in A_j. But if $w \in \Gamma_n^l$ is a nondiagonal matrix, then $w \notin Z_j(\Gamma_n^l)$. For,

$$w = h^\alpha c, \quad 0 < \alpha < n, \quad c \in A_l,$$
$$(c_l, w) = (c_l, h^\alpha) = (c_l, h)^\alpha = \lambda c_{l-1}^\alpha, \quad \lambda \in \Delta^*.$$

Since $j - 1 < l - 1$,

$$\lambda c_{l-1}^\alpha \notin A_{j-1}, \quad w \notin Z_j(\Gamma_n^l).$$

Hence (8) is true for $j < l$. Since $\Gamma_n^l : A_l = n$ and $A_l : Z_{l-1}(\Gamma_n^l) = n$, we have

$$\Gamma_n^l : Z_{l-1}(\Gamma_n^l) = n^2.$$

Now, $\Gamma_n^l = (h)(c_l)A_{l-1}$. The factor group

$$\Gamma_n^l / A_{l-1} = (h)\, A_{l-1}/A_{l-1} \cdot (c_l)\, A_{l-1}/A_{l-1}$$

is abelian, because $(h, c_l) \in A_{l-1}$. Thus $Z_l(\Gamma_n^l) = \Gamma_n^l$, so that Γ_n^l is nilpotent of class l. Obviously $A_{l-1} : A_1 = n^{l-2}$. Therefore

$$\Gamma_n^l : Z_1(\Gamma_n^l) = n^l.$$

This proves (iv).

It remains to prove (iii). Let $\Gamma_n^l \subset \Gamma$, where Γ is a nilpotent subgroup of $GL(n, \Omega)$ of class l. Then $Z_1(\Gamma) = A_1$. Let $j < l$ and suppose that $Z_i(\Gamma) \subset A_i$ for all $i < j$. We claim that

$$Z_j(\Gamma) \subset A_j. \tag{9}$$

If $g_j \in Z_j(\Gamma)$, then

$$d_{j-1} = (g_j, h) \in Z_{j-1}(\Gamma) \subset A_{j-1}, \qquad \det d_{j-1} = 1.$$

Consequently there is an element d_j in A_j such that $(d_j, h) = (g_j, h) = d_{j-1}$. Hence $g_j = d_j v$, $(v, h) = 1$. We consider (c_2, g_j):

$$(c_2, g_j) = (c_2, d_j v) = d'_{j-1} \in Z_{j-1}(\Gamma) \subset A_{j-1}.$$

Therefore the matrix $c_2 d_j v c_2^{-1} v^{-1} d_j^{-1}$ is diagonal, and so is $v c_2^{-1} v^{-1}$; and $(c_2, g_j) = (c_2, v)$. Further,

$$(h, d'_{j-1}) = (h, c_2 v c_2^{-1} v^{-1}) = h c_2 v c_2^{-1} v^{-1} h^{-1} v c_2 v^{-1} c_2^{-1}$$
$$= h c_2 v c_2^{-1} h^{-1} c_2 v^{-1} c_2^{-1}.$$

Since $c_2 h = \epsilon h c_2$, we have $h^{-1} c_2 = \epsilon c_2 h^{-1}$. Hence

$$(h, d'_{j-1}) = h c_2 v c_2^{-1} \varepsilon c_2 h^{-1} v^{-1} c_2^{-1} = \varepsilon h c_2 h^{-1} c_2^{-1} = \varepsilon \varepsilon^{-1} E_n = E_n.$$

Thus $c_2 v c_2^{-1} v^{-1}$ is a diagonal matrix commuting with h. Since $\det(c_2, v) = 1$, we have $(c_2, v) = \epsilon^\alpha E_n$, $v = \lambda h^\alpha$, $\lambda \in \Omega^*$, and $g_j = \lambda d_j h^\alpha$. We consider

$$(c_{j+1}, g_j) = (c_{j+1}, \lambda d_j h^\alpha) = c_{j+1} d_j h^\alpha c_{j+1}^{-1} h^{-\alpha} d_j^{-1}$$
$$= c_{j+1} d_j d_j^{-1} h^\alpha c_{j+1}^{-1} h^{-\alpha} = (c_{j+1}, h^\alpha) = \mu c_j^\alpha \in A_{j-1}, \quad \mu \in \Omega^*.$$

Hence we obtain $\alpha = 0$ and $g_j = \lambda d_j$; that is, (9) holds for $j < l$. In particular, $Z_{l-1}(\Gamma) \subset A_{l-1}$. Now let g be an arbitrary element of Γ. Then

$$(g, h) = d_{l-1} \in Z_{l-1}(\Gamma) \subset A_{l-1}, \qquad \det d_{l-1} = 1.$$

Thus there is a matrix d_l in A_l such that $(d_l, h) = d_{l-1}$. Hence $g = d_l v$, $(v, h) = E_n$, $v = \lambda h^\alpha$, $g = \lambda d_l h^\alpha \in \Gamma_n^l$ and $\Gamma = \Gamma_n^l$. ∎

COROLLARY 1. *For any $l > 1$ the group $GL(n, \Omega)$ contains a finite irreducible nilpotent subgroup $\mathfrak{G}$ of order n^{l+1} and of class l.*

PROOF. We set $\mathfrak{G} = \Gamma_n^l \cap SL(n, \Omega)$. Evidently $\Gamma_n^l = \mathfrak{G}\Omega^* E_n$. Hence

$\mathfrak{G}$ is an irreducible nilpotent subgroup of $GL(n, \Omega)$ of class l. Now,

$$\Gamma_n^l/\Omega^* E_n \cong \mathfrak{G}/\Omega^* E_n \cap \mathfrak{G}.$$

By Theorem 8, $\Gamma_n^l : \Omega^* E_n = n^l$. Obviously $\text{card}(\Omega^* E_n \cap \mathfrak{G}) = n$. Consequently $\text{card}\, \mathfrak{G} = n^{l+1}$. ∎

LEMMA 9. *Let Γ be an irreducible nilpotent subgroup of $GL(n, \Omega)$ of class l such that $Z_2(\Gamma)/Z_1(\Gamma)$ contains an element of order n. Then Γ is conjugate in $GL(n, \Omega)$ to a subgroup of Γ_n^l.*

PROOF. It is obvious that $Z_1(\Gamma) \subset \Omega^* E_n$. We may therefore assume that $Z_1(\Gamma) = \Omega^* E_n$. Let $aZ_1(\Gamma)$ be an element of order n in $Z_2(\Gamma)/Z_1(\Gamma)$. Then, by Lemma 3, there is an element $b \in \Gamma$ such that

$$(a, b) = \varepsilon E_n, \tag{10}$$

where ϵ is an element of order n in Ω^*. Since $\Gamma \supset \Omega^* E_n$ and Ω is algebraically closed, we can choose a and b so that

$$a^n = b^n = E_n. \tag{11}$$

It follows from (10) and (11) that in a suitable basis of Ω^n

$$a = c_2 = \text{diag}\,[1, \varepsilon, \ldots, \varepsilon^{n-1}], \quad b = h,$$

where h is given by (6). Hence, replacing Γ by a conjugate group, we may assume that $c_2 \in Z_2(\Gamma)$ and $h \in \Gamma$. Now let g be an arbitrary element of Γ. Then

$$(c_2, g) = \varepsilon^\alpha E_n = (c_2, h^\alpha), \quad (c_2, h^{-\alpha} g) = E_n.$$

Therefore $h^{-\alpha}g$ is a diagonal matrix, and $\Gamma = (h)\Phi$, where Φ is the centralizer of c_2 in Γ; that is, the set of all diagonal matrices in Γ. Since $\Gamma_n^l = (h)A_l$, the lemma will be proved if we can show that $\Phi \subset A_l$. We set $\Phi_j = \Phi \cap Z_j(\Gamma)$, $j = 1, \ldots, l$. Then $\Phi_1 = \Phi \cap Z_1(\Gamma) = A_1$. Let $j > 1$ and $\Phi_{j-1} \subset A_{j-1}$. We show that $\Phi_j \subset A_j$. If $\varphi \in \Phi_j$, then $(\varphi, h) \in \Phi$, $(\varphi, h) \in A_{j-1}$ and $(\varphi, h) \in \Phi_{j-1}$. By construction φ is an element of A_j. Hence $\Phi_j \subset A_j$; so

$$\Phi = \Phi_l = \Phi \cap Z_l(\Gamma) = \Phi \cap \Gamma \subset A_l. \blacksquare$$

THEOREM 9. *Let n be square-free. Then every irreducible nilpotent subgroup of $GL(n, \Omega)$ of class l is conjugate in $GL(n, \Omega)$ to a subgroup of Γ_n^l. Any two irreducible subgroups of $GL(n, \Omega)$ that are maximal among the nilpotent subgroups of class l are conjugate.*

PROOF. Let Γ be an irreducible nilpotent subgroup of $GL(n, \Omega)$ of class l. Since n is square-free, by Corollary 2.1 the exponent of $Z_2(\Gamma)/Z_1(\Gamma)$ is n. Hence, by Lemma 9, Γ is conjugate in $GL(n, \Omega)$ to a subgroup of Γ_n^l. And if Γ is a maximal nilpotent subgroup of class l in $GL(n, \Omega)$, then it is conjugate to Γ_n^l. ∎

Irreducible nilpotent linear groups over a finite field have been studied by Suprunenko and Apatenok [1], [2], who give a complete classification of the maximal irreducible nilpotent subgroups of $GL(n, \Delta)$ for finite Δ. In particular, they show that when Δ is finite, two maximal irreducible nilpotent subgroups of $GL(n, \Delta)$ are conjugate if their centers are conjugate.

§28. Irreducible locally nilpotent matrix groups

1. *Reduction to absolutely irreducible groups. The factor group of the center.* A group is said to be *locally nilpotent* if every finitely generated subgroup of it is nilpotent.

Throughout this subsection, Δ is an arbitrary field.

THEOREM 1. *Let G be an absolutely irreducible locally nilpotent subgroup of $GL(n, \Delta)$. Then $G/Z_1(G)$ is periodic and $\Pi(G/Z_1(G)) = \Pi(n)$.*

PROOF. Since G is absolutely irreducible, $Z_1(G) \subset \Delta^* E_n$. Every element $g \in G$ is contained in some absolutely irreducible nilpotent subgroup $\mathfrak{H}$ of G. For if $g_1, \ldots, g_{n^2}$ are linearly independent matrices in G, then the group $\mathfrak{H}$ they generate is nilpotent, absolutely irreducible and contained in G. By Corollary 27.2.3,

$$\Pi(\mathfrak{H}/Z_1(\mathfrak{H})) = \Pi(n).$$

On the other hand, $Z_1(\mathfrak{H}) \subset Z_1(G) \subset \Delta^* E_n$. Hence for any $h \in \mathfrak{H}$ the order of $hZ_1(\mathfrak{H})$ in $\mathfrak{H}/Z_1(\mathfrak{H})$ is the same as that of $hZ_1(G)$ in $G/Z_1(G)$. This implies the theorem. ∎

THEOREM 2. *Let G be an irreducible locally nilpotent subgroup of $GL(n, \Delta)$. Then $G/Z_1(G)$ is periodic, and $\Pi(G/Z_1(G)) \subset \Pi(n)$.*

PROOF. As in the proof of Lemma 27.1, it can be proved that G is contained in a locally nilpotent subgroup $\mathfrak{G}$ of $GL(n, \Delta)$ having the following properties:

(i) $\mathfrak{G}$ is an absolutely irreducible subgroup of $GL(r, \Sigma)$, with $\Sigma : \Delta = m$, $mr = n$.

(ii) $Z_1(\mathfrak{G}) = \Sigma^* E_r \supset Z_1(G)$.

By Theorem 1, $\Pi(\mathfrak{G}/Z_1(\mathfrak{G})) = \Pi(r)$. But

$$G/Z_1(G) = G/G \cap Z_1(\mathfrak{G}) \cong GZ_1(\mathfrak{G})/Z_1(\mathfrak{G}).$$

Therefore $\Pi(G/Z_1(G)) \subset \Pi(r) \subset \Pi(n)$. ∎

THEOREM 3. *Let G be an irreducible locally nilpotent subgroup of $GL(n, \Delta)$, $[G]$ the linear Δ-hull of G and $[Z_1(G)]$ the linear Δ-hull of $Z_1(G)$. Suppose that the center of $[G]$ is $[Z_1(G)]$. Then $\Pi(G/Z_1(G)) = \Pi(n/\mu)$, where $\mu = [Z_1(G)] : \Delta$.*

PROOF. We set $[Z_1(G)] = FE_{n/\mu}$, where F is the image of an irreducible component of $[Z_1(G)]$, and $n_1 = n/\mu$. Then $G \subset GL(n_1, F)$, and the center of

$[G]$ is FE_{n_1}. On the other hand, $[G] = T_l$, where T is a division ring whose center is $FE_{n_1/l}$, $n_1 = l(T:E)$. The centralizer of T_l in the full matrix algebra E_{n_1} is DE_l, where D is a division ring anti-isomorphic to T. Obviously, $D \cap T = FE_{n_1/l}$. Now let Σ be a maximal commutative subfield of D. Then

$$(\Sigma : F)^2 = T : F = D : F = n_1 l^{-1} = m^2.$$

The group $\mathfrak{G} = \Sigma^* G$ is locally nilpotent. We consider the linear F-hull $[\mathfrak{G}]$ of $\mathfrak{G}$. It is clear that $[\mathfrak{G}] = \Sigma T_l$. Furthermore, ΣT_l is an algebra over Σ, $\Sigma T_l : \Sigma = m^2 l^2$. On the other hand, the elements of ΣT_l are matrices of order ml over Σ. Hence $[\mathfrak{G}] = \Sigma_{ml}$, so that $\mathfrak{G}$ is an absolutely irreducible subgroup of $GL(ml, \Sigma)$. By Theorem 1,

$$\Pi(\mathfrak{G}/\Sigma^* E_{ml}) = \Pi(ml) = \Pi(m^2 l) = \Pi(n_1).$$

Next,

$$G/Z_1(G) \cong GZ_1(\mathfrak{G})/Z_1(\mathfrak{G}) = \mathfrak{G}/\Sigma^* E_{ml}.$$

Therefore $\Pi(G/Z_1(G)) = \Pi(n_1) = \Pi(n/\mu)$. ∎

THEOREM 4. *Let G be a maximal irreducible locally nilpotent subgroup of $GL(n, \Delta)$, and Σ the linear Δ-hull of $Z_1(G)$. Then G is a maximal absolutely irreducible locally nilpotent subgroup of $GL(r, \Sigma)$, where $r = n/(\Sigma : \Delta)$.*

This follows from Theorem 14.7.

THEOREM 5. *Let G be a maximal irreducible locally nilpotent subgroup of $GL(n, \Delta)$ and Σ the linear Δ-hull of $Z_1(G)$. Then $\Pi(G/Z_1(G)) = \Pi(r)$, where $r = n/(\Sigma : \Delta)$.*

PROOF. By Theorem 4, G is an absolutely irreducible subgroup of $GL(r, \Sigma)$. By Theorem 1 this implies that $\Pi(G/Z_1(G)) = \Pi(r)$. ∎

THEOREM 6. *$GL(n, \Delta)$ contains absolutely irreducible locally nilpotent subgroups if and only if for any $q \in \Pi(n)$ there is an element of order q in Δ^*.*

PROOF. Suppose first that $GL(n, \Delta)$ contains an absolutely irreducible locally nilpotent subgroup. Then it obviously contains an absolutely irreducible nilpotent subgroup G. By Corollary 27.2.3,

$$\Pi(G/Z_1(G)) = \Pi(n).$$

By Lemma 27.5, $\Pi(G/Z_1(G)) = \Pi(k)$, where k is the exponent of $Z_2(G)/Z_1(G)$. Consequently

$$\Pi(n) = \Pi(k). \tag{1}$$

Since G is absolutely irreducible, $Z_1(G) \subset \Delta^* E_n$. By Lemma 3, there is an element of order k in Δ^*. Hence and from (1) it follows that for any q in $\Pi(n)$ the group Δ^* contains an element of order q.

Conversely, suppose that Δ^* contains an element of order q for every prime divisor q of n. We show that $GL(n, \Delta)$ contains absolutely irreducible locally nilpotent subgroups. First suppose that $n = q^l$, where q is a prime. We consider the group $G = Q \sim N_{q^l}$, where Q is a subgroup of Δ^* of order q and N_{q^l} a Sylow subgroup of the symmetric group S_{q^l}. If Ω is any extension field of Δ, then by Lemma 15.4, G is an irreducible subgroup of $GL(q^l, \Omega)$. Hence G is a finite q-group and so is nilpotent. Thus for $n = q^l$ the assertion is proved.

Now let $n = q_1^{l_1} \cdots q_k^{l_k}$ be the canonical decomposition of n, and $G_j, j = 1, \ldots, k$, an absolutely irreducible locally nilpotent subgroup of $GL(q_j^{l_j}, \Delta)$. Then the Kronecker product $G_1 \dot{\times} \cdots \dot{\times} G_k$ is a locally nilpotent absolutely irreducible subgroup of $GL(n, \Delta)$. ∎

THEOREM 6a. *$GL(n, \Delta)$ contains irreducible locally nilpotent subgroups if and only if there is an extension Σ of Δ such that $\Sigma : \Delta$ divides n and for any $p \in \Pi(n/(\Sigma : \Delta))$ there is an element of order p in Σ^*.*

PROOF. Suppose that $GL(n, \Delta)$ contains irreducible locally nilpotent subgroups. Then it also contains irreducible nilpotent subgroups. Thus, by Lemma 27.1, if Σ is an extension of Δ such that $\Sigma : \Delta | n$ and $r = n/(\Sigma : \Delta)$ then $GL(r, \Sigma)$ contains an absolutely irreducible nilpotent subgroup $\mathfrak{G}$. By Theorem 6, for any $p \in \Pi(r)$ there is an element of order p in Σ^*. This proves the necessity.

Now for the sufficiency. Suppose that Δ has an extension Σ such that for any $p \in \Pi(n/(\Sigma : \Delta))$ there is an element of order p in Σ^*. Then, by Theorem 6, $GL(r, \Sigma)$, where $r = n/(\Sigma : \Delta)$, contains an absolutely irreducible locally nilpotent subgroup G such that $Z_1(G) = \Sigma^* E_r$. Representing the elements σ of Σ as matrices of degree $\Sigma : \Delta$ over Δ, we may assume that G is a subgroup of $GL(n, \Delta)$. By Lemma 14.4, G is an irreducible subgroup of $GL(n, \Delta)$. ∎

2. *Irreducible locally nilpotent subgroups of $GL(n, \Omega)$ over an algebraically closed field Ω.* Let Ω be *algebraically closed.* It follows from Theorem 6 that $GL(n, \Omega)$ contains irreducible locally nilpotent subgroups if and only if char $\Omega \notin \Pi(n)$.

LEMMA 1. *Let G be an irreducible locally nilpotent subgroup of $GL(n, \Omega)$ such that $Z_1(G) = \Omega^* E_n$. Then $G_0 = G \cap SL(n, \Omega)$ is locally finite, $\Pi(G_0) = \Pi(n)$, and*

$$G = G_0 \Omega^*. \tag{2}$$

PROOF. It follows from the obvious equality (2) that G_0 is an irreducible locally nilpotent subgroup of $GL(n, \Omega)$. Any finite set of elements of G_0 is contained in an irreducible nilpotent subgroup $H \subset SL(n, \Omega)$. By Corollary

27.3.1, H is finite and $\Pi(H) = \Pi(n)$. This proves the lemma. ∎

LEMMA 2. *An irreducible locally nilpotent subgroup G of $GL(n, \Omega)$ is monomial.*

PROOF. From the properties of nonrefinable decompositions into systems of imprimitivity it follows that we need only prove imprimitivity of G for $n > 1$. Suppose that G is primitive. We may assume that $Z_1(G) = \Omega^*E_n$. Since G is primitive, Ω^*E_n is its maximal abelian normal subgroup. Since G is solvable, we have $G : \Omega^*E_n < \infty$, i.e. $G_0 = G \cap SL(n, \Omega)$ is a finite group. Thus G_0 is a nilpotent primitive subgroup of $GL(n, \Omega)$. But then it follows from Lemma 27.6 that $n = 1$. ∎

We now describe the maximal irreducible locally nilpotent subgroups of $GL(n, \Omega)$. As we mentioned above, if char $\Omega \in \Pi(n)$, then $GL(n, \Omega)$ has no irreducible locally nilpotent subgroups. Let char $\Omega \notin \Pi(n)$. We first consider the case $n = p^\nu$, where p is a prime. We define an irreducible locally nilpotent subgroup Γ_{p^ν} of $GL(p^\nu, \Omega)$ by setting

$$\Gamma_{p^\nu} = (P \sim N_{p^\nu})\, \Omega^* E_n,$$

where $n = p^\nu$, P is a Sylow p-subgroup of Ω^*, and N_{p^ν} a Sylow p-subgroup of S_{p^ν}. We know that N_{p^ν} is transitive and $P \sim N_{p^\nu}$ is a Sylow p-subgroup P_{p^ν} of $GL(p^\nu, \Omega)$. By Lemma 15.4, P_{p^ν} is irreducible. Since P_{p^ν} is a locally finite p-group, $\Gamma_{p^\nu} = P_{p^\nu}\Omega^*E_n$ is an irreducible locally nilpotent subgroup of $GL(p^\nu, \Omega)$. ∎

LEMMA 3. *Every irreducible locally nilpotent subgroup of $GL(p^\nu, \Omega)$ is conjugate in $GL(p^\nu, \Omega)$ to a subgroup of Γ_{p^ν}.*

PROOF. Let G be an irreducible locally nilpotent subgroup of $GL(p^\nu, \Omega)$. We may assume that $G \supset \Omega^*E_{p^\nu}$. By Lemma 1, $G = G_0\Omega^*$, where $G_0 = G \cap SL(p^\nu, \Omega)$ and $\Pi(G_0) = \Pi(p^\nu) = \{p\}$. Thus G_0 is a p-group and is contained in a Sylow p-subgroup Q of $GL(p^\nu, \Omega)$. By Lemma 25.6, $tQt^{-1} = P_{p^\nu}$, where $t \in GL(p^\nu, \Omega)$. Hence $tGt^{-1} = tG_0t^{-1}\Omega^*E_{p^\nu} \subset P_{p^\nu}\Omega^*E_{p^\nu} = \Gamma_{p^\nu}$. ∎

COROLLARY 1. *Γ_{p^ν} is a maximal irreducible locally nilpotent subgroup of $GL(p^\nu, \Omega)$. Any two maximal irreducible locally nilpotent subgroups of $GL(p^\nu, \Omega)$ are conjugate.*

We now construct an irreducible locally nilpotent subgroup Γ_n of $GL(n, \Omega)$. Let $n = p_1^{\nu_1} \cdots p_k^{\nu_k}$ be the canonical decomposition of n. We set

$$\mathfrak{H}_n = P_{p_1^{\nu_1}} \dot{\times} \cdots \dot{\times} P_{p_k^{\nu_k}}, \qquad \Gamma_n = \mathfrak{H}_n \Omega^* E_n.$$

Since the $P_{p_i^{\nu_i}}$ are irreducible and locally nilpotent, Γ_n is an irreducible locally nilpotent subgroup of $GL(n, \Omega)$.

THEOREM 7. *Every irreducible locally nilpotent subgroup of $GL(n, \Omega)$ is conjugate in $GL(n, \Omega)$ to a subgroup of Γ_n. Every maximal irreducible locally nilpotent subgroup of $GL(n, \Omega)$ is conjugate in $GL(n, \Omega)$ to Γ_n.*

PROOF. Let G be an irreducible locally nilpotent subgroup of $GL(n, \Omega)$. We may assume that $\Omega^*E_n \subset G$. Then $G = G_0\Omega^*E_n$, where $G_0 = G \cap SL(n, \Omega)$. By Lemma 1, G_0 is periodic and $\Pi(G_0) = \Pi(n)$. Consequently G_0 is a direct product $\mathfrak{P}_1 \times \cdots \times \mathfrak{P}_k$, where $\mathfrak{P}_j$ is a Sylow p_j-subgroup of G_0, $j = 1, \ldots, k$. By Lemma 16.4, the irreducible components of $\mathfrak{P}_j$ are pairwise equivalent. By Theorem 2, the degree of an irreducible component of $\mathfrak{P}_j$ divides $p_j^{\nu_j}$, $j = 1, \ldots, k$. Hence by applying Theorem 16.3 we obtain $G_0 = \bar{\mathfrak{P}}_1 \dot{\times} \cdots \dot{\times} \bar{\mathfrak{P}}_k$, where $\bar{\mathfrak{P}}_j$ is an irreducible subgroup of $GL(p_j^{\nu_j}, \Omega)$ isomorphic to $\mathfrak{P}_j$, $j = 1, \ldots, k$, and $\dot{\times}$ denotes the Kronecker product. Now $\bar{\mathfrak{P}}_j$ is conjugate in $GL(p_j^{\nu_j}, \Omega)$ to a subgroup of $P_{p_j}^{\nu_j}$. Hence G_0 is conjugate in $GL(n, \Omega)$ to a subgroup of $\mathfrak{H}_n$ and G to a subgroup of Γ_n. ■

3. *Maximal irreducible locally nilpotent matrix groups over an arbitrary field* Δ. By Theorem 4, *in the study of maximal irreducible locally nilpotent matrix groups we may confine ourselves to the case of absolutely irreducible groups.*

THEOREM 8. *Let $n = p_1^{\alpha_1} \cdots p_k^{\alpha_k}$ be the canonical decomposition of n, and G a maximal absolutely irreducible locally nilpotent subgroup of $GL(n, \Delta)$. Then $G = G_1 \dot{\times} \cdots \dot{\times} G_k$, where G_j is a maximal absolutely irreducible locally nilpotent subgroup of $GL(p_j^{\alpha_j}, \Delta)$, $j = 1, \ldots, k$.*

PROOF. Since G is absolutely irreducible and maximal among the locally nilpotent subgroups of $GL(n, \Omega)$, we have $Z_1(G) = \Delta^*E_n$. By Theorem 1,

$$\Pi(G/\Delta^*E_n) = \Pi(n) = \{p_1, \ldots, p_k\}.$$

Therefore

$$G/\Delta^*E_n = H_1/\Delta^*E_n \times \cdots \times H_k/\Delta^*E_n, \tag{3}$$

where H_j/Δ^*E_n is a Sylow p_j-subgroup of G/Δ^*E_n. The groups H_i and H_j commute elementwise for $i \neq j$. For let $a \in H_i$ and $b \in H_j$, $i \neq j$. Then $ab = \lambda ba$, $\lambda \in \Delta^*$, $a^{p_j^\gamma} \in \Delta^*E_n$, $b^{p_j^\delta} \in \Delta^*E_n$. Hence $\lambda^{p_i^\gamma} = \lambda^{p_j^\delta} = 1$, and so $\lambda = 1$ and $ab = ba$. Now let $\mathfrak{B}_i$ be the linear Δ-hull of H_i. We claim that the center C of $\mathfrak{B}_i$ is ΔE_n. It is evident that $C \supset \Delta E_n$. On the other hand, the elements of C commute with every element of G. Consequently $C \subset \Delta E_n$, and so $C = \Delta E_n$. From (3) we find

$$G = H_1 U, \tag{4}$$

where $U = H_2 \cdots H_k$ and H_1 commutes elementwise with U. The irreducible components of H_1 are pairwise equivalent (Lemma 16.4). Hence in a suitable basis of Δ^n the matrices $h_1 \in H_1$ can be written in the form $h_1 = g_1 \dot{\times} E_{n/n_1}$, where g_1 ranges over an irreducible locally nilpotent subgroup G_1 of $GL(n_1, \Delta)$ as h_1 ranges over H_1. Since the center of $\mathfrak{B}_1$ is ΔE_n, it follows that the center of the linear Δ-hull $[G_1]$ of G_1 is ΔE_{n_1}, and the center of G_1 is $\Delta^* E_{n_1}$. Hence Theorem 3 is applicable to G_1; that is,

$$\Pi(G_1/Z_1(G_1)) = \Pi(n_1).$$

Obviously $G_1/Z_1(G_1) \cong H_1/\Delta^*E_n$. Since H_1/Δ^*E_n is a Sylow p_1-subgroup of G/Δ^*E_n, we have $n_1 = p_1^{\beta_1}, 0 < \beta_1 \leqslant \alpha_1$. We now consider U (see (4)). The center of its linear Δ-hull $[U]$ is ΔE_n, and its irreducible components are pairwise equivalent. Let n_2 be the degree of an irreducible component of $[U]$. Applying the preceding arguments, we find that

$$n_2 = p_2^{\beta_2} \dots p_k^{\beta_k}, \quad 0 < \beta_j \leqslant \alpha_j, \quad j = 2, \dots, k.$$

It is clear that $n_1 n_2 \leqslant n$. Theorem 16.3 is applicable to G_1, H_1 and U. Therefore

$$G = G_1 \times U_1, \tag{5}$$

where G_1 is an absolutely irreducible locally nilpotent subgroup of $GL(p_1^{\alpha_1}, \Delta)$, $U_1 \cong U$ an absolutely irreducible locally nilpotent subgroup of $GL(np_1^{-\alpha_1}, \Delta)$, and

$$Z_1(G_1) = \Delta^* E_{p_1^{\alpha_1}}, \quad Z_1(U_1) = \Delta^* E_{np_1^{-\alpha_1}}.$$

Since G is a maximal locally nilpotent subgroup of $GL(n, \Delta)$, it follows that G_1 is a maximal locally nilpotent subgroup of $GL(p_1^{\alpha_1}, \Delta)$ and U_1 a maximal locally nilpotent subgroup of $GL(np_1^{-\alpha_1}, \Delta)$. Applying the preceding arguments to U_1 and using (5), we obtain what is required. ■

This theorem reduces the classification of the maximal absolutely irreducible locally nilpotent matrix groups of degree n to the case when n is a prime power.

We now consider the maximal absolutely irreducible locally nilpotent subgroups of $GL(p^\alpha, \Delta)$, where p is a prime.

THEOREM 9. (i) *Let G be a maximal absolutely irreducible locally nilpotent subgroup of $GL(p^\alpha, \Delta)$. Then $G/\Delta^*E_{p^\alpha}$ is a Sylow p-subgroup of the projective group $PGL(p^\alpha, \Delta) = GL(p^\alpha, \Delta)/\Delta^*E_{p^\alpha}$.*

(ii) *Let H/Δ^*E_p be a Sylow p-subgroup of $PGL(p^\alpha, \Delta)$ and H an absolutely irreducible subgroup of $GL(p^\alpha, \Delta)$. Then H is a maximal locally nilpotent subgroup of $GL(p^\alpha, \Delta)$.*

PROOF. We first prove (ii). Let Ω be the algebraic closure of Δ. Obviously

$$GL(p^\alpha, \Delta)/\Delta^*E_{p^\alpha} \cong GL(p^\alpha, \Delta)\,\Omega^*E_{p^\alpha}/\Omega^*E_{p^\alpha} \subset GL(p^\alpha, \Omega)/\Omega^*E_{p^\alpha}.$$

Consequently $H/\Delta^*E_{p^\alpha}$ is isomorphic to a p-subgroup of $GL(p^\alpha, \Omega)/\Omega^*E_{p^\alpha}$. Now any p-subgroup of $GL(p^\alpha, \Omega)/\Omega^*E_{p^\alpha}$ is contained in a Sylow p-subgroup

$\mathcal{P}/\Omega^*E_{p^\alpha}$ of $GL(p^\alpha, \Omega)/\Omega^*E_{p^\alpha}$. We can write $\mathcal{P}$ in the form

$$\mathcal{P} = \mathcal{P}_0\Omega^*E_{p^\alpha}, \qquad \mathcal{P}_0 = SL(p^\alpha, \Omega) \cap \mathcal{P}.$$

Since $\mathcal{P}/\Omega^*E_{p^\alpha} \cong \mathcal{P}_0/(\eta E_{p^\alpha})$, where η is an element of order p^α in Ω^*, it follows that $\mathcal{P}_0$ is a p-subgroup of $GL(p^\alpha, \Omega)$. Hence $\mathcal{P}_0$ is locally finite, and $\mathcal{P}$ and H are locally nilpotent. Now let $H \subset H_1$, where H_1 is a locally nilpotent subgroup of $GL(p^\alpha, \Delta)$. Then

$$Z_1(H_1) = Z_1(H) = \Delta^*E_{p^\alpha}, \qquad \Pi(H_1/\Delta^*E_{p^\alpha}) = \{p\}.$$

Therefore $H_1/\Delta^*E_{p^\alpha}$ is a p-subgroup of $GL(p^\alpha, \Delta)/\Delta^*E_{p^\alpha}$, and

$$H_1/\Delta^*E_{p^\alpha} \supset H/\Delta^*E_{p^\alpha}.$$

Hence $H_1 = H$, and (ii) is proved.

We now prove (i). From the conditions of the theorem it follows that $\Pi(G/\Delta^*E_{p^\alpha}) = \{p\}$. Consequently $G/\Delta^*E_{p^\alpha}$ is contained in a Sylow p-subgroup $H/\Delta^*E_{p^\alpha}$ of $PGL(p^\alpha, \Delta)$. By (ii), H is locally nilpotent. Thus $G = H$. ∎

The maximal irreducible locally nilpotent subgroups of $GL(n, R)$ over the real field R were classified by Suprunenko in [9], where the following theorems are proved:

1. *If $n > 1$ is odd, $GL(n, R)$ has no irreducible locally nilpotent subgroups.*

2. *If n is even but not a power of 2, $GL(n, R)$ has up to conjugacy exactly one maximal irreducible locally nilpotent subgroup.*

3. *If $n = 2^\alpha$ $(\alpha \geq 1)$, the set of all maximal irreducible locally nilpotent subgroups of $GL(n, R)$ splits into two conjugacy classes.*

4. *Any irreducible locally nilpotent subgroup of $GL(n, R)$, $n > 2$, is imprimitive.*

§29. Reducible locally nilpotent groups

1. *The group $\Gamma_k(F)$.* Let Δ be an arbitrary field, F an irreducible subgroup of $GL(m, \Delta)$, D the centralizer of F in the algebra Δ_m, $k > 1$ an integer, and $n = km$. We define a subgroup $\Gamma_k(F)$ of $GL(n, \Delta)$ by setting

$$\Gamma_k(F) = (F \dot{\times} E_k)\, T_k(D), \tag{1}$$

where $F \dot{\times} E_k$ consists of all the matrices of the form

$$f \dot{\times} E_k = \operatorname{diag}[f, \ldots, f], \qquad f \in F,$$

and $T_k(D)$ of all the matrices of the form

$$\begin{bmatrix} E_m & d_{12} & \ldots & d_{1k} \\ 0 & E_m & \ldots & d_{2k} \\ \ldots & \ldots & \ldots & \ldots \\ 0 & 0 & \ldots & E_m \end{bmatrix}, \qquad d_{ij} \in D \subset \Delta_m. \tag{2}$$

It is obvious that $\Gamma_k(F)$ is the direct product of $F \dot{\times} E_k$ and $T_k(D)$.

THEOREM 1. *$\Gamma_k(F)$ is indecomposable.*(1)

PROOF. As is easy to see, a subset M of Δ_n is decomposable if and only if the centralizer of M in Δ_n contains an element whose minimal polynomial is $x(x-1)$. We show that every matrix c in the centralizer C of $\Gamma_k(F)$ in Δ_n is either nilpotent or invertible. This will prove the theorem. Let

$$c = \begin{bmatrix} c_{11} & \dots & c_{1k} \\ \cdot & \cdot & \cdot \\ c_{k1} & \dots & c_{kk} \end{bmatrix}, \qquad c_{ij} \in \Delta_m.$$

Since c commutes with every matrix in $F \dot{\times} E_k$, it follows that $c_{ij} \in D$. We first show that $c_{ij} = 0$ for $i > j$. It is clearly sufficient to prove that $c_{i1} = 0$ for $i = 2, \dots, k$. Let

$$u = \begin{bmatrix} E_m & 0 & \dots & d_{1\nu} & 0 \\ 0 & E_m & \dots & 0 & 0 \\ \cdot & \cdot & \cdot & \cdot & \cdot \\ 0 & 0 & \dots & 0 & E_m \end{bmatrix}, \qquad d_{1\nu} \in D^{\bullet}, \quad \nu = 2, \dots k.$$

Then from $cu = uc$ we obtain $c_{11} = c_{11} + d_{1\nu}c_{\nu 1}$, so that $c_{\nu 1} = 0$. Therefore

$$c = \begin{bmatrix} c_1 & c_{12} & \dots & c_{1k} \\ 0 & c_2 & \dots & c_{2k} \\ \cdot & \cdot & \cdot & \cdot \\ 0 & 0 & \dots & c_k \end{bmatrix}, \qquad c_i \in D.$$

Obviously

$$\begin{bmatrix} c_i & c_{i,i+1} \\ 0 & c_{i+1} \end{bmatrix} \begin{bmatrix} E_m & E_m \\ 0 & E_m \end{bmatrix} = \begin{bmatrix} E_m & E_m \\ 0 & E_m \end{bmatrix} \begin{bmatrix} c_i & c_{i,i+1} \\ 0 & c_{i+1} \end{bmatrix}.$$

Hence $c_i = c_{i+1}$, and so $c_1 = c_2 = \dots = c_k = d \in D$. Thus c is either invertible or nilpotent. ■

By Theorem 13.5, $T_k(D)$ is nilpotent of class $k-1$. Therefore $\Gamma_k(F)$ is nilpotent or locally nilpotent if and only if F is. Later we shall prove that any indecomposable maximal locally nilpotent subgroup of $GL(n, \Delta)$, where Δ is a perfect field, is conjugate in $GL(n, \Delta)$ to $\Gamma_k(F)$ for suitable F and k.

2. *Equivalence of the irreducible components of an indecomposable locally nilpotent group.* Here we prove

THEOREM 2. *The irreducible components of a locally nilpotent indecomposable subgroup of $GL(n, \Delta)$, where Δ is an arbitrary field, are pairwise equivalent.*

We precede the proof with a number of lemmas. Let Δ be an arbitrary

---

(1)That is, Δ^n cannot be expressed as a direct sum of two nontrivial $\Gamma_k(F)$-invariant subspaces.

field, and $G_{m,l}$ a nilpotent subgroup of $GL(m + l, \Delta)$ whose matrices have the form

$$g = \begin{bmatrix} a_g & b_g \\ 0 & c_g \end{bmatrix}, \tag{3}$$

where $\gamma_1 : g \longmapsto a_g$ and $\gamma_2 : g \longmapsto c_g$ are irreducible inequivalent components of $G_{m,l}$ of degrees m and l, respectively.

LEMMA 1. *Let $h \in G_{m,l}$, $\gamma_1(h) = E_m$ and $\gamma_2(h) = E_l$. Then $h = E_{m+l}$.*

PROOF. Let

$$h = \begin{bmatrix} E_m & b \\ 0 & E_l \end{bmatrix} \in G_{m,l}, \qquad b \neq 0. \tag{4}$$

If $g \in G_{m,l}$, the commutator (g, h) is either E_{m+l} or of the form (4). Hence, as $G_{m,l}$ is nilpotent, it follows that there is a matrix $h_1 \neq E_{m+1}$ of the form (4) in $G_{m,l}$ such that $(g, h_1) = E_{m+l}$ for all $g \in G_{m,l}$. Let

$$h_1 = \begin{bmatrix} E_m & b_1 \\ 0 & E_l \end{bmatrix}, \qquad b_1 \neq 0.$$

Then from (3) we find

$$(g, h_1) = \begin{bmatrix} E_m & a_g b_1 c_g^{-1} - b_1 \\ 0 & E_l \end{bmatrix} = E_{m+l}.$$

Hence $a_g b_1 c_g^{-1} - b_1 = 0$, i.e. $a_g b_1 = b_1 c_g$. By Schur's lemma (Lemma 13.1), $b_1 = 0$. This contradiction proves the lemma. ■

Lemma 1 implies

LEMMA 2. *If $h \in G_{m,l}$, $\gamma_1(h) \in Z_1(\operatorname{Im} \gamma_1)$ and $\gamma_2(h) \in Z_1(\operatorname{Im} \gamma_2)$, then $h \in Z_1(G_{m,l})$.*

LEMMA 3. *The index of $Z_1(G_{m,l})$ in $G_{m,l}$ is finite and not divisible by* char Δ.

PROOF. We set $\Gamma_1 = \operatorname{Im} \gamma_1$ and $\Gamma_2 = \operatorname{Im} \gamma_2$. By Theorem 27.1, the index of $Z_1(\Gamma_i)$ in Γ_i is finite and not divisible by char Δ ($i = 1, 2$). We define two subgroups of $G_{m,l}$:

$$H_1 = \gamma_1^{-1}(Z_1(\Gamma_1)), \qquad H_2 = \gamma_2^{-1}(Z_1(\Gamma_2)).$$

By Lemma 2, $Z_1(G_{m,l}) = H_1 \cap H_2$. Since H_1 and H_2 are subgroups of finite index, $Z_1(G_{m,l})$ is also of finite index in $G_{m,l}$. Further,

$$\operatorname{char} \Delta \notin \Pi(G_{m,l}/Z_1(G_{m,l})),$$

because char $\Delta \neq \Pi(G_{m,l}/H_i)$, $i = 1, 2$. ■

LEMMA 4. *$G_{m,l}$ is completely reducible.*

PROOF. We consider the linear Δ-hull C of $Z_1(G_{m,l})$. The elements of C have the form

$$c = \begin{bmatrix} c_1 & c_3 \\ 0 & c_2 \end{bmatrix},$$

where $c_1 \in \Sigma_1$, $c_2 \in \Sigma_2$, and Σ_1 and Σ_2 are finite extensions of Δ. Therefore, if c is a nilpotent element, then $c_1 = c_2 = 0$. But then $a_g c_3 = c_3 c_g$ for any $g \in G_{m,l}$. Hence and from Schur's lemma it follows that $c_3 = 0$, and so $c = 0$. Thus C is a completely reducible algebra and $Z_1(G)$ a completely reducible group. By Lemma 3 and Theorem 17.1, $G_{m,l}$ is completely reducible. ∎

LEMMA 5. *Lemma 4 remains valid if $G_{m,l}$ is assumed to be locally nilpotent instead of nilpotent.*

PROOF. A locally nilpotent subgroup of $GL(m + l, \Delta)$ is contained in the linear Δ-hull of a nilpotent subgroup. ∎

Observe that if a matrix t of $GL(m + l, \Delta)$ simultaneously transforms the matrices of $G_{m,l}$ to the form $\operatorname{diag}[a_g, c_g]$, it can be chosen thus:

$$t = \begin{bmatrix} E_m & b \\ 0 & E_l \end{bmatrix}. \tag{5}$$

For let

$$u_1, \ldots, u_{m+l} \tag{6}$$

be a basis of Δ^{m+l} in which the matrices of $G_{m,l}$ take the form (3). Then the subspace Q_1 spanned by the vectors $u_1, \ldots, u_m$ is invariant under $G_{m,l}$. Since $G_{m,l}$ is completely reducible, we have $\Delta^n = Q_1 \dot{+} Q_2$, where Q_2 is also invariant under $G_{m,l}$. The matrix t_1 expressing the basis $u_1, \ldots, u_m, g_1, \ldots, g_l$, where $g_1, \ldots, g_l$ is a basis of Q_2, in terms of (6), has the form

$$t_1 = \begin{bmatrix} E_m & b_1 \\ 0 & d \end{bmatrix}.$$

Hence

$$t_1^{-1} g t_1 = \operatorname{diag}[a_g, d^{-1} c_g d].$$

Now if $t_2 = \operatorname{diag}[E_m, d^{-1}]$, $t = t_1 t_2$, then t has the form (5) and

$$t^{-1} g t = \operatorname{diag}[a_g, c_g]$$

for any $g \in G_{m,l}$.

PROOF OF THEOREM 2. Let G be a locally nilpotent subgroup of $GL(n, \Delta)$. By a suitable choice of basis of Δ^n the matrices g of G can be transformed simultaneously to the form

$$g = \begin{bmatrix} a_1(g) & a_{12}(g) & \dots & a_{1k}(g) \\ 0 & a_2(g) & \dots & a_{2k}(g) \\ \cdot & \cdot & \cdot & \cdot \\ 0 & 0 & \dots & a_k(g) \end{bmatrix}, \tag{7}$$

where $\gamma_j : g \mapsto a_j(g)$ is an irreducible component of G of degree n_j, $j = 1, \dots, k$; $n_1 + \cdots + n_k = n$. We shall prove the following assertion, which is equivalent to Theorem 2:

($*$) *Suppose that among the irreducible components γ_j there are two inequivalent ones. Then G is decomposable.*

For some i, $1 \leqslant i < k$, let γ_i and γ_{i+1} be inequivalent. When g ranges over G, the matrix

$$\begin{bmatrix} a_i(g) & a_{i,i+1}(g) \\ 0 & a_{i+1}(g) \end{bmatrix}$$

ranges over a locally nilpotent subgroup of $GL(n_i + n_{i+1}, \Delta)$. By Lemma 5 we may assume that $a_{i,i+1}(g) = 0$ for any $g \in G$. Hence by going over to another basis of Δ^n we can interchange γ_i and γ_{i+1}. Thus we may assume that the equivalent irreducible components γ_j in (7) are arranged in blocks; that is, if γ_i and γ_j, $i < j$, are equivalent, then γ_ν and γ_i are equivalent for $i < \nu < j$. We now prove ($*$) by induction on k. For $k = 2$ it follows from Lemma 5. Let $k > 2$ and suppose that ($*$) is true for $k - 1$ irreducible components. Since some of the γ_j are inequivalent, by the inductive hypothesis either the group G_{k-1} of matrices

$$g_{k-1} = \begin{bmatrix} a_1(g) & \dots & a_{1,k-1}(g) \\ \cdot & \cdot & \cdot \\ 0 & \dots & a_{k-1}(g) \end{bmatrix},$$

or the group $\bar{G}_{k-1}$ of matrices

$$\bar{g}_{k-1} = \begin{bmatrix} a_2(g) & \dots & a_{2k}(g) \\ \cdot & \cdot & \cdot \\ 0 & \dots & a_k(g) \end{bmatrix}$$

is decomposable. Suppose that G_{k-1} is decomposable. Then the matrices $g \in G$ can be put in the form

$$g = \begin{bmatrix} g_{k-1}(g) & \begin{matrix} a_{1k}(g) \\ \vdots \\ a_{k-1,k}(g) \end{matrix} \\ 0 & a_k(g) \end{bmatrix} = \begin{bmatrix} A_1(g) & 0 & \begin{matrix} a_{1k}(g) \\ \vdots \\ a_{\tau k}(g) \end{matrix} \\ 0 & A_2(g) & B(g) \\ 0 & 0 & a_k(g) \end{bmatrix}, \tag{8}$$

where

$$A_1(g) = \begin{bmatrix} a_1(g) & \dots & a_{1\tau}(g) \\ 0 & \dots & a_{2\tau}(g) \\ \cdot & \cdot & \cdot \\ 0 & \dots & a_\tau(g) \end{bmatrix}, \tag{9}$$

and the irreducible components γ_i, $1 \leqslant i \leqslant \tau$, are not equivalent to γ_k. We define a mapping f_τ as follows (see (8) and (9)):

$$f_\tau\colon\ g \mapsto \begin{bmatrix} a_\tau(g) & a_{\tau k}(g) \\ 0 & a_k(g) \end{bmatrix}. \tag{10}$$

By (8) and (9) we have $f_\tau(xy) = f_\tau(x)f_\tau(y)$ for $x, y \in G$, so that f_τ is a homomorphism. Hence $\operatorname{Im} f_\tau$ is locally nilpotent. By Lemma 5 and (5), there is a matrix

$$t = \begin{bmatrix} E_{n_\tau} & b \\ 0 & E_{n_k} \end{bmatrix},$$

such that

$$t^{-1} \begin{bmatrix} a_\tau(g) & a_{\tau k}(g) \\ 0 & a_k(g) \end{bmatrix} t = \operatorname{diag}[a_\tau(g), a_k(g)] \tag{11}$$

for any $g \in G$. Now we set

$$t_\tau = \begin{bmatrix} E_{n_1 \dots n_\tau} & \begin{matrix} 0 \\ \cdot \\ \cdot \\ \cdot \\ b \end{matrix} \\ 0 & E_{n_{\tau+1} \dots n_k} \end{bmatrix}.$$

Then, by (8) and (11), the matrix $t_\tau^{-1} g t_\tau$ has the form (8), but is such that $a_{\tau k}(g) = 0$ for any $g \in G$. Next we define a mapping

$$f_{\tau-1}\colon\ g \mapsto \begin{bmatrix} a_{\tau-1}(g) & a_{\tau-1,k}(g) \\ 0 & a_k(g) \end{bmatrix}$$

similar to (10), and we eliminate $a_{\tau-1,k}(g)$ from (8). Repeating this procedure τ times, we prove that G is decomposable. If $\overline{G}_{k-1}$ is decomposable, the reasoning is analogous. ∎

3. *Indecomposable locally nilpotent groups.* Let G be an indecomposable locally nilpotent subgroup of $GL(n, \Delta)$. Then, by Theorem 2, in some basis of Δ^n the elements of G have the following form:

$$g = \begin{bmatrix} a(g) & a_{12}(g) & \dots & a_{1k}(g) \\ 0 & a(g) & \dots & a_{2k}(g) \\ \cdot & \cdot & \cdot & \cdot \\ 0 & 0 & \dots & a(g) \end{bmatrix}, \tag{12}$$

where $\gamma\colon g \longmapsto a(g)$ is an irreducible representation of G of degree m, $km = n$. We set $\operatorname{Im}\gamma = A$ and denote by D the centralizer of A in Δ_m.

THEOREM 3. *Let A be a d-group. Then G is conjugate in $GL(n, \Delta)$ to a subgroup of*

$$\Gamma_k(A) = (A \dot{\times} E_k)\, T_k(D). \tag{13}$$

In particular, if G is a maximal locally nilpotent subgroup of $GL(n, \Delta)$, then G is conjugate in $GL(n, \Delta)$ to $\Gamma_k(A)$.

In other words, *if A is a d-group, then in a suitable basis of Δ^n the matrices g of G have the form*

$$g = \begin{bmatrix} a(g) & c_{12}(g)\,a(g) & \dots & c_{1k}(g)\,a(g) \\ 0 & a(g) & \dots & c_{2k}(g)\,a(g) \\ \cdot & \cdot & \cdot & \cdot \\ 0 & 0 & \dots & a(g) \end{bmatrix}, \tag{14}$$

where $c_{ij}(g) \in D$.

By Corollary 27.1.1, an irreducible locally nilpotent linear group over a perfect field is a d-group. Therefore Theorem 3 implies

THEOREM 3a. *Let Δ be a perfect field. Then an indecomposable locally nilpotent subgroup G of $GL(n, \Delta)$ is conjugate in $GL(n, \Delta)$ to a subgroup of $\Gamma_k(A)$.*

We precede the proof of Theorem 3 with a sequence of lemmas.

LEMMA 6. *Let c be a d-matrix in $GL(m, \Delta)$, and let*

$$u = \begin{bmatrix} c & b \\ 0 & c \end{bmatrix}.$$

Then there is a polynomial $f(x)$ in $\Delta[x]$ such that

$$f(u) = \begin{bmatrix} c & b_1 \\ 0 & c \end{bmatrix} = u_1,$$

and the minimal polynomial of u_1 is the same as the minimal polynomial $\varphi(x)$ of c.

PROOF. Since c is a d-matrix, the polynomials $\varphi(x)$ and $\varphi'(x)$ are relatively prime. Hence there exist polynomials $f_1(x)$ and $f_2(x)$ in $\Delta[x]$ such that

$$\varphi(x) f_1(x) - \varphi'(x) f_2(x) = 1. \tag{15}$$

We set $f(x) = x + \varphi(x) f_2(x)$. Then

$$\varphi(x + \varphi(x) f_2(x)) = \varphi(x) + \varphi'(x)\varphi(x) f_2(x) + (\varphi(x))^2 q(x).$$

By (15), $\varphi'(x) f_2(x) = \varphi(x) f_1(x) - 1$. Therefore

$$\varphi(f(x)) = \varphi(x)^2 f_2(x) + (\varphi(x))^2 q(x) \equiv 0 \pmod{(\varphi(x))^2}.$$

Hence $\varphi(f(u)) = \varphi(u_1) = 0$. ∎

LEMMA 7. *Let $a, b \in \Delta_n$, $c = ab - ba$, $ca = ac$, and suppose that a is a d-matrix. Then c is nilpotent.*

PROOF. Since $ca = ac$, for some extension Σ of Δ there is a matrix q in $GL(n, \Sigma)$ such that

$$a_1 = q^{-1}aq = \operatorname{diag}[\alpha_1, \ldots, \alpha_n],$$

$$c_1 = q^{-1}cq = \begin{bmatrix} \gamma_1 & * & \ldots & * \\ 0 & \gamma_2 & \ldots & * \\ \cdot & \cdot & \cdot & \cdot \\ 0 & 0 & \ldots & \gamma_n \end{bmatrix}.$$

We set $b_1 = \|\beta_{ij}\| = q^{-1}bq$. Then $c_1 = a_1b_1 - b_1a_1$. Hence

$$\gamma_i = \alpha_i\beta_{ii} - \beta_{ii}\alpha_i = 0, \qquad i = 1, \ldots, n,$$

so that c is nilpotent. ∎

We first consider a nilpotent group $\Gamma \subset GL(2m, \Delta)$ whose matrices g have the form

$$g = \begin{bmatrix} a(g) & b(g) \\ 0 & a(g) \end{bmatrix}, \tag{16}$$

where $\gamma: g \longmapsto a(g)$ is an irreducible representation of Γ of degree m. Let $A = \operatorname{Im}\gamma$ and let D be the centralizer of A in Δ_m. We assume that A is a d-group. The next result is easy to check.

LEMMA 8. *The matrix*

$$h = \begin{bmatrix} E_m & b \\ 0 & E_m \end{bmatrix} \tag{17}$$

is in the centralizer of Γ in $GL(2m, \Delta)$ if and only if $b \in D$.

Henceforth we assume that a matrix h of the form (17), with $b \in D$, is in $Z_1(\Gamma)$.

LEMMA 9. *Let h be a matrix of the form* (17), *and suppose that $(g, h) \in Z_1(\Gamma)$ for any g in Γ. Then $(g, h) \in E_{2m}$ for every g in Γ.*

PROOF. By (16) and (17) we have

$$(g, h) = \begin{bmatrix} E_m & a(g)\,b\,(a(g))^{-1} - b \\ 0 & E_m \end{bmatrix}.$$

By Lemma 8, for any $g \in \Gamma$

$$a(g)\,b\,(a(g))^{-1} - b = b_1 \in D.$$

Hence

$$a(g)b - ba(g) = b_1 a(g) = c. \tag{18}$$

Obviously c commutes with $a(g)$. Since A is a d-group, $a(g)$ is a d-matrix. By Lemma 7, c is nilpotent. Since $a(g)$ is nonsingular, $b_1 \in D$ is singular. But D is a division ring, and so $b_1 = 0$, and $(g, h) = E_{2m}$. ∎

LEMMA 10. *Suppose that a matrix h of the form* (16) *belongs to* Γ. *Then*

$$h \in Z_1(\Gamma). \tag{19}$$

PROOF. Let

$$h \in \Gamma \setminus Z_1(\Gamma). \tag{20}$$

It is clear that (g, h), $g \in \Gamma$, is also of the form (16). Since Γ is nilpotent, there is an integer $k > 1$ such that

$$(x_k, (x_{k-1}, \ldots, (x_1, h) \ldots)) = E_{2m} \tag{21}$$

for any $x_1, \ldots, x_k \in \Gamma$, and also there exist $g_1, \ldots, g_{k-1} \in \Gamma$ such that

$$(g_{k-1}, (g_{k-2}, \ldots, (g_1, h) \ldots)) \neq E_{2m}. \tag{22}$$

Let $h_1 = (g_{k-2}, \ldots, (g_1, h) \cdots)$ if $k > 2$, and $h_1 = h$ if $k = 2$. Then, by (21) and Lemma 9, $h_1 \in Z_1(\Gamma)$, which contradicts (22). ∎

LEMMA 11. *Let*

$$v = \begin{bmatrix} c & b \\ 0 & c \end{bmatrix}, \quad v_1 = \begin{bmatrix} c_1 & b_1 \\ 0 & c_1 \end{bmatrix} \tag{23}$$

be matrices in $GL(2m, \Delta)$ *such that*

$$c, c_1 \in Z_1(A), \tag{24}$$

and

$$(g, v), \quad (g, v_1) \in Z_1(\Gamma) \tag{25}$$

for every $g \in \Gamma$. *Suppose that* $\alpha, \beta \in \Delta$, *and that* $u = \alpha v + \beta v_1$ *is nonsingular. Then* $(g, u) \in Z_1(\Gamma)$ *for any* $g \in \Gamma$.

PROOF. From (23) and (25) we obtain

$$(g, vv_1) = (g, v)(g, v_1) \in Z_1(\Gamma), \quad (g, vv_1^{-1}) \in Z_1(\Gamma).$$

Consequently it is sufficient to consider the case $v_1 = E_{2m}$. We set

$$(g, u) = \begin{bmatrix} E_m & x_{\alpha\beta} \\ 0 & E_m \end{bmatrix}.$$

From

$$\begin{bmatrix} a(g) & b(g) \\ 0 & a(g) \end{bmatrix} \begin{bmatrix} \alpha c + \beta E_m & \alpha b \\ 0 & \alpha c + \beta E_m \end{bmatrix}$$
$$= \begin{bmatrix} E_m & x_{\alpha\beta} \\ 0 & E_m \end{bmatrix} \begin{bmatrix} \alpha c + \beta E_m & \alpha b \\ 0 & \alpha c + \beta E_m \end{bmatrix} \begin{bmatrix} a(g) & b(g) \\ 0 & a(g) \end{bmatrix}$$

we find that

$$\alpha a(g)b + \alpha b(g)c + \beta b(g)$$
$$= (\alpha c + \beta E_m)b(g) + [\alpha b + x_{\alpha\beta}(\alpha c + \beta E_m)]a(g).$$

Hence

$$x_{\alpha\beta}(\alpha c + \beta E_m) = \alpha[a(g)b - ba(g) + b(g)c - cb(g)]a(g)^{-1}. \quad (26)$$

In particular, for $\alpha = 1$ and $\beta = 0$ we obtain

$$x_{10}c = [a(g)b - ba(g) + b(g)c - cb(g)]a(g^{-1}). \quad (27)$$

By assumption, $x_{10} \in D$. By (26) and (27),

$$x_{\alpha\beta}(\alpha c + \beta E_m) = \alpha x_{10}c.$$

Therefore $x_{\alpha\beta} \in D$ and $(g, u) \in Z_1(\Gamma)$. ∎

LEMMA 12. *Let v and v_1 be matrices of the form* (23), *and suppose that $u = \alpha v + \beta v_1$ is singular. Then u commutes with every matrix $g \in \Gamma$.*

PROOF. Since u is singular,

$$\alpha v v_1^{-1} + \beta E_n = \begin{bmatrix} 0 & b \\ 0 & 0 \end{bmatrix}.$$

For any $\gamma \neq \beta$ ($\gamma \in \Delta$), by Lemmas 11 and 9 we see that $\alpha v v_1^{-1} + \gamma E_{2m}$ commutes with every g in Γ. Clearly

$$\alpha v v_1^{-1} + \gamma E_{2m} = \begin{bmatrix} (\gamma - \beta)E_m & b \\ 0 & (\gamma - \beta)E_m \end{bmatrix}.$$

Therefore $b \in D$. Since

$$u = \begin{bmatrix} 0 & bc_1 \\ 0 & 0 \end{bmatrix}, \qquad bc_1 \in D,$$

this proves the lemma. ∎

We now set $V = \gamma^{-1}(Z_1(A))$.

LEMMA 13. *Let u be a nonsingular matrix in the linear Δ-hull $[V]$ of V. Then $(g, u) \in Z_1(\Gamma)$ for any g in Γ.*

PROOF. For any $v \in V$, by Lemma 10 we have $(g, v) \in Z_1(\Gamma)$ for every $g \in \Gamma$. For u we can write

$$u = \alpha_1 v_1 + \dots + \alpha_r v_r,$$

where $\alpha_j \in \Delta$, $\alpha_j \neq 0$, and $v_j \in V$, $j = 1, \dots, r$. We define r matrices

$$u_1 = \alpha_1 v_1, \quad u_2 = u_1 + \alpha_2 v_2, \quad \dots, \quad u = u_r = u_{r-1} + \alpha_r v_r.$$

If $u_1, \dots, u_{r-1}$ are nonsingular, Lemma 13 follows from Lemma 11.

Now suppose that $u_1, \dots, u_{k-1}$ are nonsingular but u_k is singular. Since u_{k-1} is nonsingular, by Lemma 12

$$u_k = \begin{bmatrix} 0 & t \\ 0 & 0 \end{bmatrix}, \qquad t \in D.$$

For u_{k+1} we have

$$u_{k+1} = u_k + \alpha_{k+1} v_{k+1}.$$

If

$$v_{k+1} = \begin{bmatrix} \lambda E_m & b \\ 0 & \lambda E_m \end{bmatrix}, \qquad \lambda \in \Delta^*,$$

then, by Lemma 10, u_{k+1} commutes with every $g \in \Gamma$. But if

$$v_{k+1} = \begin{bmatrix} c & b \\ 0 & c \end{bmatrix}, \qquad c \in Z_1(A), \qquad c \notin \Delta^* E_m,$$

then

$$u_{k+1} = u_k + E_{2m} + \alpha_{k+1}(v_{k+1} - \alpha_{k+1}^{-1} E_{2m}).$$

It is obvious that for $g \in \Gamma$

$$(g,\ u_k + E_{2m}) = E_{2m}, \qquad (g,\ v_{k+1} - \alpha_{k+1}^{-1} E_{2m}) \in Z_1(\Gamma).$$

Therefore $v_{k+1} - \alpha_{k+1}^{-1} E_{2m}$ and $u_k + E_{2m}$ are matrices of the form (23). By Lemma 11,

$$(g,\ u_{k+1}) \in Z_1(\Gamma). \ \blacksquare$$

We continue our study of the algebra $[V]$. Suppose that the linear Δ-hull of $Z_1(A)$ is a field P. Since A is a d-group, $Z_1(A)$ consists of d-matrices. Hence the elements of $Z_1(A) \subset P$ are separable over Δ_m, and P is a separable extension of Δ. We define an epimorphism

$$\psi\colon [V] \longrightarrow P, \qquad \begin{bmatrix} d & b \\ 0 & d \end{bmatrix} \mapsto d \in P.$$

Since P is separable over Δ, it follows that $P = \Delta[c]$, where c is a primitive element of P. Let $u \in \psi^{-1}(c)$, and let U be the multiplicative group of the commutative ring $\Delta[u] \subset [V]$. By Lemma 13, every element of U commutes with every matrix of Γ, up to a factor in $Z_1(\Gamma)$. It is clear that $Z_1(\Gamma)$ and U commute elementwise. Hence $\Gamma U = U\Gamma$ is also a nilpotent subgroup of $GL(2m, \Delta)$. We may thus assume that $U \subset \Gamma$ and $U \subset Z_2(\Gamma)$. By Lemma 6, there is a matrix u_1 in U whose minimal polynomial is the same as that of c. We may assume that $u_1 = \operatorname{diag}[c, c]$, since u_1 can be brought to this form by means of the matrix

$$\begin{bmatrix} E_m & S \\ 0 & E_m \end{bmatrix}.$$

LEMMA 14. $u_1 \in Z_1(\Gamma)$.

PROOF. Since $u \in U \subset Z_2(\Gamma)$, we have

$$(g, u_1) = \begin{bmatrix} E_m & t \\ 0 & E_m \end{bmatrix}, \quad t \in D.$$

On the other hand, $tca(g) = b(g)c - cb(g)$, and the d-matrix c commutes with $b(g)c - cb(g)$. By Lemma 7, $tca(g)$ is nilpotent. Hence $t = 0$ and $u_1 \in Z_1(\Gamma)$. ∎

LEMMA 15. *Let*

$$h = \begin{bmatrix} z & b \\ 0 & z \end{bmatrix} \in \gamma^{-1}(Z_1(A)).$$

Then $h \in Z_1(\Gamma)$.

PROOF. Obviously, if $g \in \Gamma$, then

$$(g, h) = \begin{bmatrix} E_m & x \\ 0 & E_m \end{bmatrix}.$$

By Lemma 10, $(g, h) \in Z_1(\Gamma)$, $x \in D$. Since $gh = (g, h)hg$, we have

$$xza(g) = a(g)b - ba(g) + b(g)z - zb(g).$$

By Lemma 14, $b(g)z = zb(g)$. Hence

$$xza(g) = a(g)b - ba(g).$$

By Lemma 7, $xza(g)$ is nilpotent. Therefore $x = 0$ and $h \in Z_1(\Gamma)$. ∎

We may thus assume that $Z_1(\Gamma)$ consists of all the matrices of the form

$$\begin{bmatrix} z & t \\ 0 & z \end{bmatrix}, \quad t \in D, \quad z \in Z_1(A) = P^*, \quad P = \Delta[c]. \tag{28}$$

Lemma 15 and Theorem 27.1 imply

LEMMA 16. $Z_1(\Gamma)$ *has finite index in* Γ:

$$\Gamma : Z_1(\Gamma) = A : Z_1(A). \tag{29}$$

Now let $M = P^*E_2$ and $\bar{\Gamma} = \Gamma/M$. We consider the cosets in $\bar{\Gamma}$ of the normal subgroup $\bar{Z}_1 = Z_1(\Gamma)/M$.

LEMMA 17. *Every coset of* $\bar{Z}_1$ *in* $\bar{\Gamma}$ *contains one and only one element* $\bar{g}_1 \in \bar{\Gamma}$ *whose order is finite and divisible by* char Δ.

PROOF. Let

$$g = \begin{bmatrix} a(g) & b(g) \\ 0 & a(g) \end{bmatrix} \in \Gamma,$$

and let the order of the coset $a(g)Z_1(A)$ in $A/Z_1(A)$ be l. By Theorem 27.1, char $\Delta \nmid l$. By Lemma 15,

$$g^l = \begin{bmatrix} z & b \\ 0 & z \end{bmatrix} \in Z_1(\Gamma), \quad z \in Z_1(A), \quad b \in D.$$

Next,

$$h = \begin{bmatrix} E_m & -l^{-1}z^{-1}b \\ 0 & E_m \end{bmatrix} \in Z_1(\Gamma)$$

and

$$(gh)^l = \begin{bmatrix} z & b \\ 0 & z \end{bmatrix} \begin{bmatrix} E_m & -z^{-1}b \\ 0 & E_m \end{bmatrix} = \operatorname{diag}[z, z] \in M.$$

We set $g_1 = gh$ and $\bar{g}_1 = g_1 M$. Then $\bar{g}_1 \in \bar{\Gamma}$ and $g_1 M \in \bar{g}_1 \bar{Z}_1$. The order of $\bar{g}_1$ is obviously l. Since $\bar{Z}_1$ is isomorphic to the group H of all matrices of the form

$$h = \begin{bmatrix} E_m & t \\ 0 & E_m \end{bmatrix}, \quad t \in D, \tag{30}$$

it follows that $\bar{Z}_1$ is torsion-free when char $\Delta = 0$ and a p-group when char $\Delta = p$. Consequently the order of any element of $\bar{g}_1 \bar{Z}_1$ other than g_1 is either infinite or divisible by char Δ. ■

LEMMA 18. *$GL(2m, \Delta)$ contains an element*

$$q = \begin{bmatrix} E_m & b \\ 0 & E_m \end{bmatrix},$$

such that the matrices g of $q^{-1}\Gamma q$ have the form

$$g = \begin{bmatrix} a(g) & t(g)\, a(g) \\ 0 & a(g) \end{bmatrix},$$

where $t(g) \in D$.

PROOF. The set $\bar{N}$ of all elements of $\bar{\Gamma}$ whose orders are finite and not divisible by char Δ is a normal subgroup of $\bar{\Gamma}$. By Lemma 17, $\bar{\Gamma} = \overline{NZ_1}$. Hence, if $\bar{N} = N/M$, then

$$\Gamma = NZ_1(\Gamma) = NH, \tag{31}$$

where H is the group of all matrices of the form (30). M is completely reducible, and $N:M$ is finite and not divisible by char Δ. Therefore, by Theorem 17.1, N is completely reducible. Thus the matrices g_0 of N can be transformed by means of a matrix q to the form

$$g_0 = \operatorname{diag}[a(g_0), a(g_0)].$$

Now q commutes with every matrix in H, Together with (31), this proves the lemma. ■

We recall that in Lemmas 8–18 Γ is a nilpotent group of matrices of the form (16).

MAIN LEMMA. *Lemma* 18 *remains valid if Γ is assumed to be locally nilpotent instead of nilpotent.*

PROOF. Let $a_1, \ldots, a_p$ be a maximal system of matrices in A that are linearly independent over Δ, and let

$$g_j = \begin{bmatrix} a_j & b_j \\ 0 & a_j \end{bmatrix}, \quad j = 1, \ldots, p, \tag{32}$$

be elements of Γ. The group B generated by the matrices (32) is nilpotent and has two inequivalent irreducible components. By Lemma 18 we may assume that the elements of B are all of the form

$$b = \begin{bmatrix} a(b) & t(b)a(b) \\ 0 & a(b) \end{bmatrix}, \tag{33}$$

where $t(b) \in D$; that is, $B \subset \Gamma_2(A)$. We claim that $\Gamma \subset \Gamma_2(A)$; and this will prove the lemma. Let g be an arbitrary element of Γ and U the subgroup of Γ generated by B and g. Then, by Lemma 18, $GL(2m, \Delta)$ contains a matrix

$$q = \begin{bmatrix} E_m & W \\ 0 & E_m \end{bmatrix},$$

such that for any $u \in U$

$$q^{-1}uq = \begin{bmatrix} a(u) & f(u)a(u) \\ 0 & a(u) \end{bmatrix}, \quad f(u) \in D.$$

In particular, this is true for $b \in B$:

$$\begin{bmatrix} a(b) & t(b)a(b) \\ 0 & a(b) \end{bmatrix}\begin{bmatrix} E_m & W \\ 0 & E_m \end{bmatrix} = \begin{bmatrix} E_m & W \\ 0 & E_m \end{bmatrix}\begin{bmatrix} a(b) & f(b)a(b) \\ 0 & a(b) \end{bmatrix}.$$

Hence $a(b)W - Wa(b) = (f(b) - t(b))a(b)$. By Lemma 7, $f(b) - t(b) = 0$, and $W \in D$. Consequently

$$qg = gq, \quad g \in \Gamma_2(A), \quad \Gamma \subset \Gamma_2(A). \blacksquare$$

LEMMA 19. *Suppose that a matrix g of a locally nilpotent subgroup G_1 of $GL(ms, \Delta)$ is of the form*

$$g = \begin{bmatrix} a(g) & c_{12}(g)a(g) & \ldots & c_{1,s-1}(g)a(g) & a_{1s}(g) \\ 0 & a(g) & \ldots & c_{2,s-1}(g)a(g) & c_{2s}(g)a(g) \\ \cdot & \cdot & \cdot & \cdot & \cdot \\ 0 & 0 & \ldots & a(g) & c_{s-1,s}(g)a(g) \\ 0 & 0 & \ldots & 0 & a(g) \end{bmatrix}, \tag{35}$$

where $a(g)$ ranges over an irreducible locally nilpotent subgroup A of $GL(m, \Delta)$, $c_{ij}(g) \in D$, and D is the centralizer of A in Δ_m. Then the group $G_1T_s(D)$ is locally nilpotent.

PROOF. It is sufficient to show that the group $\mathfrak{G} = T_s(D)\mathfrak{H}$, where $\mathfrak{H}$ is any nilpotent subgroup of G_1, is nilpotent. A direct computation shows that matrices in $T_s(D)$ of the form

$$\begin{bmatrix} E_m & 0 & \dots & 0 & d_{1s} \\ 0 & E_m & \dots & 0 & 0 \\ \cdot & \cdot & \cdot & \cdot & \cdot \\ 0 & 0 & \dots & 0 & E_m \end{bmatrix}, \quad d_{1s} \in D, \tag{36}$$

commute with every matrix of G_1 and are therefore elements of $Z_1(\mathfrak{G})$. Next, the commutator of a matrix

$$\begin{bmatrix} E_m & 0 & \dots & d_{1,s-1} & d_{1s} \\ 0 & E_m & \dots & 0 & d_{2s} \\ 0 & 0 & \dots & 0 & 0 \\ \cdot & \cdot & \cdot & \cdot & \cdot \\ 0 & 0 & \dots & 0 & E_m \end{bmatrix}, \quad d_{ij} \in D, \tag{37}$$

with any matrix in $\mathfrak{G}$ has the form (36), and so all matrices of the form (37) occur in $Z_2(\mathfrak{G})$. Similarly, $T_s(D) \subset Z_{s-1}(\mathfrak{G})$. This completes the proof. ∎

LEMMA 20. *Let G_1 be a locally nilpotent subgroup of $GL(ms, \Delta)$ consisting of matrices of the form* (35). *Then the matrices g in G_1 can be brought simultaneously to the form*

$$\begin{bmatrix} a(g) & c_{12}(g)a(g) & \dots & c_{1,s-1}(g)a(g) & c_{1s}(g)a(g) \\ 0 & a(g) & \dots & c_{2,s-1}(g)a(g) & c_{2s}(g)a(g) \\ \cdot & \cdot & \cdot & \cdot & \cdot \\ 0 & 0 & \dots & 0 & a(g) \end{bmatrix} \tag{38}$$

by means of a matrix

$$\begin{bmatrix} E_m & 0 & \dots & 0 & k \\ 0 & E_m & \dots & 0 & 0 \\ \cdot & \cdot & \cdot & \cdot & \cdot \\ 0 & 0 & \dots & 0 & E_m \end{bmatrix}, \quad k \in \Delta_m.$$

PROOF. By Lemma 19 we may assume that $T_s(D) \subset G_1$. Hence G_1 contains, together with a matrix g of the form (35), also the matrix

$$g_2 = g \begin{bmatrix} E_m & c_{12}(g) & \dots & c_{1s}(g) \\ 0 & E_m & \dots & c_{2s}(g) \\ \cdot & \cdot & \cdot & \cdot \\ 0 & 0 & \dots & E_m \end{bmatrix}^{-1} = \begin{bmatrix} a(g) & 0 & \dots & 0 & b(g) \\ 0 & a(g) & \dots & 0 & 0 \\ \cdot & \cdot & \cdot & \cdot & \cdot \\ 0 & 0 & \dots & 0 & a(g) \end{bmatrix}.$$

From (38) it follows that

$$G_1 = T_s(D)\,\mathfrak{H}_1, \tag{39}$$

where the matrices g_2 of $\mathfrak{H}_1$ have the form

$$g_2 = \begin{bmatrix} a(g) & 0 & \dots & 0 & b(g) \\ 0 & a(g) & \dots & 0 & 0 \\ \cdot & \cdot & \cdot & \cdot & \cdot \\ 0 & 0 & \dots & 0 & a(g) \end{bmatrix}.$$

Now $\mathfrak{H}_1 \subset G_1$ is locally nilpotent. Since the mapping

$$\gamma_2\colon g_2 \mapsto \begin{bmatrix} a(g) & b(g) \\ 0 & a(g) \end{bmatrix}$$

is an isomorphism, Im γ_2 is also locally nilpotent. Therefore, by the Main Lemma, the matrices of Im γ_2 are simultaneously brought to the form

$$\begin{bmatrix} a(g) & c(g)a(g) \\ 0 & a(g) \end{bmatrix}, \quad c(g) \in D,$$

by means of a matrix

$$\begin{bmatrix} E_m & W \\ 0 & E_m \end{bmatrix}.$$

We now set

$$t = \begin{bmatrix} E_m & 0 & \dots & 0 & W \\ 0 & E_m & \dots & 0 & 0 \\ \cdot & \cdot & \cdot & \cdot & \cdot \\ 0 & 0 & \dots & 0 & E_m \end{bmatrix} \in GL(ms, \Delta).$$

Then

$$t^{-1}g_2t = \begin{bmatrix} a(g) & 0 & \dots & 0 & c(g)a(g) \\ 0 & a(g) & \dots & 0 & 0 \\ \cdot & \cdot & \cdot & \cdot & \cdot \\ 0 & 0 & \dots & 0 & a(g) \end{bmatrix},$$

Obviously t commutes with every matrix in $T_s(D)$. Hence, by (39), t transforms the matrices of G_1 to the form (38). ∎

PROOF OF THEOREM 3. By assumption, the matrices g of the locally nilpotent group G have the form (12). By the Main Lemma, $GL(2m, \Delta)$ contains matrices

$$t_j = \begin{bmatrix} E_m & W_{j,j+1} \\ 0 & E_m \end{bmatrix}, \quad j = 1, \dots, k-1,$$

such that

$$t_j^{-1}\begin{bmatrix} a(g) & a_{j,j+1}(g) \\ 0 & a(g) \end{bmatrix} t_j = \begin{bmatrix} a(g) & c_{j,j+1}(g)a(g) \\ 0 & a(g) \end{bmatrix},$$

where $c_{j,j+1}(g) \in D$ for any $g \in G$. We set

$$q_1=\begin{bmatrix} E_m & W_{12} & 0 & \dots & 0 \\ 0 & E_m & W_{23} & \dots & 0 \\ \cdot & \cdot & \cdot & \cdot & \cdot \\ 0 & 0 & 0 & \dots & W_{k-1,k} \\ 0 & 0 & 0 & \dots & E_m \end{bmatrix}.$$

Then the submatrix of $q_1^{-1}gq_1$ in the $(j, j+1)$ position is

$$c_{j,j+1}(g)\,a(g), \quad j=1, \dots, k-1,$$

where $c_{j,j+1}(g) \in D$ for any $g \in G$. By Lemma 20 there is a matrix q_2 of the form

$$q_2=\begin{bmatrix} E_m & 0 & W_{13} & 0 & \dots & 0 & 0 \\ 0 & E_m & 0 & W_{24} & \dots & 0 & 0 \\ \cdot & \cdot & \cdot & \cdot & \cdot & \cdot & \cdot \\ 0 & 0 & 0 & 0 & \dots & 0 & W_{k-2,k} \\ 0 & 0 & 0 & 0 & \dots & E_m & 0 \\ 0 & 0 & 0 & 0 & \dots & 0 & E_m \end{bmatrix}$$

such that the submatrix of $q_2^{-1}q_1^{-1}gq_1q_2$ in the (i, j) position, $0 < j - i \leqslant 2$, is $c_{ij}(g)a(g)$, $c_{ij}(g) \in D$. By repeated application of Lemma 20 we finally bring the matrices g of G to the form (14). ∎

It follows, in particular, from Theorem 3 that *up to conjugacy in* $GL(n, \Delta)$ *a maximal indecomposable locally nilpotent subgroup* G *of* $GL(n, \Delta)$ *is uniquely determined by the group* $\operatorname{Im} \gamma = A$, *where* γ *is an irreducible component of* G, *provided that* A *is a d-group.*

Neither Theorem 3 nor the corollary just stated can be extended to the case when $A = \operatorname{Im} \gamma$ is not a d-group. Here is an example of a commutative indecomposable group whose matrices cannot be transformed simultaneously to the form (14). Let $\Sigma : \Delta = m$, Σ a simple inseparable extension of Δ, $\Sigma = \Delta(\omega)$ and $\Sigma \subset \Delta_m$. Let $R \subset \Delta_{2m}$ be the set of matrices

$$r=\begin{bmatrix} 0 & \sigma \\ 0 & 0 \end{bmatrix},$$

where $\sigma \in \Sigma \subset \Delta_m$. We set

$$u=\begin{bmatrix} \sigma & d \\ 0 & \sigma \end{bmatrix},$$

where

$$d=\begin{bmatrix} 1 & 0 & \dots & 0 \\ 0 & 0 & \dots & 0 \\ \cdot & \cdot & \cdot & \cdot \\ 0 & 0 & \dots & 0 \end{bmatrix} \in \Delta_m, \qquad \mathfrak{A} = \Delta[u] + R, \ \mathfrak{A}^* = G.$$

Then G is a commutative indecomposable subgroup of $GL(2m, \Delta)$. By Theorem 1 in Suprunenko [21], the matrices of G cannot be brought to the form (14).

THEOREM 4 (ZASSENHAUS [1]). *A locally nilpotent subgroup G of $GL(n, \Delta)$ is nilpotent if and only if* Im γ *is nilpotent for every irreducible component γ of G.*

PROOF. The necessity is obvious, so we need prove only the sufficiency. Suppose that Im γ is nilpotent for every irreducible component γ of G. It is clear that this property of G is preserved if we replace Δ by its algebraic closure Ω. Hence we may assume that Im γ is a d-group. Thus Theorem 3 is applicable to the indecomposable components of G. Since $\Gamma_k(A)$ is nilpotent whenever A is, this implies that G is nilpotent. ∎

4. *Other results on locally nilpotent groups.* A locally nilpotent subgroup of $GL(n, \Delta)$, where Δ is an arbitrary field, is always solvable; therefore, by Theorem 22.5, any locally nilpotent d-group in $GL(n, \Delta)$ is completely reducible. By Corollary 27.1.1, if the ground field Δ is perfect, the converse is also true: a locally nilpotent completely reducible matrix group over Δ is a d-group. It is readily seen that if the ground field is not perfect, there are even abelian irreducible matrix groups that are not d-groups.

THEOREM 5. *Let Δ be an arbitrary field and G a completely reducible locally nilpotent subgroup of $GL(n, \Delta)$. Then any subgroup of G is completely reducible.*

PROOF. Let H be an arbitrary subgroup of G, $h_1, \ldots, h_r$ a maximal system of elements of H linearly independent over Δ, and H_1 the group generated by $h_1, \ldots, h_r$. The theorem will be proved if we can show that H_1 is completely reducible. We supplement $h_1, \ldots, h_r$ to a maximal linearly independent system of elements of G:

$$h_1, h_2, \ldots, h_r, g_{r+1}, \ldots, g_s. \tag{40}$$

Since G is completely reducible, so is the group G_1 generated by the system (40). G_1 is nilpotent and contains H_1; therefore (see Kuroš [1], Russian p. 389; English vol. II, p. 215) there exists a normal series

$$G_1 \supset H_t \supset H_{t-1} \supset \ldots \supset H_1 \supset (E_n). \tag{41}$$

Since G_1 is completely reducible, it clearly follows from Theorem 16.2 that every group in (41) is completely reducible. ∎

COROLLARY 1. *Let G be a completely reducible locally nilpotent matrix group over an arbitrary field. Then every element of G is a semisimple matrix.*

We now return to the groups $\Gamma_k(F)$ of §29.1.

LEMMA 21. *Let F be an irreducible d-subgroup of $GL(m, \Delta)$. Then $F \dot{\times} E_k$ is the set of all d-matrices in $\Gamma_k(F)$, and $T_k(D)$ is the set of all unipotent matrices in $\Gamma_k(F)$.*

PROOF. Let $a = (f \dot{\times} E_k)u \in \Gamma_k(F), f \in F, u \in T_k(D)$. Obviously

$$\det(xE_n - a) = (\det(xE_m - f))^k.$$

Hence, if a is unipotent, then $\det(xE_m - f) = (x-1)^m$. Since F is a d-group, for a unipotent matrix a we have $f = E_m$ and $a = u \in T_k(D)$. Thus $T_k(D)$ is the set of all unipotent matrices of $\Gamma_k(F)$. Now let $a = (f \dot{\times} E_k)u$ be a d-matrix. The d-matrix $b = f^{-1} \dot{\times} E_k$ commutes with a. Therefore, by Lemma 17.1, $u = ab$ is also a d-matrix. Hence $u = E_n$ and $a = b \dot{\times} E_k \in F \dot{\times} E_k$. ∎

THEOREM 6. *Let Δ be an arbitrary field, G a locally nilpotent subgroup of $GL(n, \Delta)$, $d(G)$ the set of all d-matrices in G, and $u(G)$ the set of all unipotent matrices in G. Then $d(G)$ and $u(G)$ are normal subgroups of G.*

PROOF. It is sufficient to prove that $d(G)$ and $u(G)$ are subgroups of G. Since the property of being a d-matrix or a unipotent matrix is preserved under extension of the ground field, we may regard G as a subgroup of $GL(n, P)$, where P is the algebraic closure of Δ.

We may clearly assume that G consists of matrices of the form

$$g = \operatorname{diag}[g_1, g_2, \ldots, g_t], \tag{42}$$

where g_j ranges over a locally nilpotent indecomposable subgroup G_j of $GL(n_j, P)$, $j = 1, \ldots, t$, and $n_1 + \cdots + n_t = n$. Since an algebraically closed field is perfect, Theorem 3a is valid for G_j. Hence we may assume that $G_j \subset \Gamma_{k_j}(F_j)$, where F_j is an irreducible locally nilpotent subgroup of $GL(m_j, P)$, $m_j k_j = n_j$.

By Lemma 21, the theorem is true for the group $G_j \subset \Gamma_{k_j}(F_j)$. This and (42) imply the validity of the theorem for G. ∎

LEMMA 22. *Let F and H be irreducible subgroups of $GL(m, \Delta)$ such that $[F] = [H]$. If F is a proper subgroup of H, then $\Gamma_k(F)$ is a proper subgroup of $\Gamma_k(H)$.*

PROOF. Since $[F] = [H]$, the centralizer of H in Δ_m is the same as the centralizer D of F in Δ_m. Hence

$$\Gamma_k(H) = (H \dot{\times} E_k)T_k(D) \supset (F \dot{\times} E_k)T_k(D) = \Gamma_k(F). \quad ∎$$

COROLLARY 1. *Let F be an absolutely irreducible locally nilpotent subgroup of $GL(m, \Delta)$. If $\Gamma_k(F)$ is a maximal locally nilpotent subgroup of $GL(n, \Delta)$, then F is a maximal locally nilpotent subgroup of $GL(m, \Delta)$.*

We mention without proof the following property of $\Gamma_k(F)$. *Let* Δ *be a perfect field,* F *an irreducible locally nilpotent subgroup of* $GL(m, \Delta)$, M *the multiplicative group of the ring* $[F]$, $k > 1$ *and* $mk = n$. *Then* $\Gamma_k(F)$ *is a maximal locally nilpotent subgroup of* $GL(n, \Delta)$ *if and only if* F *is a maximal locally nilpotent subgroup of* M (Suprunenko [20]).

THEOREM 7. *Let* Δ *be an algebraically closed field and* $\mathfrak{M}$ *the set of all maximal locally nilpotent subgroups of* $GL(n, \Delta)$. *Then the number of conjugacy classes of subgroups of* $GL(n, \Delta)$ *into which* $\mathfrak{M}$ *splits cannot exceed a bound* $k(n)$ *depending only on* n.

PROOF. Let $G \in \mathfrak{M}$. Then, by Theorem 3a, G is conjugate in $GL(n, \Delta)$ to the group of all matrices g of the form

$$g = \operatorname{diag}[g_1, g_2, \ldots, g_\nu],$$

where $g_j \in \Gamma_{k_j}(F_j)$, F_j is an irreducible locally nilpotent subgroup of $GL(m_j, \Delta)$, $m_j k_j = n_j$, and $n_1 + \cdots + n_\nu = n$. By Lemma 22, F_j is a maximal locally nilpotent subgroup of $GL(m_j, \Delta)$. By Theorem 28.7, F_j is determined up to conjugacy in $GL(m_j, \Delta)$ by the number m_j. Thus G is determined up to conjugacy in $GL(n, \Delta)$ by a representation of n in the form

$$n = k_1 m_1 + k_2 m_2 + \ldots + k_\nu m_\nu, \tag{43}$$

where $k_j > 0$ and $m_j > 0$.(2) Hence the required number of classes cannot exceed the number of such representations of n in the form (43). ■

In conclusion, we quote one more theorem without proof.

THEOREM 8. *Suppose that a group* G *has a faithful representation over some field. Then the following four assertions are equivalent*:

1. G *is locally nilpotent.*
2. G *satisfies the normalizer condition*
3. G *is an Engel group.*
4. G *is a ZA-group.*

For a proof, see Garaščuk [1] or Garaščuk and Suprunenko [1].*

---

(2)We can add that, by Theorem 28.6, char $\Delta \notin \Pi(m_j)$.

**Editor's note.* For recent developments in the theory of nilpotent and locally nilpotent linear groups, the reader should consult the book by Wehrfritz [3].

BIBLIOGRAPHY

E. Artin

[1] *Geometric algebra*, Interscience, New York, 1957. MR **18**, 553.

L. Auslander

[1] *On a problem of Philip Hall*, Ann. of Math. (2) **86** (1967), 112–116. MR **36** #1540.

R. Baer

[1] *Linear algebra and projective geometry*, Academic Press, New York, 1952. MR **14**, 675.

H. Bass

[1] *K-theory and stable algebra*, Inst. Hautes Études Sci. Publ. Math. No. 22 (1964), 5–60. MR **30** #4805.

H. Bass, M. Lazard and J.-P. Serre

[1] *Sous-groupes d'indice fini dans SL(n, Z)*, Bull. Amer. Math. Soc. **70** (1964), 385–392. MR **28** #5117.

H. Bass, J. Milnor and J.-P. Serre

[1] *Solution of the congruence subgroup problem for* SL_n $(n \geq 3)$ *and* Sp_{2n} $(n \geq 2)$, Inst. Hautes Études Sci. Publ. Math. No. 33 (1967), 59–137. MR **39** #5574.

A. Borel

[1] *Groupes linéaires algébriques*, Ann. of Math. (2) **64** (1956), 20–82. MR **19**, 1195.

N. Bourbaki

[1] *Algèbre.* Chaps. 1–3, 2nd ed., Actualités Sci. Indust., nos. 1144, 1236, 1044, Hermann, Paris, 1951, 1955, 1958; English transl. of 1970 ed., Hermann, Paris; Addison-Wesley, Reading, Mass. 1973. MR **6**, 113; **9**, 406; **30** #3104; **43** #2.

R. Brauer and W. Feit

[1] *An analogue of Jordan's theorem in characteristic p*, Ann. of Math. (2) **84** (1966), 119–131. MR **34** #246.

J. L. Brenner

[1] *The linear homogeneous group.* III, Ann. of Math. (2) **71** (1960), 210–223. MR **22** #1622.

W. Burnside

[1] *On an unsettled question in the theory of discontinuous groups*, Quart. J. Pure Appl. Math. **33** (1902), 230–238.

[2] *On criteria for the finiteness of the order of a group of linear substitutions*, Proc. London Math. Soc. (2) **3** (1905), 345–440.

[3] *On the arithmetical nature of the coefficients in a group of linear substitutions*, Proc. London Math. Soc. (2) **7** (1908), 8–13.

A. H. Clifford

[1] *Representations induced in an invariant subgroup*, Ann. of Math. (2) 38 (1937), 533–550.

C. W. Curtis and I. Reiner

[1] *Representation theory of finite groups and associative algebras*, Pure and Appl. Math., vol. 11, Interscience, New York, 1962. MR 26 #2519.

L. E. Dickson

[1] *Linear groups*, Teubner, Leipzig, 1901.

J. Dieudonné

[1] *Les déterminants sur un corps non commutatif*, Bull. Soc. Math. France 71 (1943), 27–45. MR 7, 3.

[2] *La géométrie des groupes classiques*, Ergebnisse der Mathematik und ihrer Grenzgebiete, Heft 5, Springer-Verlag, Berlin, 1955. MR 17, 236.

[3] *Sur les groupes classiques*, Actualités Sci. Indust., no. 1040, Hermann, Paris, 1948. MR 9, 494.

P. G. L. Dirichlet

[1] *Vorlesungen über Zahlentheorie*, 4th ed., Vieweg, Braunschweig, 1894.

J. D. Dixon

[1] *Complete reducibility of infinite groups*, Canad. J. Math. 16 (1964), 267–274. MR 28 #5120.

[2] *The solvable length of a solvable linear group*, Math. Z. 107 (1968), 151–158. MR 38 #4569.

W. Feit and J. Thompson

[1] *Solvability of groups of odd order*, Pacific J. Math. 13 (1963), 775–1029. MR 29 #3538.

M. S. Garaščuk

[1] *On the theory of generalized nilpotent linear groups*, Dokl. Akad. Nauk BSSR 4 (1960), 276–277. (Russian) MR 24 #A1954.

M. S. Garaščuk and D. A. Suprunenko

[1] *Linear nil-groups*, Dokl. Akad. Nauk BSSR 4 (1960), 407–408. (Russian) MR 24 #A1955.

E. S. Golod

[1] *On nil-algebras and finitely approximable p-groups*, Izv. Akad. Nauk SSSR Ser. Mat. 28 (1964), 273–276; English transl., Amer. Math. Soc. Transl. (2) 48 (1965), 103–106. MR 28 #5082.

O. Grün

[1] *Über eine Faktorgruppe freier Gruppen.* I, Deutsche Math. 1 (1936), 772–782.

M. Hall

[1] *The theory of groups*, Macmillan, New York, 1959. MR 21 #1996.

P. Hall

[1] *Nilpotent groups*, Canad. Math. Congress Summer Seminar, Univ. of Alberta, Edmonton, Alberta, 1957; republished as: *The Edmonton notes on nilpotent groups*, Queen Mary College Math. Notes, London, Ontario, 1969. MR 44 #316.

B. Huppert

[1] *Lineare auflösbare Gruppen*, Math. Z. 67 (1957), 479–518. MR 19, 729.

I. D. Ivanjuta

[1] *Sylow p-subgroups of the countable symmetric group*, Ukrain. Mat. Ž. **15** (1963), 240–249. (Russian) MR **28** #2153.

[2] *On complete sets of Sylow subgroups of the countable symmetric group*, Ukrain. Mat. Ž. **18** (1966), no. 3, 112–115. (Russian) MR **34** #2673.

C. Jordan

[1] *Sur la résolution algébrique des équations primitives de degré* p^2 *(p étant premier impair)*, J. Math. Pures Appl. (2) **13** (1868), 124–126.

[2] *Traité des substitutions et des équations algébriques*, Gauthier-Villars, Paris, 1870.

[3] *Mémoire sur les équations différentielles linéaires à intégral algébrique*, J. Reine Angew. Math. **84** (1878), 89–215.

L. Kaloujnine [L. A. Kalužnin]

[1] *Sur les p-groupes de Sylow du groupe symétrique de degré* p^m, C. R. Acad. Sci. Paris **221** (1945), 222–224. MR **7**, 239.

[2] *Sur les p-groupes de Sylow du groupe symétrique de degré* p^m *(Suite centrale ascendante et descendante)*, C. R. Acad. Sci. Paris **223** (1946), 703–705. MR **8**, 251.

[3] *La structure des p-groupes de Sylow des groupes symétriques finis*, Ann. Sci. École Norm. Sup. (3) **65** (1948), 239–276. MR **10**, 505.

I. Kaplansky

[1] *An introduction to differential algebra*, Actualités Sci. Indust., no. 1251, Hermann, Paris, 1957. MR **20** #177.

M. I. Kargapolov

[1] *On periodic matrix groups*, Sibirsk. Mat. Ž. **3** (1962), 834–838. (Russian) MR **26** #2517.

[2] *Finitely generated linear groups*, Algebra i Logika **6** (1967), no. 5, 17–20. (Russian) MR **38** #5942.

O. H. Kegel

[1] *Lectures on locally finite groups*, Mathematical Institute, Oxford, 1969.

O. H. Kegel and B. A. F. Wehrfritz

[1] *Locally finite groups*, North-Holland Math. Library, North-Holland, Amsterdam, 1973.

W. Klingenberg

[1] *Die Struktur der linearen Gruppen über einem nichtkommutativen lokalen Ring*, Arch. Math. **13** (1962), 73–81. MR **26** #1367.

E. R. Kolchin

[1] *On certain concepts in the theory of algebraic matric groups*, Ann. of Math. (2) **49** (1948), 774–789. MR **10**, 232.

A. G. Kuroš

[1] *Theory of groups*, 3rd ed., "Nauka", Moscow, 1967; English transl. of 2nd ed., Chelsea, New York, 1960. MR **22** #727; **40** #2740.

[2] *Lectures on general algebra*, Fizmatgiz, Moscow, 1962; English transl., Chelsea, New York, 1963; Internat. Ser. Monographs Pure Appl. Math., vol. 70, Pergamon Press, Oxford, 1965. MR **25** #5097; **28** #1228; **31** #3483.

S. Lang

[1] *Algebra*, Addison-Wesley, Reading, Mass., 1965. MR **33** #5416.

A. I. Mal'cev

[1] *On faithful representation of infinite groups by matrices*, Mat. Sb. 8 (50) (1940), 405–422; English transl., Amer. Math. Soc. Transl. (2) 45 (1965), 1–18. MR 2, 216.

[2] *On a general method for obtaining local theorems in group theory*, Ivanov. Gos. Ped. Inst. Učen. Zap. Fiz.-Mat. Fak. 1 (1941), 3–9; English Transl., Chapter II in A. I. Mal'cev, *The metamathematics of algebraic systems. Collected papers*: 1936–1967, Studies in Logic and Foundations of Math., vol. 66, North-Holland, Amsterdam, 1971. MR 17, 823.

[3] *On certain classes of infinite solvable groups*, Mat. Sb. 28 (70) (1951), 567–588; English transl., Amer. Math. Soc. Transl. (2) 2 (1956), 1–21. MR 13, 203; 17, 824.

[4] *Foundations of linear algebra*, 3rd ed., "Nauka", Moscow, 1970; English transl. of 2nd ed., Freeman, San Francisco, Calif., 1963. MR 29 #3477.

[5] *Algebraic systems*, "Nauka", Moscow, 1970; English transl., Die Grundlehren der math. Wissenschaften, Band 192, Springer-Verlag, Berlin and New York, 1973. MR 44 #142.

G. Maxwell

[1] *Infinite general linear groups over rings*, Trans. Amer. Math. Soc. 151 (1970), 371–375. MR 41 #8531.

J. L. Mennicke

[1] *Finite factor groups of the unimodular group*, Ann. of Math. (2) 81 (1965), 31–37. MR 30 #2083.

Ju. I. Merzljakov

[1] *Central series and commutator series of matrix groups*, Algebra i Logika 3 (1964), no. 4, 49–59. (Russian) MR 30 #170.

[2] *On linear groups with bounded cyclic subgroups*, Sibirsk. Mat. Ž. 7 (1966), 318–322 = Siberian Math. J. 7 (1966), 257–259. MR 33 #5742.

[3] *Integer representation of holomorphs of polycyclic groups*, Algebra i Logika 9 (1970), 539–558 = Algebra and Logic 9 (1970), 326–337. MR 43 #6298.

J. Milnor

[1] *Whitehead torsion*, Bull. Amer. Math. Soc. 72 (1966), 358–426. MR 33 #4922.

V. L. Nisnevič

[1] *Über Gruppen, die durch Matrizen über einem kommutativen Feld isomorph darstellbar sind*, Mat. Sb. 8 (50) (1940), 395–404. (Russian; German summary) MR 2, 309.

P. S. Novikov and S. I. Adjan

[1] *On finite periodic groups*. I, II, III, Izv. Akad. Nauk SSSR Ser. Mat. 32 (1968), 212–244, 251–524, 709–731 = Math. USSR Izv. 2 (1968), 209–237, 241–479, 665–685. MR 39 #1532a,b,c.

V. P. Platonov

[1] *On the theory of algebraic linear groups*, Dokl. Akad. Nauk SSSR 146 (1962), 1025–1026 = Soviet Math. Dokl. 3 (1962), 1448–1449. MR 26 #1370.

[2] *Solvable algebraic groups*, Dokl. Akad. Nauk SSSR 151 (1963), 48–51 = Soviet Math. Dokl. 4 (1963), 924–927. MR 27 #2508.

[3] *Splittable linear groups*, Dokl. Akad. Nauk SSSR 151 (1963), 286–289 = Soviet Math. Dokl. 4 (1963), 1002–1005. MR 29 #4806.

[4] *Periodic subgroups of algebraic groups*, Dokl. Akad. Nauk SSSR 153 (1963), 270–272 = Soviet Math. Dokl. 4 (1963), 1653–1655. MR 28 #2152.

[5] *Solvable algebraic groups*, Dokl. Akad. Nauk BSSR 7 (1963), 298–300. (Russian) MR 27 #3639.

[6] *On solvable algebraic linear groups*, Dokl. Akad. Nauk BSSR 7 (1963), 439–442. (Russian) MR 27 #5825.

[7] *On complete reducibility of linear groups*, Vescī Akad. Navuk BSSR Ser. Fīz.-Tėhn. Navuk 1964, no. 3, 126–128. (Russian) MR 30 #171.

[8] *Structure of periodic linear groups and algebraic groups*, Dokl. Akad. Nauk SSSR 160 (1965), 541–544 = Soviet Math. Dokl. 6 (1965), 144–147. MR 30 #3924.

[9] *Algebraic d-groups*, Dokl. Akad. Nauk BSSR 7 (1963), 729–731. (Russian) MR 28 #1260.

[10] *Automorphisms of algebraic groups*, Dokl. Akad. Nauk SSSR 168 (1966), 1257–1260 = Soviet Math. Dokl. 7 (1966), 825–828. MR 35 #4230.

[11] *Theory of algebraic linear groups and periodic groups*, Izv. Akad. Nauk SSSR Ser. Mat. 30 (1966), 573–620; English transl., Amer. Math. Soc. Transl. (2) 69 (1968), 61–110. MR 33 #7428.

[12] *The Frattini subgroup of linear groups and finite approximability*, Dokl. Akad. Nauk SSSR 171 (1966), 798–801 = Soviet Math. Dokl. 7 (1966), 1557–1560. MR 35 #2971.

[13] *Linear groups with identical relations*, Dokl. Akad. Nauk BSSR 11 (1967), 581–582. (Russian) MR 35 #4314.

[14] *Proof of the finiteness hypothesis for solvable subgroups of algebraic groups*, Sibirsk. Mat. Ž. 10 (1969), 1084–1090 = Siberian Math. J. 10 (1968), 800–804. MR 40 #7355.

V. P. Platonov and A. E. Zalesskiĭ

[1] *The Auerbach problem*, Dokl. Akad. Nauk BSSR 10 (1966), 5–6. (Russian) MR 33 #224.

B. I. Plotkin

[1] *On some criteria for locally nilpotent groups*, Uspehi Mat. Nauk 9 (1954), no. 3 (61), 181–186; English transl., Amer. Math. Soc. Transl. (2) 17 (1961), 1–7. MR 16, 444; 23 #A1711.

[2] *On nil-groups*, Dokl. Akad. Nauk SSSR 94 (1954), 999–1001. (Russian) MR 15, 682.

[3] *Infinite-dimensional linear groups*, Dokl. Akad. Nauk SSSR 153 (1963), 42–45 = Soviet Math. Dokl. 4 (1963), 1617–1620. MR 28 #141.

[4] *Groups of automorphisms of algebraic systems*, "Nauka", Moscow, 1966; English transl., Wolters-Noordhoff, Groningen, 1972. MR 40 #217.

C. Procesi

[1] *The Burnside problem*, J. Algebra 4 (1966), 421–425. MR 35 #2956.

I. Reiner

[1] *Normal subgroups of the unimodular group*, Illinois J. Math. 2 (1958), 142–144. MR 20 #2379.

A. Rosenberg

[1] *The structure of the infinite general linear group*, Ann. of Math. (2) 68 (1958), 278–294. MR 21 #1319.

I. Schur

[1] *Über Gruppen periodischer linearer Substitutionen*, S.-B. Preuss. Akad. Wiss. 1911, 619–627.

A. Selberg

[1] *On discontinuous groups in higher-dimensional symmetric spaces*, Contributions to Function Theory (Internat. Colloq. Function Theory, Bombay, 1960), Tata Institute of Fundamental Research, Bombay, 1960, pp. 147–164. MR 24 #A188.

J.-P. Serre

[1] *Représentations linéaires des groupes finis*, Hermann, Paris, 1967. MR 38 #1190.

O. Ju. Šmidt

[1] *Selected works. Mathematics*, Izdat. Akad. Nauk SSSR, Moscow, 1959. (Russian) MR 21 #1259.

D. A. Suprunenko

[1] *Primitive solvable substitution groups*, Mat. Sb. 20 (62) (1947), 331–350. (Russian) MR 8, 562; 14, 1277.

[2] *Solvable matrix groups*, Učen. Zap. Leningrad. Gos. Univ. No. 12 (1951), 74–112. (Russian)

[3] *Irreducible nilpotent matrix groups of prime degree*, Mat. Sb. 31 (73) (1952), 353–358. (Russian) MR 14, 447.

[4] *On nilpotent groups of matrices*, Belorussk. Gos. Univ. Učen. Zap. Ser. Fiz.-Mat. 1953, no. 15, 3–6. (Russian) MR 17, 456.

[5] *On nilpotent transitive subgroups of the symmetric group*, Dokl. Akad. Nauk SSSR 99 (1954), 23–25. (Russian) MR 16, 793.

[6] *On irreducible nilpotent matrix groups*, Mat. Sb. 35 (77) (1954), 501–512. (Russian) MR 16, 793.

[7] *Locally nilpotent irreducible subgroups of the full linear group*, Dokl. Akad. Nauk SSSR 102 (1955), 41–44. (Russian) MR 17, 456.

[8] *Soluble and nilpotent linear groups*, Ministerstvo Vysšego Obrazovanija SSSR, Izdat. Belgosuniv. im V. I. Lenina, Minsk, 1958; English transl., Transl. Math. Monographs, vol. 9, Amer. Math. Soc., Providence, R. I., 1963. MR 27 #4861a,b.

[9] *Real linear locally nilpotent groups*, Mat. Sb. 50 (92) (1960), 59–66; English transl., Amer. Math. Soc. Transl. (2) 45 (1965), 23–31. MR 22 #2653.

[10] *On periodic linear groups*, Sibirsk. Mat. Ž. 3 (1962), 87–94. (Russian) MR 27 #2560.

[11] *Periodic subgroups of solvable matrix groups*, Dokl. Akad. Nauk SSSR 147 (1962), 310–312 = Soviet Math. Dokl. 3 (1962), 1612–1614. MR 26 #220.

[12] *On the Minkowski homomorphism*, In Memory of N. G. Čebotarev, Izdat. Kazan. Univ., Kazan, 1964, pp. 89–92. (Russian) MR 33 #4151.

[13] *The kernel of a homomorphism*, Sibirsk. Mat. Ž. 6 (1965), 199–206. (Russian) MR 30 #3893.

[14] *Locally nilpotent matrix groups over an arbitrary field*, Mat. Sb. 68 (110) (1965), 614–622. (Russian) MR 33 #198.

[15] *Locally nilpotent subgroups of the infinite symmetric group*, Dokl. Akad. Nauk SSSR **167** (1966), 302–304 = Soviet Math. Dokl. **7** (1966), 392–394. MR 33 #2709.

[16] *A condition for complete reducibility of a matrix group*, Izv. Akad. Nauk SSSR Ser. Mat. **27** (1963), 435–438. (Russian) MR 26 #6268.

[17] *On solvable subgroups of the multiplicative group of a field*, Izv. Akad. Nauk SSSR Ser. Mat. **26** (1962), 631–638; English transl., Amer. Math. Soc. Transl. (2) **46** (1965), 153–161. MR 25 #5111.

[18] *Two theorems on matrix groups*, Dokl. Akad. Nauk BSSR **8** (1964), 491–494. (Russian) MR 30 #172.

[19] *On maximal nilpotent subgroups of the full linear group over a factor ring* $Z/(p^m)$, Mat. Sb. **66** (108) (1965), 598–607. (Russian) MR 30 #4839.

[20] *Irreducible components of a reducible maximal locally nilpotent matrix group*, Izv. Akad. Nauk SSSR Ser. Mat. **32** (1968), 527–533 = Math. USSR Izv. **2** (1968), 481–486. MR 37 #1487.

[21] *The theory of commutative matrix algebras over an imperfect field*, Vescī Akad. Navuk BSSR Ser. Fiz.-Mat. Navuk **1967**, no. 3, 9–19. (Russian) MR 36 #208.

[22] *On the theory of solvable linear groups*, Sibirsk. Mat. Ž. **10** (1969), 1161–1172 = Siberian Math. J. **10** (1969), 859–867. MR 40 #4381.

[23] *Subgroups of the full linear group over a sfield D that contain the group U(n, D) of all special triangular matrices*, Dokl. Akad. Nauk BSSR **14** (1970), 305–308. (Russian) MR 43 #3354.

D. A. Suprunenko and R. F. Apatenok

[1] *Nilpotent irreducible groups over a finite field*, Dokl. Akad. Nauk BSSR **3** (1959), 475–478. (Russian)

[2] *Nilpotent irreducible matrix groups over a finite field*, Dokl. Akad. Nauk BSSR **5** (1961), 535–537. (Russian) MR 27 #5834.

D. A. Suprunenko and M. S. Garaščuk

[1] *Linear groups with a category*, Dokl. Akad. Nauk BSSR **6** (1962), 411–414. (Russian) MR 26 #1369.

[2] *Linear groups with Engel's condition*, Dokl. Akad. Nauk BSSR **6** (1962), 277–280. (Russian) MR 26 #1368.

D. A. Suprunenko and V. P. Platonov

[1] *On a theorem of Schur*, Dokl. Akad. Nauk BSSR **7** (1963), 510–512. (Russian) MR 27 #4860.

D. A. Suprunenko and R. I. Tyškevič

[1] *Commutative matrices*, Nauka i Tehnika, Minsk, 1966; English transl., Academic Press, New York, 1968. MR 34 #1356.

[2] *Reducible locally nilpotent linear groups*, Izv. Akad. Nauk SSSR Ser. Mat. **24** (1960), 787–806. (Russian) MR 25 #124.

D. A. Suprunenko and G. P. Žavrid

[1] *Intransitive nilpotent subgroups of the symmetric group*, Dokl. Akad. Nauk BSSR **2** (1958), 320–322. (Russian) MR 41 #313.

R. G. Swan

[1] *Representation of polycyclic groups*, Proc. Amer. Math. Soc. **18** (1967), 573–574. MR 35 #4306.

R. T. Vol'vačev

[1] *Sylow p-subgroups of the general linear group*, Izv. Akad. Nauk SSSR Ser. Mat. 27 (1963), 1031–1054; English transl., Amer. Math. Soc. Transl. (2) 64 (1967), 216–243. MR 28 #4037.

[2] *Periodic nilpotent linear groups over the field of rational numbers*, Dokl. Akad. Nauk SSSR 167 (1966), 499–502 = Soviet Math. Dokl. 7 (1966), 403–406. MR 36 #2712.

B. L. van der Waerden

[1] *Moderne Algebra.* Vol. I, 2nd ed., Springer, Berlin, 1937; English transl., Ungar, New York, 1949. MR 10, 587.

[2] *Moderne Algebra.* Vol. II, 2nd ed., Springer, Berlin, 1940; English transl., Ungar, New York, 1950. MR 10, 587.

[3] *Gruppen von linearen Transformationen*, Springer, Berlin, 1935.

B. A. F. Wehrfritz

[1] *Sylow theorems for periodic linear groups*, Proc. London Math. Soc. (3) 18 (1968), 125–140. MR 36 #3893.

[2] *Soluble periodic linear groups*, Proc. London Math. Soc. (3) 18 (1968), 141–157. MR 36 #3894.

[3]* *Infinite linear groups. An account of the group-theoretic properties of infinite groups of matrices*, Ergebnisse der Math. und ihrer Grenzgebiete, Band 76, Springer-Verlag, Berlin and New York, 1973. MR 49 #436.

A. J. Weir

[1] *Sylow p-subgroups of the classical groups over finite fields with characteristic prime to p*, Proc. Amer. Math. Soc. 6 (1955), 529–533. MR 17, 235.

G. P. Žavrid

[1] *On the theory of solvable matrix groups*, Dokl. Akad. Nauk BSSR 11 (1967), 873–877. (Russian) MR 36 #2713.

A. E. Zalesskiĭ

[1] *Solvable subgroups of the multiplicative group of a locally finite algebra*, Mat. Sb. 61 (103) (1963), 408–417. (Russian) MR 28 #134.

[2] *Solvable subgroups of the multiplicative group of a simple algebra*, Dokl. Akad. Nauk BSSR 7 (1963), 80–82. (Russian) MR 26 #6254.

[3] *Sylow p-subgroups of the general linear group over a skew field*, Izv. Akad. Nauk SSSR Ser. Mat. 31 (1967), 1149–1158 = Math. USSR Izv. 1 (1967), 1099–1108. MR 36 #2710.

H. Zassenhaus

[1] *Beweis eines Satzes über diskrete Gruppen*, Abh. Math. Sem. Univ. Hamburg 12 (1938), 289–312.

---

*Added by the editor.

Subject Index